Applied Reliability-Centered Maintenance

Applied Reliability-Centered Maintenance

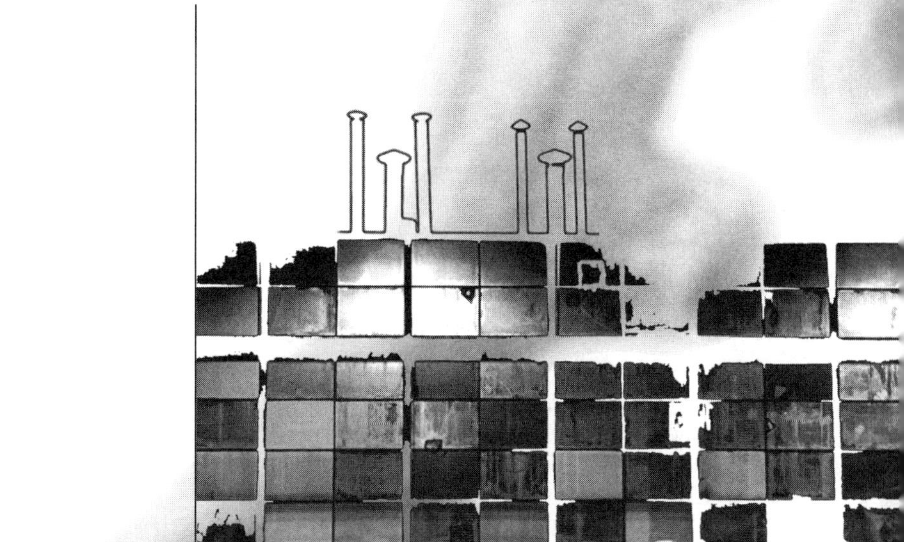

by Jim August, PE
OME, Inc.

PennWell

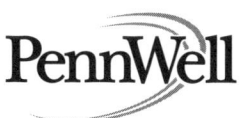

Copyright © 1999 by
PennWell
1421 S. Sheridan/P.O. Box 1260
Tulsa, Oklahoma 74101

Library of Congress Cataloging-in-Publication Data
August, Jim
 Applied reliability centered maintenance / by Jim August.
 p.cm.
 ISBN 0-87814-746-2
 1. Plant maintenance. 2. Reliability (Engineering) 3. Maintainability
 (Engineering)
 I.Title.

TS192 .A94 1999
658.2'02--dc21
99-050100

Cover Design: Shanon Garvin and Brian Firth
Layout: Brian Firth

Printed in the United States of America

03 02 01 00 99 1 2 3 4 5

Table of Contents

Figures

Tables

Acronyms List

AB	A or B (with respect to train, or piece of equipment)
A-E	Architect-engineer
ALARA	As low as reasonably achievable
ANI	American Nuclear Insurers
ANSI	American National Standards Institute
ARCM	Applied reliability centered maintenance
ASCE	American Society of Civil Engineers
ASME	American Society of Mechanical Engineers
ASQC	American Society for Quality Control
B/C	Benefit/cost
BFP	Boiler feed pump
BPV	Boiler and pressure vessel
B&W	Babcock and Wilcox
BWR	Boiling water reactor
Cal.	Calibration
CBM	Condition-based maintenance
CC	Channel checks
CCF	Common cause failure
CDM	Condition-directed maintenance
CDM (FF)	Condition-directed maintenance (failure finding)
CE	Combustion Engineering
CEM	Continuous emissions monitor
CFR	Code of federal regulations
CIC	Component identification codes
CM	Corrective maintenance
CNMM	Condition-monitoring (based) maintenance
CMMS (Legacy)	Computerized maintenance management systems
CNM	Condition monitoring
CO	Conditional overhaul
CRT	Cathode ray tube
CT	Combustion turbine
CWP	Circulating water pump
CWT	Circulating water tower
DC	Design change
DCS	Distributed-control system
DOE	Department of Energy

DOT	Department of Transportation
DP	Differential pressure
E	Emergency
EEI	Edison Electric Institute
EFOR	Equivalent forced outage rate
EG	Equipment group
"E" MWR	Emergency maintenance work request
EO	Equipment operator
EPA	Environmental Protection Agency
EPRI	Electric Power Research Institute
EQ	Environmentally qualified
FAA	Federal Aviation Administration
FAI	Failed as is
FC	Fails closed
FD	Forced draft
FERC	Federal Energy Regulatory Commission
FF	Functional failure
FMEA	Failure modes and effects analysis
FMECA	Failure modes and effects criticality analysis
FO	Fails open
FOR	Forced outage rate
FT	Functional tests
FTA	Fault tree analysis
FTM	Fixed-time maintenance
GADS	Generation availability data system
GE	General Electric
GPA	Grade point average
GUI	Graphical user interface
HEU	Hydraulic equipment units
HP	Horse power
HRSG	Heat recovery steam generator
HTGR	High temperature gas reactor
I&C	Instrumentation and control
IA	Instrument air
ID	Induction draft
IEEE	Institute of Electrical and Electronic Engineers
INPO	Institute of Nuclear Plant Operations
IPP	Independent power producer
KISS	Keep it simple, stupid!
LAN	Local area network
LCM	Life-cycle maintenance

LCO	Limiting conditions for operations
LTA	Logic tree analysis
LWR	Light water reactor
MIS	Maintenance information system
MORT	Management oversight risk tree
MOV	Motor operated valve
MPFF	Maintenance preventable functional failure
mREM	One thousandth of a REM
MS	Microsoft
MSG-3	Maintenance Steering Group-3
MTBF	Mean time between failures
MTTR	Mean time to repair
MW	Megawatt
MWO	Maintenance work order
MWR	Maintenance work request
NCE	New century energies
NDE	Non-destructive examination
NEC	National electric code
NERC	North American Electric Reliability Council
NFPA	National Fire Protection Association
NPM	No planned maintenance
NPPD	Nebraska Public Power District (utility)
NRC	Nuclear Regulatory Commission
NSM	No scheduled maintenance
O&M	Operation and maintenance
OCM	On-condition maintenance
OCMFF	On-condition maintenance (failure-finding)
OEM	Original equipment manufacturer
OOS	Out-of-service
OM	Operator manual
OSHA	Occupational Safety and Health Administration
OTF	Operate to failure
P&ID	Process and instrumentation drawings
PA	Primary air
PC	Primary containment
PCRV	Pre-stressed concrete reactor vessel
PdM	Predictive maintenance
PG&E	Pacific Gas & Electric
PM	Preventive maintenance
PMO	Preventive maintenance optimization
PRB	Powder River Basin

PUC	Public utility commissions
PV	Present value
PWR	Pressurized water reactors
R	Reliability
RAM	Reliability, availability and maintainability
RCA	Root cause analysis
RCFA	Root cause failure analysis
RCM	Reliability centered maintenance
RD	Response-driven
REM	Roentgen equivalent, man
RL	Random limit
RO	Reverse osmosis
RPS	Reactor protective system
RTF	Run to failure
SAE	Society of Automotive Engineers
SBAC	Soot-blowing air compressor
SL	Straight line
SNAFU	Situation normal all fouled up
SOA	Society of Actuaries
SP	Surveillance programs
SPC	Statistical process control
SRCM	Streamlined reliability centered maintenance
SSC	Structures, systems and components
SWOT	Strength weakness opportunity threat (analysis)
TBM	Time-based maintenance
TC	Thermocouple
TMI	Three Mile Island
TPM	Total productive maintenance
TQM	Total quality management
TRCM	Traditional reliability centered maintenance
UAL	United Airlines
VAR	Volt amp reactive
VM	Vibration monitoring
VOM	Volt ohm meter
VWO	Valves wide open
WO	Work order
WSSC	Western States Coordinating Council
Y2K	Year 2000

Acknowledgements

...a chaise breaks down, but doesn't wear out

-Oliver Wendell Holmes, "The Deacon's Masterpiece"
(with credit to Stan Nolan & Howard Heap)

Elements of reliability centered maintenance (RCM) aren't new; time based maintenance, make-or-buy, re-work, performance testing and corrective maintenance (CM) have been terms used traditionally to describe aspects of scheduled maintenance programs. On another level, RCM brought order to a confused and complex subject. Like Inuit (Eskimo) language's many terms for snow, maintenance has many descriptions. The beauty of RCM is the order that it brings to these terms in the context of a strategy.

Strategy is an underlying theme of RCM. As in chess or war, strategy requires supporting tactics. These are mastered as preliminaries. Without appreciating the tactics of maintenance, the need and value for strategy can be missed. Even with a strategy, tactical battles can be lost. Yet with a strategy comes a comprehensive vision for managing short, intermediate, and long-term maintenance. A strategy provides a resource map that uniquely identifies the multiple roles that must be played by various work groups.

Experts come into play in several ways: (1) developing the strategy and supporting tactics for the existing plant, (2) identifying the paths to future goals, and (3) managing the emerging maintenance requirements of the plant. By whatever name it's given, understanding on-condition/condition-directed maintenance complementary tasks is the general lesson of RCM. Too often in my experience, failures weren't recognized because limits weren't defined. Redundancy or sheer nerve were adequate support decision-making. At some point, otherwise failed equipment can be operated. Perhaps not economically, but operated.

To assist the reader with the terminology, I offer a rough term equivalence guide. What we used to call preventative maintenance (PM) is now recognized as mainly rework and restore time-based maintenance (TBM). What we used to call corrective maintenance is a combination of condition-directed, condition-based, and failure maintenance. True failures are infrequent in even reactive environments. They virtually don't occur in some. Surveillance and operating tests now have formal roles as special types of "on-condition" maintenance tasks. These tasks trigger maintenance, of course. The contribution of RCM is:

- to put order into the maintenance lexicon and tool kit
- identify how to reduce operating costs using condition monitoring and on-condition maintenance (OCM) tasks while maintaining operating goals

I'd like to recognize all those who helped me with this work—friends, family, and peers. My friends patiently tolerated my probing questions and helped review materials. My family tolerated my long hours and absence. My reviewers tolerated my long-winded initial drafts! (I recommend writing a book to everyone.) Perhaps foremost though, I'd like to recognize the giants who led the way, including the commercial aviation pioneers, the Federal Aviation Administration (FAA), and the Airline Transport Association. The risk of putting together a list is leaving someone out. Published works capture the ideas of many people. The Department of Defense deserves special merit for noting a promising technology, and chartering its documentation in published form (*Reliability Centered Maintenance*). Like Newton, we stand on the shoulders of giants. But like Einstein, they painstakingly put together pieces of a puzzle that were already available in various nebulous forms.

For my reviewers I'd like to cite those who provided specific comments and assistance. First, I'd like to recognize Alan Bern, my colleague at TU Electric. Alan painstakingly plowed through the first roughs with exuberance and helpful comments. On the technical level, Alan's detailed review was invaluable. Second, my long-time pal

Joe Hunter, who shared an operator's perspective. Next, my friend and former boss who provided crucial support in times of need—Frank Novachek. (We haven't always seen things the same, but our friendship has endured.) Next, my colleague Jon Anderson, who knows more practical maintenance than I could ever dream of. Lastly, I want to thank the many fine crafts who have been so helpful sharing their insights over the years. The fixers are a crusty crew, difficult in the extreme to converse with, but a truly fine bunch of people who've made this career worthwhile. Corporate supporters—those who provide work and a place to practice the craft—are also most welcome. These include Jim Love, manager at New Century Energies (NCE), along with Dick Chuvarsky, Jim Stevens, Richard Roe, Mike Blossom, at Arapahoe (NCE), and Mike Young, Chuck Fidler and Chuck Gaines from Cooper (Nebraska Public Power District [NPPD]). They supported the pragmatic work side to allow this to happen.

A former manager once exclaimed that he knew of no one else who made so much effort to understand, and fundamentally grasp things. I take that in a complimentary way. Make no mistake, low cost maintenance ultimately boils down to technical expertise. If you have technically competent staff you're lucky. Some companies treat it superficially. But like chess, the overall strategy ultimately comes down to understanding the roles and selections of the pieces. Equipment and diagnostic technology must be understood, and this understanding comes only with years of experience.

In many ways RCM is nothing but a framework. To achieve the program requires finishing, so to speak. At this age I can finally admit I know very little of the useful knowledge in this world, and learn more every day. The beauty of well-designed equipment is that you really need to know very little to operate it successfully. The corollary is that you need to know useful knowledge very well to diagnose and repair equipment effectively. While most people are reasonably successful with modest levels of understanding, those who know equipment well and have the best maintenance model turn in superior performance. RCM provides a model to point out our weaknesses so we all learn more quickly.

Finally, what can I say about applied RCM (ARCM)? Why the

new term? RCM has been given a black eye by analysis. Some would make it a religion or take it into non-measurable, esoteric philosophy. I believe that RCM's most appropriate application is fundamentally as a technology. There's little new except complexity, failure behavior, numbers, exploration, and RCM's integrating perspective. Since no one has offered a completely comprehensive text emphasizing maintenance technology, I offer this work.

The most tedious and difficult aspects of RCM are selecting task limits and intervals. Fortunately, RCM also provides techniques to provide answers—even with incomplete information—with powerful alternatives that help us to manage risk. Technical competence is taken for granted; it may or may not be available.

ARCM is fundamentally about using RCM for value, to understand which paths to pursue, what ax to grind, and where to focus. ARCM is finding those things that provide value in a specific setting, and doing them. Just do it!

Know your equipment. Know how it ages. Know that it is aging, and how to restore it. And when it's broken, just fix it! This last missive is the hardest. A chaise breaks down but never wears out!

Preface

Things that matter most must never be at the mercy of those that matter least.

-Goerthe

There hasn't been a "maintenance" best seller since *Zen and the Art of Motorcycle Maintenance*—a book on philosophy. The author's theme was his love/hate relationship with technology, expressed as his motorcycle.

Maintenance is a tough subject to write about. It's so dry! Yet it's a subject we all immediately relate to, both professionally and as people.

The last significant new book on technical maintenance introduced us to RCM, in 1978. Since then, a few new maintenance terms have been added, such as total productive maintenance (TPM) and total quality management (TQM). Asset management is the latest twist on the subject, as I write this. The original United Airlines (UAL) work, *Reliability Centered Maintenance*, by Nolan and Heap, published in 1978, has led to 20 years of implementation history in the nuclear generation industry. Several re-interpretations have been published. My purpose in this book is to provide new fundamental interpretations of ARCM based upon the original theory, and to discuss them in terms of real-world problems and experience. These problems at one time or another demanded hours of analytical thought, planning and performance—their learning was in some instances bitterly bought.

RCM offered a fresh perspective on maintenance, focusing on the theory and methodology of traditional, non-aeronautics RCM. However, the original RCM field—aeronautics—differs radically from power generation. RCM texts currently available could easily leave the impression that applying RCM requires the skills of engineers and mathematicians.

In fact, RCM summarizes practical experience, putting it on a firm engineering and mathematical foundation. RCM provides a fresh perspective on maintenance to enable us to make better, more informed maintenance decisions. Practically, RCM principles can be learned and implemented in day-to-day work situations. Organizations that want to apply simple RCM lessons can do so quickly. While engineers and managers need RCM involvement, it is more important to achieve shop-floor RCM recognition and application. This involves reducing traditional RCM (TRCM) materials to simple catechisms and applying them to daily, routine maintenance problems. RCM offers a powerful tool to operationalize maintenance.

For maintenance managers, RCM offers a powerful tool to justify effective programs—especially when working under distant, obscure, or conservative vendor, government, or corporate guidelines. In today's environment—with its ever-more prescriptive regulation and competitive pressures—RCM is a tool that organizations can use to recapture the maintenance initiative.

But to have any value RCM must focus on implementation—operationalizing many maintenance activities that the very best organizations have implicitly practiced all along. Such an emphasis on implementation is fundamentally different from that addressed by TRCM approaches. Organizations like the Electric Power Research Institute (EPRI) have developed simplified methods, termed streamlined RCM (SRCM). A fossil version has been coined—preventative maintenance optimization (PMO). These differ in fundamental ways from the approach presented here—an approach I call ARCM.

Taken together, I believe that these methods—unlike those now accepted as traditional generation RCM and documented in several published texts—represent a simpler route to RCM benefits and return to the original RCM methods developed by Stan Nolan and Howard Heap. The distinctions may appear arbitrary to those who believe they are versed in the subject because they differ from utility RCM in several subtle but important ways. But these minor distinctions allow maintenance organizations to capture the value and effectiveness discovered by the commercial air industry more than 20 years ago, when RCM was first put to the test—the methods that still work best.

My ARCM approach is based on experience and based upon the supposition that the primary barriers to RCM implementation are not in the maintenance work force but in the prevailing state of maintenance. What is "maintenance?" Who administers it? How should it be managed and performed?

Maintenance is democratic and relies upon autonomy in utility environments. Few traditional engineering staffs are fluent in this world. Most of the new tools and products were introduced by contractors. Most non-military, non-aerospace maintenance departments practice in the traditional "maintenance environment." Utilities with whom I have worked fit this model. For successful implementation, it is essential that RCM materials and methods are suited to capital "M" maintenance.

Yet, where formal RCM studies precede formal implementation, results have been mixed. Management unfamiliarity with RCM principles and the absence of formal project management methods and measures hurt. Shop-floor readiness for change has influenced success but utilities traditionally have had low learning rates (supported by regulation.) Sometimes changes (even necessary ones) overwhelm an organization until they put the RCM effort on the shelves. Implementation demands too much too fast.

RCM has been described as "the right maintenance at the right time," but who knows the right maintenance and the right time? How do they find this out and inform the organization? When the front office calls the shots, a planned maintenance effort succeeds to the degree the shops follow its lead. Operations, maintenance, and technical support staffers often feel they intuitively know "the right maintenance and the right times," but cannot connect with those hardheaded guys who schedule outages, select parts, or establish practical expenditures. They feel swept aside by maintenance infrastructure.

If the perception is that RCM is theory—irrelevant to field maintenance—ARCM provides techniques and simple, applicable insights that improve maintenance performance. RCM theoretically provides complete maintenance solutions (and regulatory support bases) but TRCM-based maintenance requires complete, closed-form system studies. My experience as a practicing engineer is that RCM can be implemented on a daily basis for most routine maintenance at industrial facilities. Through time, properly documented and retained, RCM measures support complete equipment and system maintenance programs. On another level, there are only a few unique and new dimensions in RCM. One of these is inherent reliability. The single most important maintenance lesson is that when you have a maintenance problem—fix it. No benefit results from deferring or ignoring maintenance. The likely results are complications—compounded problems, secondary failures, and greater expenses.

This book encourages small, simple steps to help implement RCM and to help guide big projects in organizations with formal maintenance foundations. These methods counter prevailing wisdom and I eagerly look forward to discussions I hope to generate.

Examples cited here all have basis in fact and are taken from an historical perspective. Some were faced as long as 20 years ago! In some instances the plant and even the companies no longer exist. All are provided to stimulate thought and provoke reader reflection. I believe that had we known of these methods at the time, and had broader recognition of their validity, we would have been more effective and may have avoided expensive consequences. On the other hand, where we decided to be effective at the time, we typically were.

So, what key points of ARCM are covered in this text?

- The fundamental RCM strategy classifications
- Simple, neat documentation
- Measurement
- "Factual basis" maintenance planning
- Explicit failure resistance limits specified for "on-condition" maintenance
- "Partial" solutions, including:
  - outage reviews
  - PM reviews
  - "safety modification" reviews
  - Strategy

A fast-track approach to RCM described here can be applied at most facilities. These differ across industries and facilities, but my general experience is there are two high-value RCM areas:

- Instrumentation and Controls (I&C)
- Implementation

I cannot over-emphasize that implementation is where the value is, and that benefits come from ARCM discipline introduced at the shop-floor level. ARCM can deliver the results for those who have the stomach to go after them. Organizations operated with maintenance backlogs provide a worker security blanket. To substantially change these organizations will threaten equilibrium! ARCM can and will change your maintenance processes. It may turn out that you discover things you suspected all along, but could never prove!

Regulated utilities had the luxury of operating in the cost-plus world for many years. The need for greater (and documented) productivity is only dawning on these work forces. Maintenance workers need tools that can help them make these changes, and make significant contributions. ARCM provides just that.

This book is directed towards providing practical, useful material to assist generation companies with process reliability. And while many reliability aspects are inherently known, the subject has been taught on the job with the sink-or-swim approach. This book offers specific operations, maintenance, and engineering tools to improve reliability and availability and reduce costs at the plant level. It shares new ways of viewing maintenance so users can make more informed decisions. Most readers will be able to follow this text knowing little about TRCM theory. Although we bring it all together at the end, for those wishing to firm up their theoretical RCM understanding, I recommend browsing any of the RCM texts cited in the references. Most are in print, a few are excellent, and all are worthwhile.

Chapter 1

Applied RCM: An Overview

The problem with twin-engine planes is that they double your chance of engine failure.

-(Aviation anecdote)

RCM is a maintenance perspective in an operational context—understanding plant goals, needs, and equipment (e.g., how equipment serves, ages, and fails), and then developing a maintenance strategy to optimize outcomes in the context of your goals. It opens an operations-maintenance dialog. It recognizes the joint roles of operators in maintenance and maintenance in operations. When we understand and quantify roles in the maintenance process and understand maintenance limitations—where and when to involve design engineering—we get better in all ways.

These words are obvious but, sadly, operations and maintenance have been overlooked in recent industrial research in the U.S. It wasn't always this way.

Precursors to RCM

Early in the 20th century, Frederick Taylor, William Shewhart, W.E.

Deming, Joseph Juran, and other Americans developed models for improving production, management, and organizational theory. Western Electric, Ford, General Electric, United States Steel, and Westinghouse led the world as manufacturing powerhouses using these models. Americans prided themselves for producing the best products with the best methods. Understanding how things worked, inside and out, was taken for granted as American "know-how." Profound process knowledge—the foundation of American industry—was an accepted standard.

As skill levels and standards of living changed, the workforce changed as well. Production prowess became less important as the post-World War II era progressed. American goods were still in demand but inflation of the 70s was paced by cultural and social change. Capital availability dropped. The energy crisis developed. Global competitors arose. American industries were put on the defensive, their industrial hegemony ended. We still "had what it takes," but as the century progressed, it became time to rethink some processes, to sharpen the competitive edge. This was the backdrop for development of RCM.

Development of RCM

The term comes from the title of the work, *Reliability Centered Maintenance* by Stanley Nowlan and Howard Heap. It was published by United Airlines in conjunction with the U.S. Defense Department. It remains available through the National Technical Information Agency and the Department of Commerce.

RCM fills a void between reliability (R) engineering—focusing on the theory and mathematics of R—and the workplace, where maintaining production is key. Applied conscientiously over time, RCM provides production focus. While there are other tools (and no single one is perfect), and although tools and processes overlap in approaches—and adjunct tools include training, technology, and software—RCM is particularly suited to American culture and needs.

Consultants "sell" versions of RCM. At least 10 different software packages purport to allow users to "perform RCM." Two-to-four page magazine ads in maintenance periodicals promise to teach RCM in three days. (I wish these guys had been around when I took integral calculus. Perhaps I could have learned that in three days!) Some companies practice more than one version. If for no other reason than to engage small talk at industry conferences, it's useful to know what RCM is— and what is it not.

RCM has other names. PMO is one. "Common sense" is another. There are certainly competing versions of RCM, as well. An RCM process standard has been drafted. Questions outnumber answers:

- Are there fundamental RCM attributes? What are they?
- What key factors characterize a program, person, or company as RCM-based?
- What key factors demonstrate the degrees to which different programs achieve RCM?

The answers incorporate the best elements of the original aerospace maintenance and R developments of the 50s and 60s—failure analysis theory, work performance consistency, and quality. Learning these answers requires an intense awareness of maintenance by participation and benefits from experience implementing and improving maintenance programs. How well RCM elements are implemented into work practice determines the degree to which an organization embraces RCM. What work practices indicate that implementation has been achieved? They include:

- dynamic maintenance programs
- ongoing maintenance dialog among operations, maintenance, engineering, and support staffs
- awareness of operating and work strategies at the performer level
- active, effective maintenance with frequent design engineering interactions
- cost-performance information at the system and equipment level buttressed by failure statistics
- focus on improvement and improvement ideas
- continual cost reduction
- improved availability
- the ability to identify and eliminate low value work at all organizational levels

- personnel competence in areas of expertise; awareness of other's competencies (cross training)
- obsession with continuous assessment and interpretation of plant condition and health
- the ability to take organizational action based on observed performance trends
- questioning attitudes

RCM-minded organizations don't just operate plants—they improve plant operations. Before you conclude this objective is obvious, ask yourself:

- How many organizations truly focus on plant operations?
- How many "presume" operations will follow?
- How many support plant operations and production from all perspectives?

High-performing maintenance organizations share attributes with high performers identified in other fields.(Fig. 1-1)

Origin

"Maintenance" came into its own as a concept with the industrial revolution. Before that time, "machines" were designed, built, and maintained by their users. Watt, Edison, Westinghouse, the Wrights, Sikorsky, and a long list of other brilliant people conceptualized, developed, tested, and debugged their own designs. They had few peers, for design is the realm of sheer genius. "Design-build-operate" information exchange wasn't necessary-they were integrated in one and the same person.

The industrial revolution differentiated processes. Product users became separate from product makers. As production became dependent upon machines, specialties—operators and maintainers—emerged and evolved into different jobs. "Scientific work analysis" (espoused early in this century by Frederick Taylor) found that there were benefits in specialization. The assembly line—dedicating low-skilled workers to specific assembly tasks—took this position to an extreme degree. Operations diverged from maintenance. Managers didn't want opera-

Figure 1-1: Idarado Ball and Rod Mill, Pandora, Co. Informal on-the-job training, remote owners, and lack of operating strategy lead to sporadic operations, high costs, and eventual shut down. This plant employing 350 people kept the otherwise nondescript town of Telluride from becoming yet another western ghost town in the 1960's.

tors to think about maintenance—just do the job. New technology—and social developments such as unionization—enabled workforce differentiation. Electricians evolved into crafts, as had boilermakers, millwrights, and other trades. Engineering became a profession. Instrumentation sprouted as a craft, evolved to encompass controls, and then added software in modern distributed-control system (DCS) plants.

Specialization helped to create "maintenance" and other separate work groups early in this century. This has been part of our problem.

RCM

When Stan Nolan and Rowland Heap coined the term R-centered maintenance in their 1978 publication they summarized early jet engine R development by the commercial airline industry and the FAA. Ultimately, RCM was applied to jumbo airliners (beginning with the Boeing 747) to capture practical R lessons in a highly visible field. This work provides many of the concise RCM terms:

- condition-monitoring
- maintenance task
- hard-time
- logic tree analysis
- on-condition
- effectiveness
- age exploration
- failure-finding
- time-based

R studies in the late 50s were driven by the large lead the Soviets apparently held in missile technology. Spurred on by congressional funding, R studies in defense and aerospace took many paths. Spin-off benefits included development of:

- failure modes and effects analysis (FMEA) theory
- fault tree analysis (FTA)
- general failure analysis
- management system failure analysis and management oversight risk tree (MORT)

along with other R techniques.

Failure study provided lessons that were counter-intuitive. Large statistical populations were examined and preconceived notions were put aside. Some surprising results developed. One, for example, was the general "thumb rule" about wearout. (Wearout is the notion that every part has a finite lifetime, after which it will deteriorate quickly to failure. This in turn necessitates replacement.) Wearout couldn't be proven in many real applications because most products never reached the wearout stage in service. Yet, it's one of the most fundamental assumptions of any maintenance program (Fig. 1-2).

Using systems analysis, it was recognized that in complex designs, individual components and their functions could be replicated in ways that reduced or eliminated the consequence of their failure. In this way, individual component failures had little or no consequence by design.

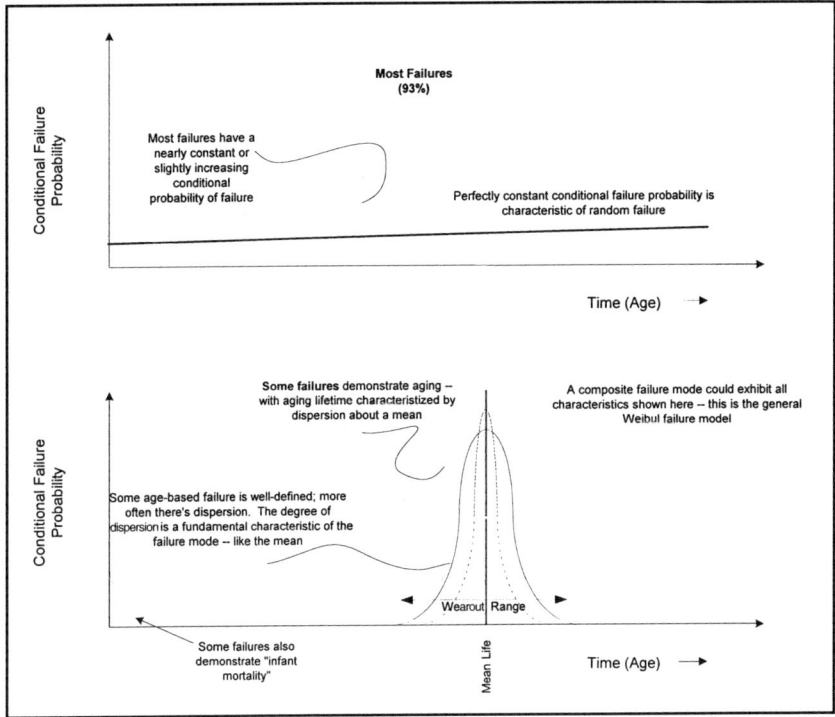

Figure 1-2: Failure Distributions

Systems failed—not individual components. Design could manage component failures. This shift in perspective to a system approach was profound. On the other hand, one couldn't assume anything about failures. Systematic study of equipment component failure modes and their roles in overall functionality was necessary. Failures that couldn't be controlled had to be addressed by redesign.

Cold War defense applications pushed design envelopes. There were mistakes—the Thresher sinking, the Apollo 7 fire, and the B-58 release problem—each representing significant design oversights. There were also many successes—the Boeing B-52, the Navy Polaris program, the F-4 Phantom, and the SR-71 "Blackbird." From the confusion and lessons of this period what we now call RCM evolved into a discernible strategy.

By the late 1950s, Americans—like people in other industrialized nations—had a cultural concept of "maintaining" things. From this we get the paradigm of preventive maintenance—PM. It's based on the notion that things shouldn't break—or wouldn't, if properly maintained—and when they did, "someone was at fault." With proper PMs, equipment would last indefinitely. PM went hand-in-hand with the concept of a TBM routine and complemented the maturing technology of the day.

Americans developed other significant concepts: preplanned obsolescence and operate to failure (OTF). Both play significant roles in the development and applications of RCM. It's said that Adolph Hitler once dismissed American industrial prowess, acknowledging only that "Americans build a good refrigerator." Intending to trivialize American efforts, he missed his mark. Building a good refrigerator was a very worthy endeavor—but the point is that the process can be transferred to other products (some of which—tanks, fighters, bombers, destroyers, and a host of other war products—helped to defeat him in World War II). It also suggests that a complex appliance—the refrigerator—had evolved to be so reliable, it required virtually no maintenance whatsoever! That a complex product could operate over its entire economic life with virtually no maintenance was an industrial milestone.

Preplanned obsolescence—replacement of a serviceable item by a technically superior product—had arrived. Rapid advances in production and technology and lowering of life-cycle costs led to products that could be replaced before the end of their functionally useful life—a uniquely American milestone. By the end of the 30s—on the eve of World War II—the stage was set for what would become RCM. Precepts included:

- PM-based maintenance strategies (TBM performance)
- OTF: products that required virtually no service
  for their entire useful lives
- technological obsolescence: products which would be retired
  prior to wearout failure
- increasingly complex products in industrial and private use
- a belief in technology and our collective ability to manage and
  control it

Post-World War II

Soon after the war, television added a whole new dimension to American life. Jet engines, rockets, and nuclear reactors were introduced. Designs were refined and matured. The steam locomotive—benefactor of a hundred years of evolution—was outflanked by submarine diesel engines that had been modified for locomotive use. Post-war production shifted to consumer products. By the 1950s, a new paradigm presented "American" products and technology as the best in the world, though new products provided new problems amid the technical advances. Technology growth led to a second "preventive" maintenance paradigm—predictive maintenance (PdM).

PdM suited the rapid advances in diagnostics and equipment taking place at the time. Using our insight into the mechanics of failures, we would be able to predict when things were going awry, and then head them off before they did. The Department of Defense applied PdM on F-105s, fast-attack submarines, and M-100 Abram tanks. Maintenance practitioners and managers embraced PdM applications such as vibration monitoring, oil sample analysis, multi-channel analyzers, and remote telemetered data. Regulators also saw the appeal in these philosophies—so much so, they sometimes mandated their use. Areas of vital public interest, such as nuclear power and air transportation, were early PdM proponents. Military procurement contracts specified PdM use. Industrial safety and environmental protection followed.

Over time, however, requirements became more prescriptive. Computerized maintenance management systems (CMMS) delivered information with ease; suddenly, organizations were buried in maintenance demands. More parties took interest in the maintenance process and had resources to pursue their interests. The vast resources of the federal government could be applied where the public interest was concerned. The PdM experience bogged down and stalled.

PdM acknowledged time-based PM but emphasized that you couldn't prevent all failures with TBM. You could do something nearly as good and possibly more useful, however—you could know when things were starting to fail. All you needed was the right diagnostic tools and the ability to interpret them. All it took was a little savvy and the right technology—and Americans had both! The model held great appeal. So much so, that thousands of predictive maintenance programs

9

were set up.

The development of the jet engine had shown shortcomings of the PM model. PM did not always work—time-based overhauls had actually made things worse in case studies. Objective reviews uncovered the not-so-obvious problem that maintenance did not always improve equipment performance. The problem was not the maintenance performers. Rather, it could be best explained in fundamental statistics—intrusive overhaul of previously satisfactory engines resulted in higher failure rates.

Enter Traditional RCM

Commercial jet engines in the 1950s posed a dilemma. Under the supervision of the FAA, and with competitive pressure plaguing airlines, jet engine R couldn't be guaranteed within the 1950s regulatory rulebook. The technology was on the forefront of a lengthy product learning cycle but low engine R had to be improved immediately. The FAA applied the accepted maintenance standards of the day and prescribed increased PM in the form of reduced hard-time overhauls. (It halved mandatory time-based engine overhauls, in fact.) Yet, statistically, many jet engine failures reflected a kind of infant mortality-with more frequent overhauls, engineers suspected total failures would increase, not decrease, as desired. A better means had to be found.

In 1959, R engineers offered a plan to systematically collect and analyze actual jet engine failure data (targeting the Wright R-2800 CA-15 and Pratt & Whitney JT-4 engines). R data they analyzed—Air Force statistical parts failure studies, as well as United Airlines parts usage records, the best available data at that time—suggested re-examination of existing failure data assumptions and interpretation. The idea of exploring aging and wearout performance of equipment in-service was born.

The embryonic space program was also studied for failure processes in its man/machine systems. Failure "modes," their effects, and processes evolved to the recognition that processes as well as products and systems failed. Root-cause failure analysis was developed and thrived. Failure evaluation, categorization, and oversight methods such as MORT were initiated. The data exhaustively collected by the United

Airlines and Air Force statistical failure studies supported new and fundamentally different interpretation of failures.

Advancing rapidly along several paths, failure pattern recognition developed into the identification and study of failure modes and their effects. Emphasis shifted from performing repairs—the historical focus of maintenance—to understanding the causes of failure. The assumption that maintenance was always effective was challenged. Systems theory and evaluation of the Pratt & Whitney JT-4 engine maintenance results laid the foundations of what has come to be called RCM.

Key aspects of the initial findings included:

- systems focus
- recognition of complexity as an important attribute in modern equipment
- failure classification by modes
- assessment of failure mode effects on systems
- numerical and statistical data evaluation of large equipment populations

Theoretical assessment of what maintenance can (and can't) do, the completeness of maintenance plans, and options for equipment assessment identified this new approach. Spanning a period of 15 years, these theories developed, through application, into RCM.

RCM brought together many loose maintenance ends under a common umbrella, covering the full maintenance spectrum to identify the range of possible solutions. It brought closure to maintenance theory. If we can identify the failure modes of interest to us—e.g., those whose costs we wish to impact—then, we can identify options. These procedures help us to generate a closed solution set around our options, or indicate that the equipment can't be operated within our criteria.

RCM provides a standard, common methodology for assessing, ranking and evaluating any maintenance environment. It encompasses previous methods and then enlarges the spectrum to include testing. RCM extends maintenance processes by providing a standard method for the development and application of any maintenance program with certain, objective results. RCM provides the structural glue that holds together the three professions of operations, maintenance, and engineering (Fig. 1-3).

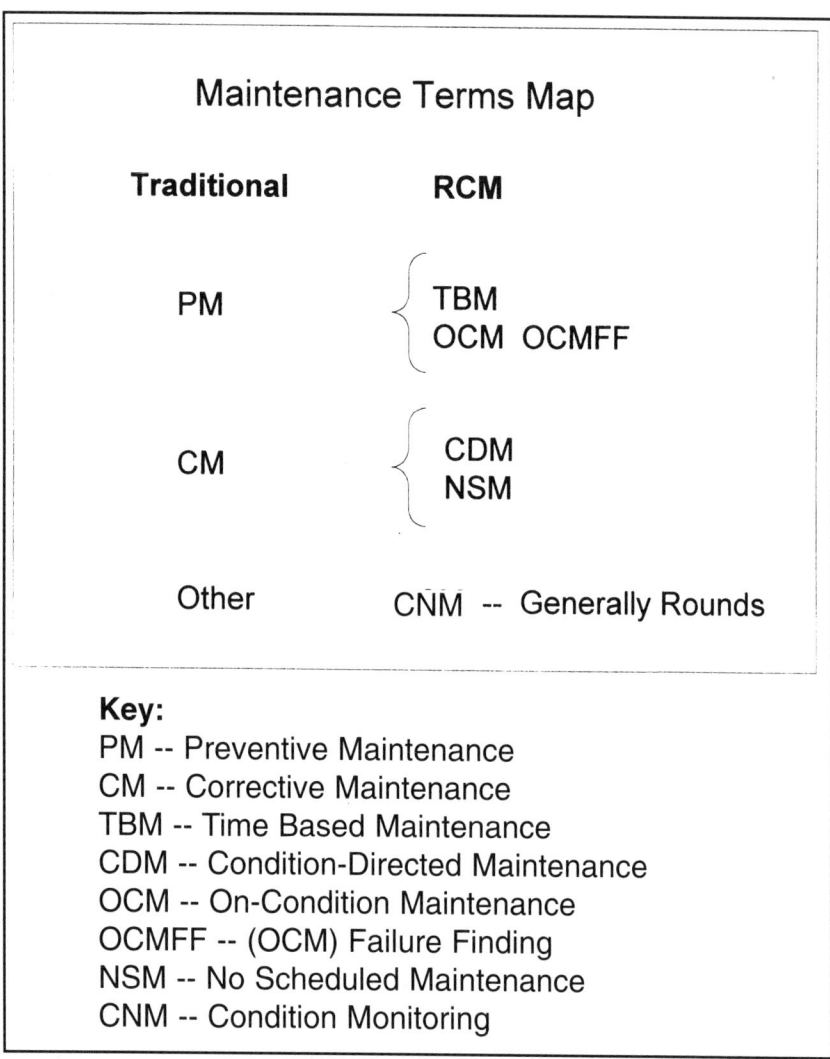

Figure 1-3: Maintenance Terms Map

Applied RCM

Because it evolved in a highly regulated environment, traditional RCM (TRCM) includes a rigorous task selection methodology that follows detailed flow paths needed to document decision-making. This is summarized in a "logic tree analysis (LTA)" process that works down a systematic hierarchy to classify component decision logic.

Simplifying and summarizing the common results of this process is reflected in ARCM.

RCM benefits come from applications. ARCM accelerates the tedious and laborious traditional process to provide immediate applications. If the key RCM points include:

- strategic mission-oriented thinking
- systems equipment approach
- function understanding
- technology assessment
- fact-based decision processes
- failure understanding (especially root causes)
- statistical failure analysis
- profound process understanding
- continuous improvement
- completeness
- functional failure focus
- risk management orientation
- benefit/cost (B/C) consideration
- failure modes identification, classification, and study

then an ARCM perspective simplifies maintenance processes, supports standardization, and identifies general strategies. It suggests where and how to focus improvement efforts. It can also be as problematic as traditional RCM.

ARCM requires that information be applied to basic processes. ARCM identifies the "what"—the work required. Maintenance work processes provide the "how" to accomplish the "what." So, ARCM requires a two-part process:

- identifying the best solutions, using information, and simplified RCM analysis
- implementing the results

While RCM doesn't directly tell us how to improve actual maintenance performance, it will tell us whether an improvement effort was effective or not. High infant mortality rates point towards better maintenance practices for improvement efforts; high random failures, on the other hand, point to operations processes. The numbers suggest where to selectively concentrate.

ARCM helps develop a general philosophy of how operations and maintenance best work together. It prevents functional losses by managing failures. It leads to process improvement. The overall objective is meeting mission goals—usually in the form of costs, safety, and risk—so the benefits of ARCM applications extend beyond the facility owners to the operating and maintenance staffs, the community, and even to the general public.

The "R" in RCM

Reliability defined

Mathematically defined, reliability (R) is a conditional probability—the ratio of acceptable outcomes to total trials. More exactly, R is the probability that components, equipment, and systems will perform their design functions without failure. It's based upon:

- definition of successful outcome(s)
- mission (including intended use and environment)
- period of interest
- conditions at onset of the mission period

While it's not the purpose of this book to develop R theory, we need to understand basic R concepts to appreciate the R in RCM. Intuitively, we should have some benchmark R numbers in mind when we look at any equipment. For example, is a feedpump R of 0.99995 satisfactory? In what context? How about overall feedwater system R? Two 50% pump combinations? Three? What are the benchmark comparison standards? How can we relate these numbers to conditions that utility managers more closely follow, such as equivalent availability, capacity, and cost?

Basic R is noticeable mainly in its absence. Reliable products, equipment, and plants establish benchmark comparisons that contrast sharply with under-performing competitors. R has value, though developing supporting analysis is complex. For instance:

R = 1 - Unreliability

Often, unreliability is provided. In those cases, R is given by the following:

0.99995 = 1 - 0.00005

The R—0.99995—can be viewed as a 99.995% probability of a successful event outcome. Let's briefly look at R in more depth.

R engineering

R engineering applies R theory to solve engineering problems. This is done by projecting a system's overall R and applying engineering methods to assure those goals are achieved. When R is allocated among constituent components, successful "mission completion" can be established for new designs with relative confidence. For existing facilities, sources of unreliability can be identified and traced back to causes— design, operation, maintenance, or a combination thereof.

Unlike military R applications that focus on individual mission events, power plants look at operating periods. These could be:

• periods between scheduled outages
• calendar periods
• budget periods
• peak production periods

"Scheduled outages" vary greatly from application to application with many possible issues of R engineering in play. Major scheduled outages occur on an interval of 12 months (or longer); a general benchmark is 18 months. Most boilers and nuclear reactors adhere to this interval for major inspections and rework. Special outages run on longer intervals—turbines on 5-to-12 year intervals, for instance. In today's economic climate, operators push design envelopes to extend outage periods. Some common outage intervals are listed in Table 1-1.

Plant	Outage	Interval
Fossil Boiler	Boiler Inspection	18 month
Combustion Turbine	Combustor Inspection	12 month
Hydro	Intake Inspection	Spring
Turbine	Stage Inspection	5 year
Nuclear	Refueling	18 month

Table 1-1: Common Outage Intervals

Not surprisingly, a great deal of R assessment data comes to the power industry from aerospace and military applications. In those cases, R theory was used to assess small production runs, low volumes, single-use components, and systems that usually involved specific, one-of-a-kind missions. Analysis was mainly government funded. By contrast, R studies funded in the private sector focus on high value, high volume products (such as computers and peripherals) supporting hardware, software, and telecommunications support. Devices may function in hundreds of thousands of operations daily with a high cost for failures. R assessment and "benchmarking" are less common in the power industry today as deregulation focuses everyone on lower production costs, and fewer companies willingly share their information.

Overall, R engineering is usually applied to high value product manufacturing, and so most R work has supported weapons, space, computers, and commercial air transport applications, where R in a product can be apportioned with a "R budget" allocated among constituent parts. Traditional heavy manufacturing is less frequently evaluated in R terms, where products have typically been in productions for tens of years. "Generation" is not traditionally thought of in manufacturing terms—and it's not rocket science, even though R concepts apply.

Within power generation, R assessment has not penetrated non-nuclear generation to a significant degree. With tight project budgets, it's easy to cut back on capital investments that have little or no short-term payback, even if they could affect "mean-time-to-repair" for essential equipment. Utilities often establish a project budget to manage costs without considering long-term operational consequences. This provides an opportunity to apply R engineering.

That being said, in commercial generation designs, R engineering uses general "thumb rules"—standards and guidelines—to achieve client contract goals. R is built upon incremental advances in production methods and facilities, standardized redundancy, layout planning, and common design packages. Designers use experience and similar designs to project plant R. Probability risk assessments are reserved for nuclear plants, where special requirements (such as the NRC's maintenance rule) override simple economics.

There are two ways to evaluate R—*a priori* (before the facts) and *a posteriori* (afterwards). Production R engineering looks at a facility's a posteriori performance, examining sources of unreliability and their causes. By allocating unreliability downward to systems, equipment, and components, engineers identify those areas with the greatest opportunity for improvement. They can then allocate resources to where they will do the most good.

A priori calculations require the use of probability theory and assumptions. This can be illustrated by tossing a coin 1,000 times. If you get 493 "heads," the *a posteriori* probability of a "head" is 0.493 or 49.3%. Probability theory tells us that for the toss of a fair coin, the *a priori* probability of a head is 50%, exactly. (Strictly speaking, the mean value probability approaches 0.50 after many tosses.)

The key assumption is that we have a fair coin. Overlooking or failing to appreciate such a simple, common assumption in a real-world problem can be painful to the owner of a manufacturing plant stuck with a costly retrofit, significantly different production costs, or both. When they assess facility R projections, owners must carefully evaluate how they were developed—their basis. Numerical results providing R— whether casual R estimates or formal failure modes and effects criticality assessment (FMECA)—are rarely provided with designs. In their absence, the owner must rely on:

• the reputation of the architectural/engineering firm
• design proposal experience
• innovative context of the design

A reputable designer with a proven track record, an existing design, and incremental enhancements supports low risk. Sometimes these factors can't all be met. Occasionally none of them can. In these cases, R analysis can even help quantify and evaluate risk.

R theory doesn't specifically tie R outcomes to unreliability sources. R engineering can measure system R, relate that to subsystems and components, and reveal which individual reliabilities are needed to achieve a given overall R. It can project R based upon supporting processes, systems, and components. It won't improve basic processes that determine overall R and it can't assure that operations meet assumptions for availability of standby or backup systems. (This issue led, in part, to the NRC's maintenance rule. If a site license assumes, for safety calculations, that a standby system is available 99.5 % of the time, then that level of readiness should be maintained in the operating plan!)

Process R

TQM and statistical process control (SPC) address production process R. Each process has different inherent design capabilities. This concept of process capability has been thoroughly developed by manufacturing process engineers and statisticians. In addition, Deming, Stewhart, Juran, Gryna and others provided many insights into the statistical basis for production process improvement. While some companies are very capable at improving production processes-most find it a struggle.

Yet, a goal of this book is to provide tools for generation engineers seeking to improve plant process R to support higher unit, plant, and system R goals. Like a body-builder developing muscle mass, however, building intrinsic process R is a laboriously slow process.

Initially, there's lots of training and other investment with no immediate payback. It takes time to generate results and earnings. Fast-track methods can provide quicker paybacks but once advocates and supporters of a process improvement project move on, it's often back to business as usual. Achieving fundamental change requires sustained commitment.

A first step is the basic measurement process. R engineering builds on the basic theory that identifies and quantifies the benefit of a two-pump versus three-pump (one-redundant) feedwater supply system—basic configurations, series and parallel, mathematical models, and where they are best suited. Operators should understand the relative cost benefit of a two versus three pump configuration. Theory tells us whether our strategies fit with the designer's intent and how likely we are to be successful. An out-of-service, functionally abandoned spare pump fundamentally changes the designer's intent. Plant staff may not appreciate this—even in an obvious case, such as this one.

A critical bearing-water system serviced large reactor cooling compressors (circulators), in an advanced nuclear plant. A spare bearing-water pump was intended to be always on standby, ready to run. In virtually every bearing-water pump trip, the standby pump either kicked in immediately or a plant shutdown quickly followed. If bearing-water was injected into the circulated gas coolant, shutdowns were lengthy. Yet, maintenance crews worked on these pumps while they were on-line "since operations could do without them." Every outage, it seemed, had more high-priority work, more pressing problems than pump maintenance. Practically, then, the pumps were out-of-service on-line. Many times, a demand for the out-of-service "standby" pump failed and water went into the gas coolant. Armed plant trips detected water in the coolant gas and brought the unit down.

This is a classic example of what can happen when maintenance is staffed initially from another plant. Personnel's appreciation of the importance of the bearing-water pumps was fundamentally out of sync with the plant's needs. They could have changed the prevailing maintenance and operations culture. The absence of a working dialog between engineering and maintenance crippled the plant.

The lesson is clear: R theory should drive maintenance—not company practices nor plant maintenance preferences. In practice, company and even industry culture are powerful change impediments.

Implementation

Implementing RCM process results is tough, and the reasons why provide insight into generation production challenges. Some are simple, others complex, but RCM analysis without implementation has no value.

An organization considering ARCM should first examine its basic work management processes, to uncover implicit "processes" that may be understood but not well defined. Managers and other parties to existing processes may not know how work is actually performed; those who perceive that potential gains would cause them to lose in any way could block RCM applications.

Maintenance has traditionally been a "craft" process whose workers usually have had great latitude to work flexibly, using their own methods, standards, and pace. For this reason, maintenance culture and practices should be reviewed for RCM alignment. Some organizational features align more naturally with RCM processes than others. They must be discussed and emphasized to support the RCM effort and so avoid later implementation pitfalls.

Many organizations find maintenance process commitments substantial, and (understandably) are reluctant to take them on. However, once an RCM-based maintenance paradigm takes hold, RCM thinking can provide compound returns. Simplified projects can achieve RCM benefits quickly—even within the budget year. The discipline can greatly focus efforts. This offers the added benefit of demonstrating change success. If value can be demonstrated, most organizations have powerful incentives to improve.

As the pace of industry deregulation and reorganization continues around competitive structures, companies will have to invest in maintenance infrastructure to remain competitive. New emphasis will be placed on R and process improvement offered through RCM. Operations, maintenance, and engineering support will all benefit as companies discover this fertile area of improvement.

Value added

If the megawatts aren't available when the buyer demands them, then no sale occurs. Availability creates the sales opportunity, since electricity "storage" is limited. Producing at a cost lower than a competitor assures sales in a competitive market. This means that availability is a significant "value-adder" and R is the practical indicator of availability performance.

Cost is another factor. A plant with a high product cost—even though its product is available—will not be as attractive in an "economic dispatch" model. Because total generation costs include operating and maintenance production costs, plants that turn to RCM favorably influence availability and cost to benefit their bottom lines.

Maintenance strategy

Nine times out of ten, operators initiate maintenance but their operating success rests on the maintenance product delivered. Maintenance can't correct fundamental design problems. In the past, only combined group efforts identified design as the fundamental problem and eliminated maintenance as a solution. RCM enables us to identify misapplied maintenance quickly—facilitating designer involvement more effectively. This helps maintenance focus on things it can correct and stick to a winning strategy.

"Design change (DC) maintenance"—design changes initiated where a maintenance solution is available—is expensive. Maintenance organizations request and implement design changes either for non-problems or for problems that have simple maintenance solutions, because:

- workers didn't understand the maintenance needs of the equipment
- problems were patently maintenance issues but persistent root-cause analysis didn't occur before initiating the DC support request cost/benefit routine process
- organizations fail to recognize the true cost of modifications

Great savings are realized when maintenance solutions are recognized, design change requests canceled, and workers roll up their sleeves—when maintenance strategy is followed, in other words!

Daily maintenance strategies should be explicit. Reviewing implicit ones often unveils substantial opportunities—the potential for more online maintenance, the elimination of ineffective maintenance, the opportunity to extend service intervals. Strategy development should include maintenance-planning managers and the rank-and-file workers, who have great insights on equipment performance in-service but who don't always understand how to change and improve maintenance decision-making.

For fossil generators, an absence of plans can mean that work is performed because equipment is available—not because it's needed. For nuclear generators, complex, multiple path maintenance approaches, extreme conservatism, and overbearing regulation cause productivity losses. Each environment can learn from the other.

When companies find cost management so important, a strategy can provide a clear road map of what is possible, and paths that must be taken.

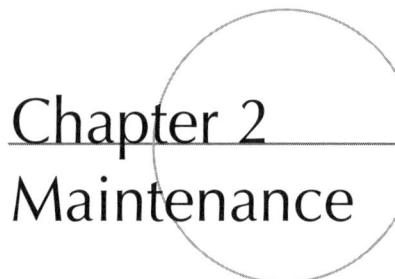

Chapter 2
Maintenance

How come dumb stuff seems so smart when you're doing it?

-Dennis the Menace

If it ain't broke, don't fix it.

-Farmer's adage

Maintenance practiced in North American power plants is crisis-oriented. Crisis—the day-to-day emergence of random events and directive management—is what the American industrial culture seems to need to manage on a daily basis. Crises energize companies. When business orientation is towards crisis, crisis is inevitable and structural.

Yet, lack of preparation for predictable events is what provides a crisis orientation. While work will always be a dynamic environment, proactive maintenance—as can be derived from a failure managent—based strategy—can manage or remove a great deal of stress for everyone. Isn't that a better way to operate?

The point is a simple one : O&M are harder to perform under adverse circumstances—especially when some plant managers wear cri-

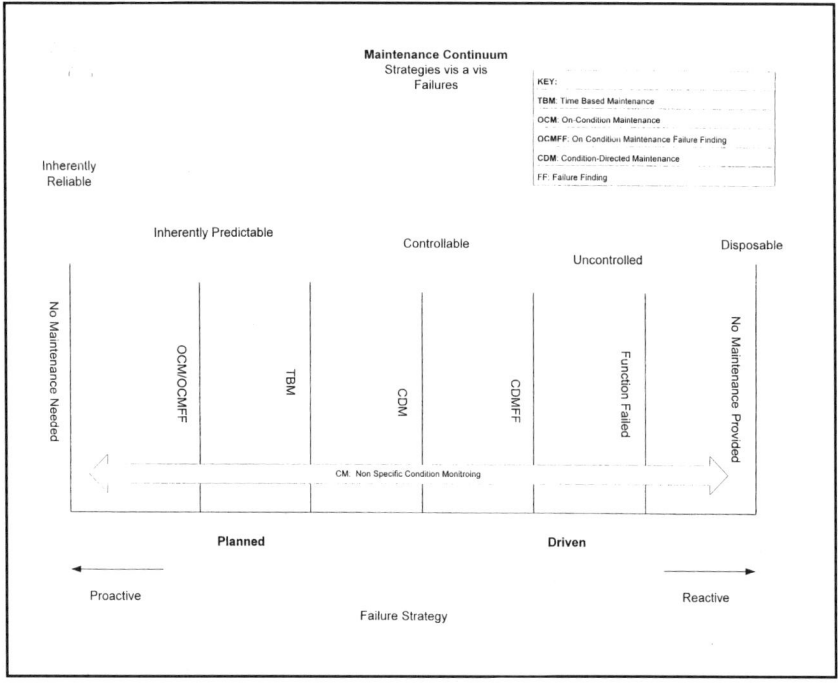

Figure 2-1: Maintenance Continuum

sis events like purple hearts and some corporate cultures reward those who promote and manage crisis, rather than stable productive workplaces. I advocate stable, predictable operations. We need to get the job done, minimizing crisis responses! And everyone needs to go home at the end of the day.

Maintenance Options

On a continuum, maintenance varies from purely reactive—failure response—to purely preventive—time-based. (Fig 2-1) Looking at maintenance across such a spectrum, there's less tendency to view any particular maintenance approach as either "good" or "bad." They're just approaches.

I don't come to this discussion totally unbiased—I believe in planned maintenance. Competence stems from knowing which method is most effective, and when. Even response-based maintenance can be planned! Different equipment with different design capabilities optimizes costs at different points on the maintenance spectrum. By the maintenance strategy we develop, we can choose where to place equipment on the spectrum based on overall risk, cost, and operational objectives. Equipment falls naturally into niches. To optimally place equipment on the maintenance spectrum requires an understanding of the equipment, its context, and the operating organization's culture, risk tolerance, and goals. Is any one place better to be than another? That depends on your operating goals—or perhaps your regulator's.

With design evolution moving steadily towards complexity, redundancies are incorporated into most designs. These provide opportunities to lower costs. Maintenance strategy(s) can use redundancy features differently to lower costs while maintaining functionality.

Environment also influences component strategies. Lubrication requirements in a dusty, dirty, or wet environment vary from those in a clean, dry one, given the same level of equipment sealing. A "dirty" environment requires frequent lubrication to purge contaminants. This function is unnecessary or greatly diminished in a clean environment.

Constituent component capability and "quality" levels also influence how quickly a component ages. For example, a high-quality lubricant with superior base stock and additives outlasts a simple mineral oil. High-quality electrical insulating materials outlive simple, inexpensive ones. Constituent material variations influence where a functional item—oil, cable, or other material—falls on the failure spectrum. Understanding the capability of constituents, and how they influence overall part capability, influence maintenance strategy. It may be more cost-effective to use high-quality lubricating oils that possess predictable aging characteristics than to condition-monitor frequently for degradation.

We usually associate low quality with product unpredictability. High-quality components have longer mean time between failures (MTBF), a more predictable "lifetime," or (typically) both. Random failure requires more efforts to:

- mitigate the failure (e.g., introduce redundancy)
- detect and correct the failures that ultimately will occur

• suffer the operational losses when uncontrollable random failure occurs

The uncertainty of randomness in failure inherently raises costs! This increases initial product cost, operating cost, or both. Ultimately, life cycle costs are higher. Predictable failure mechanisms add value. Quality manufacturers know this, and strive to deliver predictable products—predictable operations, predictable failures.

Manufacturers' engineering staffs develop "predictable" products that command higher

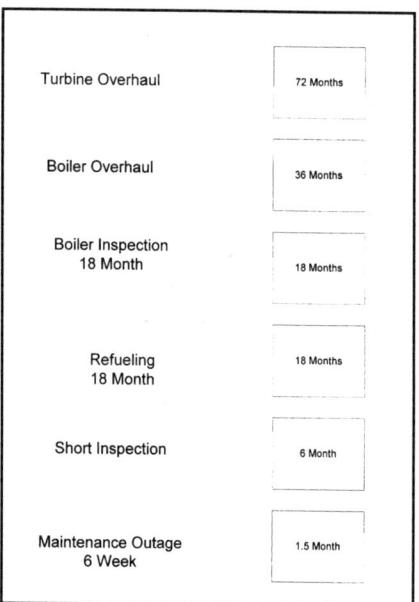

Figure 2-2: Practical PM Alignment

returns. The more reliable the product—the lower its cost—the greater the demand. When quality is perceived in a reliable product, that product commands higher prices that an informed purchaser—the guy maintaining the equipment—is willing to pay. However, buyers indirectly related to the work, who can't see or measure the costs of low quality, may purchase on cost, not value. Most craft workers are painfully aware of the cost of quality and since they have reworked many jobs due to faulty parts.

The idea is to extend maintenance intervals through the use of superior parts, engineering, and processes. (Fig. 2-2) The old saw that "all parts are created equal" is all wet. Anyone in business eventually learns that "parts are not created equal." If analysis shows that lower quality substitution is adequate—buy 'em. Until you have this profound piece of knowledge, be careful! The unknown substitute probably isn't a bargain. For a PM program:

• use long-lived parts
• align PMs
• lasso the "cowboys"

Establish a process with rules and then ensure that everyone (even cowboys) plays by the rules. There's almost never a good reason to shorten a PM interval just to get a price break on parts. The opposite should occur: if you become aware of a premium part, analyze its cost/benefit; if you find it's cost-effective to use it, buy it. Only extend the service-life interval based on the better part.

Doing part-lifetime analysis work is not trivial. Unfortunately, many people think that it is, which is why there's such a large market for low-quality parts. In a more rigorous, informed cost environment, many cost-based part suppliers couldn't survive.

When you've analyzed, compared, and tested your components, you're ready to build them into sub-assemblies, skids, and systems. The overall integration determines the failures that ultimately cause overall functional failure. Two things can happen—equipment, with a life-limiting part, can fail. It can also last indefinitely, with internal failures, while preserving function. (This is the complexity principle.) If there is a predominant age-based failure, it establishes an aging and failure profile. The composite of all component failure modes over the expected life of the equipment or assembly, and their redundancy in design, establishes the overall composite failure characteristic and behavior. This locates the equipment on a failure spectrum (Fig. 2-3).

Thus, the failure spectrum enables you to consider alternative strategies and how the maintenance strategy must change when components change. Ideally, only those changes that increase product lifetimes would occur but, unfortunately, low-quality parts and/or services compromise lifetimes with the opposite effect. In some cases the systematic downgrade of constituent parts leads to equipment capability loss—that is, as equipment becomes less capable, useful, and maintainable, utility decreases and aging accelerates.

Between major replacement programs, our ability to maintain equipment drops as many small problems gradually sap the overall equipment utility, its capability to be maintained, and its operating margins. When the operating margin is gone, failures occur. Taken together, all across a plant, they raise overall costs. This is why planned maintenance programs can maintain nearly complete performance capacity.

But, how? A policy of conscious age-exploration and learning—a

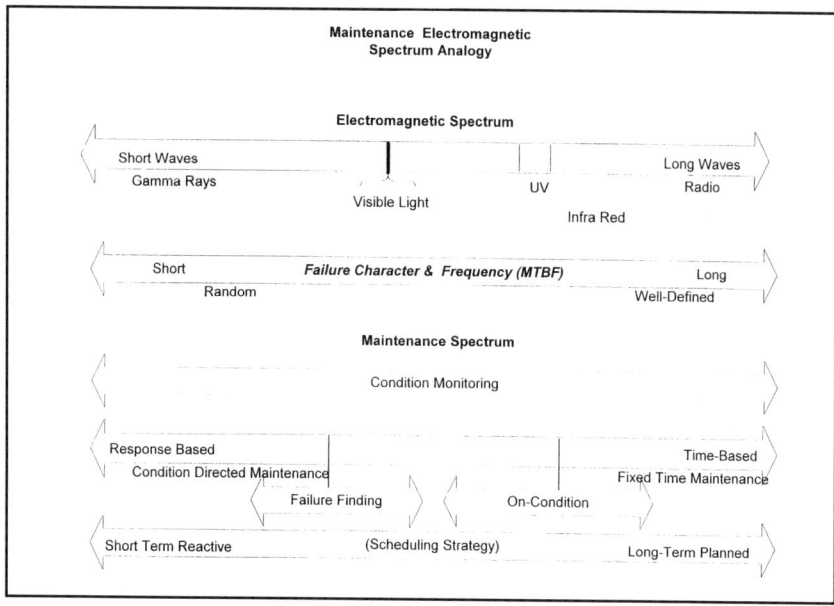

Figure 2-3: Maintenance Electromagnetic Spectrum Analogy

continuous-improvement environment—can help assure that performance loss doesn't happen. Equipment condition deteriorates from ignorance, not conscious abuse. Some "aging" can be defined more accurately as "the systematic extraction of capital value from equipment by compromising the original equipment's specifications through lower quality substitutions." A skilled, committed workforce can maintain capital equipment capabilities to nearly original specifications and control the rate of deterioration. And because original specifications include design margins, motivated plant personnel can maintain margins over part life.

The failure spectrum explains why composite operator diagnostic skills are needed, and why operators are useful in maintenance. The natural differentiation between operators and maintenance workers comes from the failures each addresses. If you have:

- low MTBF
- random failures

then only operator monitoring will be effective. The more the maintenance strategy is oriented towards "no scheduled maintenance" (NSM), the more dependent the strategy is on operators to identify failing equipment. The best operators are literally integrated into the man-machine process. Experienced, skilled operators can compensate for most failures and identify developing problems through CNM. They require little guidance. A facility with such operators who make well-designed rounds in a well-designed plant with a responsive maintenance process is an ideal plant—functional failures are rare to non-existent! Many plants meet this implemented ARCM definition today.

The failure spectrum suggests that to be effective we must really manage risk. The effectiveness depends on plant design, combined R factors and redundancies, and how equipment is operated. The key to managing risk is education. We must master knowledge of:

- design failure "capacity"
- maintenance support
- monitoring level
- maintenance response timeliness

We can place ourselves anywhere on the risk management spectrum. Favorable outcomes—including low costs—don't result by accident. An uninformed acceptance of a maintenance strategy doesn't optimize overall performance. Consistency does.

Consistency

Fossil generating-station maintenance processes and strategies are implicit; nuclear plant processes are defined (though nuclear plant processes are functionally similar to fossil). For effective RCM applications, information exchange must occur on several levels, no matter what kind of plant is involved. These processes are unique to maintenance optimization, continuous cost reduction, and performance improvement but are not routine for many reasons.

Corporate cost information is often unavailable or inaccurate. Many utilities are just learning cost management. Predictable costs are a challenge to achieve. Spontaneous, undocumented problem solving may be commonplace, and standards absent. Predictability does not happen by chance. Random approaches lead to higher costs. Developing consistent information processes that support cost management requires practical experience and time.

Consistent, integrated measurement processes provide the necessary feedback information that enables staffs to tune the maintenance plan. Available generation-oriented CMMS software and user-friendly feedback mechanisms, accessible to any employee, provide effective feedback. Twenty years ago, engineers had secretaries to type reports; today they have PCs and word processors. They file their own reports—faster! The same transition can be projected for mainframe CMMS software systems.

Consistency can be achieved! Several elements are involved.

Statistics

Informality—the lack of a maintenance strategy that is universally understood and applied—introduces random factors into work performance. This dilutes planned maintenance effectiveness and increases the frequency and dispersion of failures. Maintenance plans must address equipment to control failures, but this has been difficult to do, except at the worker level.

Statistics tell us that around 85% of the tasks in a typical large generating facility are CNM and CDM, a large fraction of which should (or needs to) be implemented by operators. Because the monitoring interval for operator tasks is short—hours to days—what does this say for plants that are characterized by:

- informal (or no) operating procedures?
- informal (or informally developed) rounds?
- informal (or no) current system operating guides?

This traditional electric generation scenario helps explain why random failure rates—in spite of mature plant designs with years of design and operation development—are high.

While each unique design requires a distinct interpretation, simple

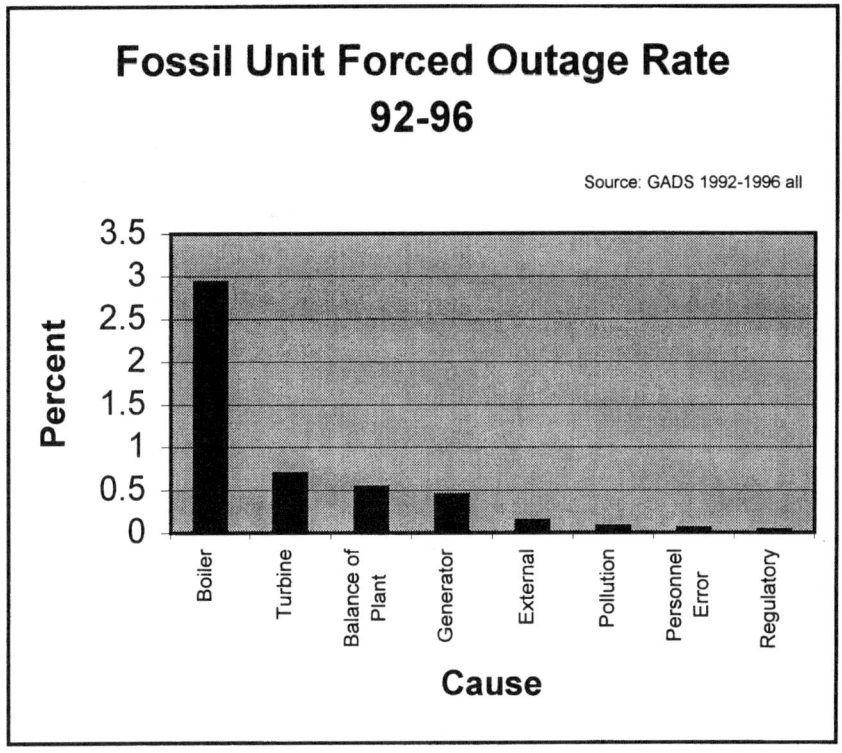

Figures 2-4: Fossil Unit Forced Outage Rate (from NERC GADS)

procedures could help to assure consistent, timely performance of plant operation tasks in fossil units. This statement is based on North American Electric Reliability Council (NERC)-reported planned and forced outage rates of 1.26 and 8.73, respectively, for all fossil units for 1996-1998. (Contrast these with figures of 1.23 and 2.96 for nuclear units.) Both plant types plan for the same level of outages. Forced outages at fossil units are just comparatively high (Fig. 2-4).

Maintenance assumptions, revisited

Applicability. Maintenance effectiveness based on periodic maintenance cannot be assumed. Complex electronic systems developed during and after World War II, the advent of complex mechanical systems (such as jet turbines) and statistical life-failure studies first suggested that PM activity effectiveness varied.

For example: The Weibull model mathematically characterized

31

infant mortality—but applications to electron tube aging quantified it. (Infant mortality—derived from mortality studies of human populations—was a general attribute of new "unburned" electronic equipment. Burn-in significantly reduced early life failures.)

This lesson formally challenged the prevailing notion that PM was automatically effective. Weibull generalized the mathematical model to be useful to test proposed PM activity. Activity must be technically effective to be considered valid PM. Since time-based jet engine overhauls actually increased in-service failures, technology should be carefully scrutinized with the applicability test.

New PM must be successful on two levels to be considered appropriate. First, is the activity technically appropriate? That is; does it really achieve its stated, intended failure-prevention purpose? We assume here that we are evaluating state-of-the-art application of the technology and that the analysis is performed by trained, qualified craft in a production environment with production equipment—not under lab conditions. "Does it actually work?" is what we're trying to answer.

Once this is assured, we can go on to the next, broader level—"is the intended work cost-effective in the production environment?" In practice, there are cut-off points where PM is no longer cost-effective—where equipment requires NSM. Formally tagging an equipment maintenance task or plan as NSM places it on a CNM basis and forces any new PM task to pass the cost criterion to be applied.

Testing applicability and cost-effectiveness—objectively, statistically—is demanding work. For this reason, few do it. They use other, less rigorous techniques—or gut feelings! However, validating PM effectiveness in the field is an ongoing chore that has high payback. The key concept is to assure and validate that any PM task meets applicability and effectiveness criteria.

Effectiveness. Technical applicability is necessary for any proposed PM. But does it make money? Is it cost-effective to do? Cost-effectiveness is a higher hurdle than applicability for new proposed PM tasks.

Before this test was generally applied, utilities routinely purchased the latest test equipment as it came along and trained a person who then became the application's advocate. The person promoted its use widely often ignoring cost-effectiveness. As a result, programs built around technology exploded. VM, non destructive examination (NDE), oil analysis, acoustics, leak detection, and other test specialties were developed extensively for many components and problems. How can sampling a 4 pint bearing sump be effective? The time involved to pull a sample matches what it would take to replace the entire contents. What's the point in performing VM on a 10 HP motor with no (historical or conceivable) impact on plant availability? Unfortunately, these techniques have been used in applications where there could never be a significant return—or any return at all. Reviewed with an ARCM effectiveness test, many of these tests simply cannot pass the cost-effectiveness criteria.

Controlling PM program scope by ruthless application of the effectiveness test is essential for program credibility.

"Add a PM." Managers often respond to failures with the cry, "do a PM." This is especially true in response to regulatory pressure. PMs provide a tool to keep an inquisitive inspector at bay. The assumption is that it's quick, simple, demonstrates action, and doesn't cost anything—right? Wrong. Time-based PMs—any single, simple PM performed through a typical maintenance organization—are expensive. In reviewing bloated programs over the years, I can only conclude that many of them had an insurance or regulatory origin, or originated with management. In many "management-derived" PMs objectives are vague, task-failure correspondence missing, crew input is absent, and planning or understanding is missing. When these gaps are formally addressed, there's concurrence that such PMs reflect "no value added"—and are dropped—or a much simpler solution is found (operator monitoring, redesign, or NSM.)

Strict adherence to failure analysis and R engineering assures applicable, effective PMs. As the primary source of operational R, what goes into the routine maintenance system must be managed with great care—much like a checkbook. A program of "controlled PMs" is superior to an informally managed program in which anyone can originate one.

The right PMs, performed consistently, add R and availability, and they lower operations cost. A credible PM system can be a prime contribution made by a conscientious plant engineering group.

Statistics and regulators. Regulators and insurers (in my experi-

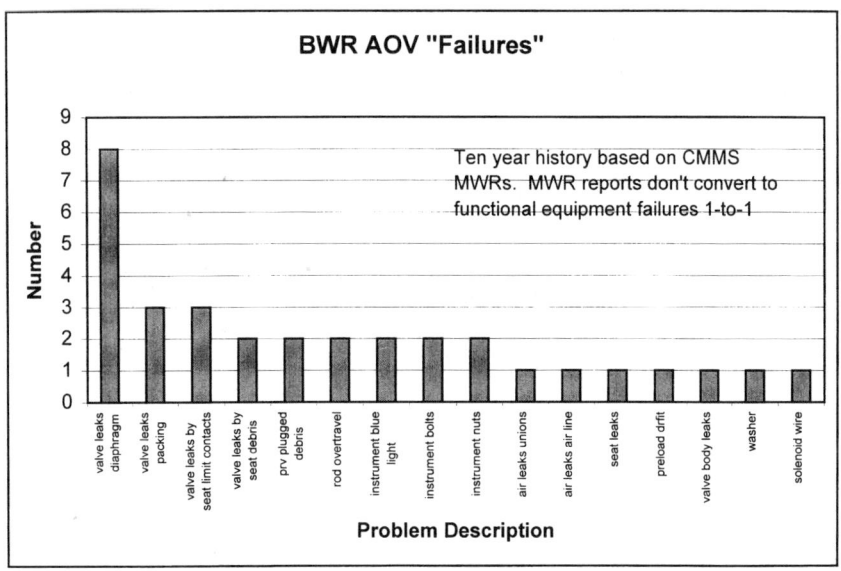

Figures 2-5: Boiling Water Reactor Air Operator Valve "Failures"

ence) ignore statistics, concentrating instead on rare, improbable, yet significant events. They track these with zeal. True risk-based regulation considers the probability of occurrence before initiating expensive corrective actions. This benefits the public interest when it's applied in the commercial aviation industry, but I don't expect to see it in electric power generation. If anything, Environmental Protection Agency (EPA)/environmental and Occupational Safety and Health Administration (OSHA)/workplace focus on fossil-fuel generation appears to be destined for more intervention.

RCM is an effective strategy to counter these trends. Truly managing failures—and having the statistics and programs to prove it—will not only keep regulators more accountable, but will ultimately improve regulation and enforcement. (Fig. 2-5) In many incidents in which regulatory action was threatened or carried out, in my estimation, regulators were justified. In too many of these cases, as I've said, PM was a palliative substitute designed to keep inspectors at bay.

RCM studies suggest that regulators consider levels of redundancy and backup before issuing citations. For example, in the nuclear industry, violations can be issued for non-compliance on vague, ill-defined, contentious, or indistinct secondary issues where no actual failures (of people or physical hardware) can be identified. Violations are based on what could have happened, or for breakdowns in management and support processes. Nuclear requirements are so exceedingly complex and numerous, that 100% compliance is unrealistic. Many of the regulations are subject to interpretation. Worse, an inspector makes his/her mark based on citations. My opinion is that minor failures that leave the functional redundancy of support systems intact, are acceptable—unless "for cause" or other compliance issues arise. Because nuclear power plant designs today are mature with few new technical issues (plant aging excepted), hypothetical failures should not be the issue. Such secondary guesswork represents an unproductive use of engineering resources. The selective application of RCM-based statistics can clarify cases.

In fossil plants, a three-barrier, defense-in-depth standard works well. Design is an inherent barrier. Finding anything other than code-specified materials and welds on high-pressure piping requires immediate correction including shutdown. The basis is that a fundamental design assumption has been broken, putting a real hazard in place.

The second barrier is the redundancy for the primary failure cause (usually instrumentation but perhaps other operating limits). Consider the case of catastrophic blade separation failure in a high-speed fan due to imbalance. Based on experience, it occurs with a warning period—if you monitor vibration. If monitoring occurs, failure will be detected, even if we break the first barrier. To assure that monitoring is in place (with necessary trips) means we must have an instrument PM program. If an operator is expected to initiate the trip (instead of an armed, programmed trip on excessive vibration), monitoring of operator performance is also needed.

The third barrier is general O&M equipment monitoring. Though non-specific, most practical failures have predecessors, be they alarms, warnings, limits, noises, vibrations, and even smells. An efficient monitoring program conducted by motivated and skilled operators effectively identifies evidence of failure. Combine them and you have defense in depth.

Application of these three barriers in fossil generation has been effective in selecting appropriate CNM. By taking appropriate actions while following the generally accepted environmental standards, we placate regulators and plant managers. Taken together, these three barriers effectively control risk posed by 99% of failures. At plants that use all three measures effectively, significant events—accidents, unit trips, and major equipment losses—are very rare.

Instrumentation has been a sticky issue at some fossil units. Fossil plants typically don't use armed trips when options to monitor high turbine vibration levels, induction draft (ID)/forced draft (FD) fan vibrations, and other faults, such as electrical faults are available. "Jumpering trips" for startups is accepted practice. (Standards are informal.) Fossil operators have great discretion to sustain operations in the face of conditions outside normal limits. Expectations may be unclear or unknown. Training emphasis is haphazard—or absent. Yet, when limits are exceeded, it's important that operators act. From an RCM perspective, knowing the essential functional instrumentation limits that relate to basic safety and equipment performance is essential. In almost all cases, the penalty is economic. Economic penalty is ever more detrimental to competitive health though.

The philosophy often is, "Let the operator run the plant—we'll cross bridges when we come to them." My experience has been that in many cases, an operator made a spontaneous call—the bridge was crossed—and someone then decided there could have been a better response. An ARCM-based maintenance program would have assured that the expected response is known before the bridge is crossed! With no planned maintenance program for essential instrumentation, it's easy to see why an operator discounts an alarm. An essential alarm that is not maintained is more than a nuisance—at best, it's a trip waiting to happen in a fossil operating program with ambiguous instrumentation guidelines.

I&C programs can greatly influence "maintenance-controllable" station performance. Nuclear plant I&C is largely controlled by technical specifications, which are comparatively constant. Fossil plant operators have the discretion to establish I&C calibration intervals, alarm check frequencies, and many other test intervals. However, because so much I&C equipment is available, important instrument calibrations (cals) can easily become buried beneath "the trivial many." Screening fossil I&C for unnecessary cals and other work is highly effective in improving overall program results and assuring completion of those cals that do make a difference.

Maintenance Process

Overview

Engineers appreciate highly complex chemical, mechanical, and other engineering processes that can be analyzed objectively. This often stands in contrast with organizational process awareness. Most managers of production facilities are engineers, but there has been less recognition of the "soft" processes as they apply to operating efficiency. After years in the utility industry—as both engineer and manager—I attribute this to a combination of "cost-plus mentality" and lack of profound maintenance process awareness endemic to American industry.

Maintenance is not static. The constant introduction and improvement of materials and processes has transformed the maintenance process. Like other processes, the environment has influenced the pace of change. Where 40 years ago, small simple-cycle plants and diesel operations were replaced by huge vertically integrated utilities, today the opposite occurs (Fig. 2-6).

Maintenance is one of many complex organizational processes that benefit greatly from process improvement techniques. For example, quality process theories found in manufacturing can be applied to maintenance performance. Maintenance can be viewed as a process that "delivers" available equipment ("products") in an operating facility on a budget. (Fig. 2-7) Traditional maintenance organizations have done an outstanding job "delivering" maintenance but rules are changing. Maintenance organizations need to deliver operating equipment more of the time at lower cost and take on more than the old "maintenance department" has done. Some independent power producers (IPPs) have replaced traditional maintenance staffs and annual-unit outages with flexibility scheduled overhauls at lower cost. Workers literally wear all hats—operator, mechanic, and technician—to develop the jack-of-all-trades "utility worker."

Figure 2-6: Change! Although this engine captured 100 years of steam design learinng, engineers could not overcome the inherent advantages of diesel locomotives. Infrastucture requirements, high operating costs, and labor agreements made steam no match for simpler, reliable diesels. Anachronisms lingered for 40 more years but operating steam locomotives disappeared forever on Amcrica's railroads between 1955 and 1960. No tear has turned back the inevitable march of technology progress.

With the modern emphasis on process—and the competitive pressure in the utility industry—maintenance processes are too ripe an improvement opportunity to not change. This challenges the corporate culture in most maintenance organizations, of course! Few floor-level maintenance managers attend workshops and skills-improvement classes. Maintenance is a highly crafted product in many companies but too many of them have historically tolerated a high level of rework. Organizations with low training and high rework often convey an implicit message that performing the job right the first time has little or no organizational value. Emphasis on new, innovative work practices means that more work performers need to see others' methods.

Maintenance improvement throughout other industries includes total quality maintenance (TQM) and TPM. While each represents a good start at defining and advancing the maintenance profession, each has intangible aspects that bring TQM to mind. TQM skills relate to the

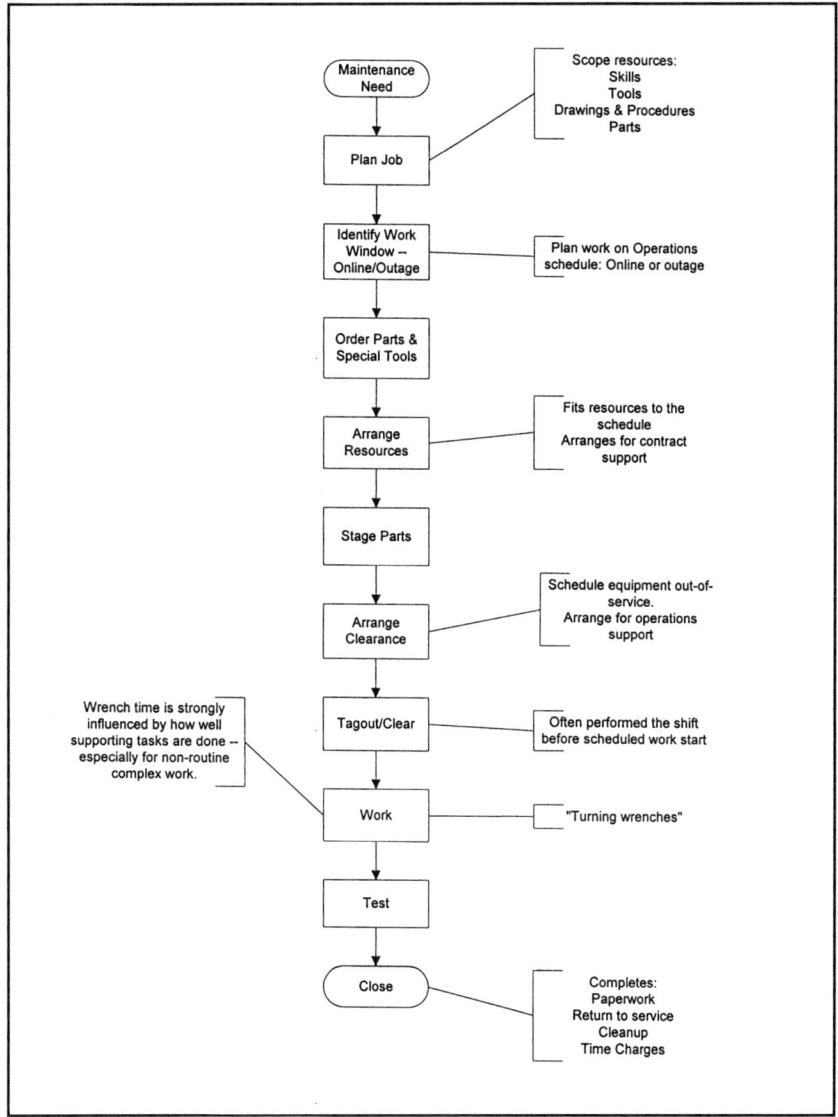

Figure 2-7: Maintenance

work but aren't recognized as critical organizational processes by traditional maintenance workgroups. While TQM has its place, tangible techniques will have greater success in the North American power environment. Successful maintenance hinges on bringing fundamental

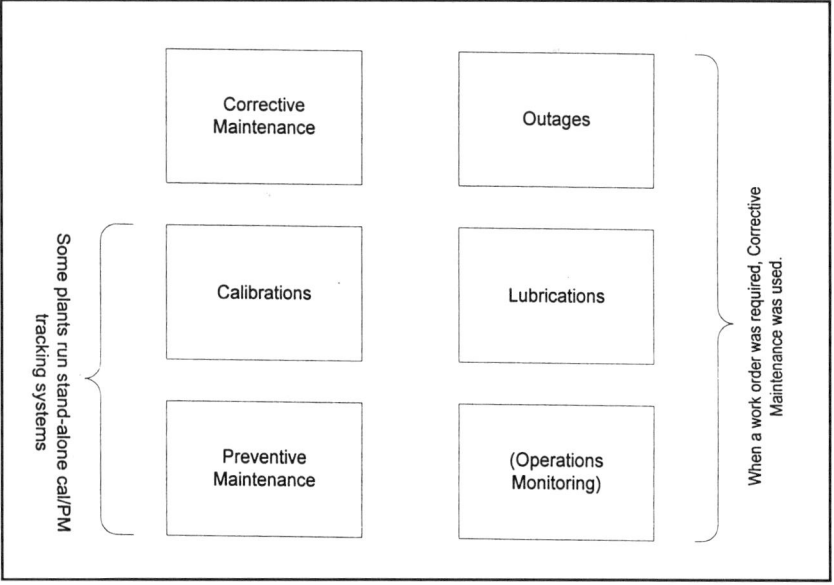

Figures 2-8: Specific Maintenance Processes

organizational skills together with work performance.

W.E. Deming identified competence levels he called "profound business knowledge." These key processes, inherent to any business, are exceedingly difficult to learn and perform and are usually developed through experience inside the business. Employees learn these techniques and processes in a work environment over many years, not necessarily even understanding why they work. Such proprietary competencies present entrance barriers in competitive markets because of this. Businesses won't openly divulge trade secrets or process details that strongly influence cost or how a process works. (Fig. 2-8)

In this context, what are the profound elements that shape an organization's maintenance performance?

Key maintenance processes

Plan. Maintenance planning identifies necessary work and decides how it should be performed. Repetitive work is standardized. Simple methods increase work performance consistency. Step-by-step work development supports the work itself. Competent crafts follow optimized work plans to minimize work, trips, and parts usage, while controlling key aspects of the work.

Weak maintenance organizations don't plan or adhere to planned work. Maintenance unit managers and craft plan their own work. Work plan standards are few or absent. Planners are not a specialized, skilled group. "Anyone can plan" characterizes the approach. Yet unplanned work is slower, has higher failure rates, and suffers from greater rework. Statistically, most maintenance work is highly repetitive over a long-time perspective. This supports planning.

Schedule. Weak maintenance organizations lack scheduling processes. They maintain incomplete database work lists, work tracking and control measures, or measurement capability in their CMMS. They can expedite work when they must but the routine work horizon is short—perhaps two or three days. They've learned to "work maintenance" within a short horizon tuned for crisis management, but less supportive of long term work plans.

Performance. RCM presumes maintenance performance is available. Plant and equipment failures due to inadequate performance show up statistically as infant mortality failures with random causes. (Consistent failure causes can be attributed to processes. Random failures must be attributable to lack of process control.) Maintenance performance is a rich subject, but one that is not within the scope of this book.

RCM (or other maintenance selection technology) helps identify the maintenance repertoire an organization chooses to support. It can identify weak maintenance processes based on statistical and proximate failure analysis. It greatly increases the craft awareness of critical equipment components, and increases their sensitivity in rework processes. It cannot (at least, not by itself) establish or compensate for deficient craft skills, processes, environments, materials, or other factors that are a direct part of the maintenance process.

Many factors contribute to excellent maintenance performance. Consistency comes from knowing the job, following standard practices, working with close engineering support, and adequate training. These elements are never perfect in any maintenance organization. They need continuous focus to assure maintenance performance is consistent, and constantly improving. Sadly, some organizations are only vaguely aware

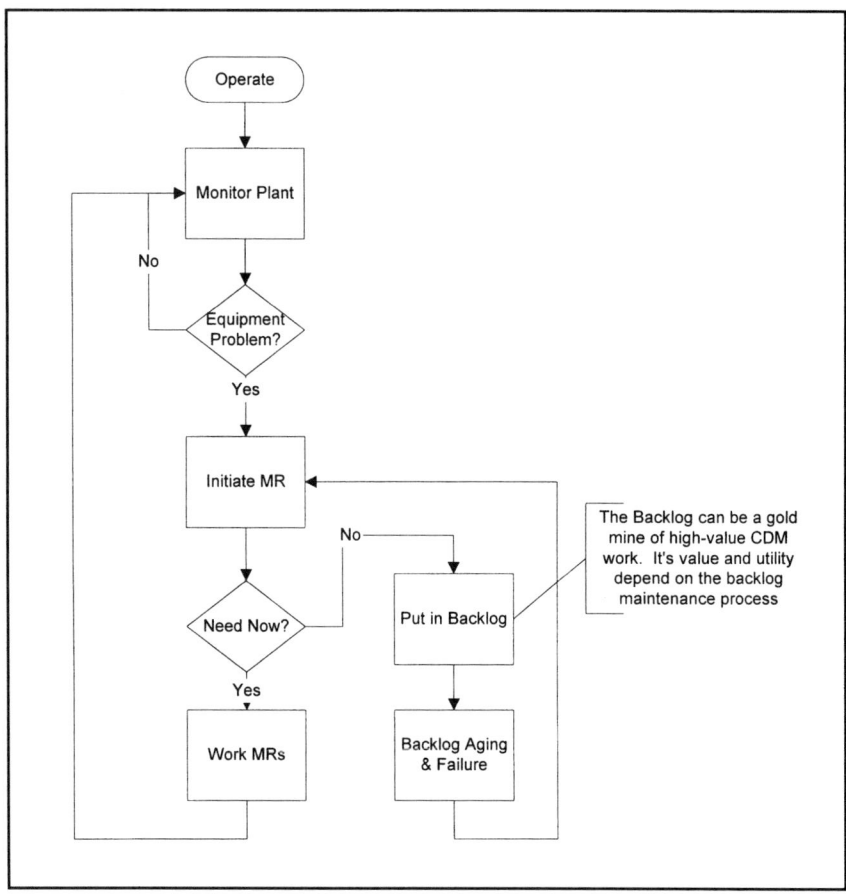

Figure 2-9: Condition-based Maintenance

of factors contributing to maintenance excellence.

Training. Utility maintenance programs were traditionally centered on the apprenticeship and journeyman steps that qualified an employee to perform any work of that skill in the plant. Seniority and positions were determined when the transition occurred. Apprenticeship programs provided a basic training but lacked engineering coordination and involvement. Instead, union and management—through a joint apprenticeship committee—determined standards, periods, curriculum, and other elements in managed apprentice programs that were grandfathered through previous union contracts.

Smart utilities have initiated pre-selection testing to assure that apprentices (and new operators, for that matter) possess the basic aptitude and knowledge to be successful in their work. If R and failure study were added to apprenticeship training, it would help maintenance restructuring towards "age exploration" and other strategic initiatives. The administrative aspects of maintenance—what information management is, and what it tells us, from engineering and management perspectives—needs to be conveyed to the craft in the field.

A mechanic who knows that a particular part gives poor service must convey that on a work order (WO) at the time of premature replacement, if R engineer is to know he found the part unsatisfactory, and trigger an engineering assessment of part performance. Although some assessments get quite detailed, many identify easily solved engineering problems, once action is triggered.

Engineering. Maintenance and engineering are not traditionally partners. Systematic support of maintenance is not the focus of engineering. Where maintenance departments can justify an engineering presence, engineers often focus on design. Few maintenance groups having engineering support dedicated to R improvement, process improvement, or cost reduction, because these responsibilities never functionally flowed down to performance-level engineers. Traditional utility engineering departments should redirect their efforts towards operational and maintenance support to help achieve R improvement and cost reduction. This redirection is fundamental, but difficult for many engineering groups.

The maintenance process defined

Maintenance is initiated either by defect—exception and deteriorated performance—or it derives from continuous monitoring, restoration, and performance. The former is corrective, or responsive maintenance; the latter, CDM. Organizational maintenance, therefore, can be thought of as either fundamentally proactive or reactive. Either can be effective.

The responsive maintenance process—traditionally used for CM—starts with an identified problem. An "out-of-spec" condition, a suspicious noise, an unresponsive machine, a failed alarm—all of these con-

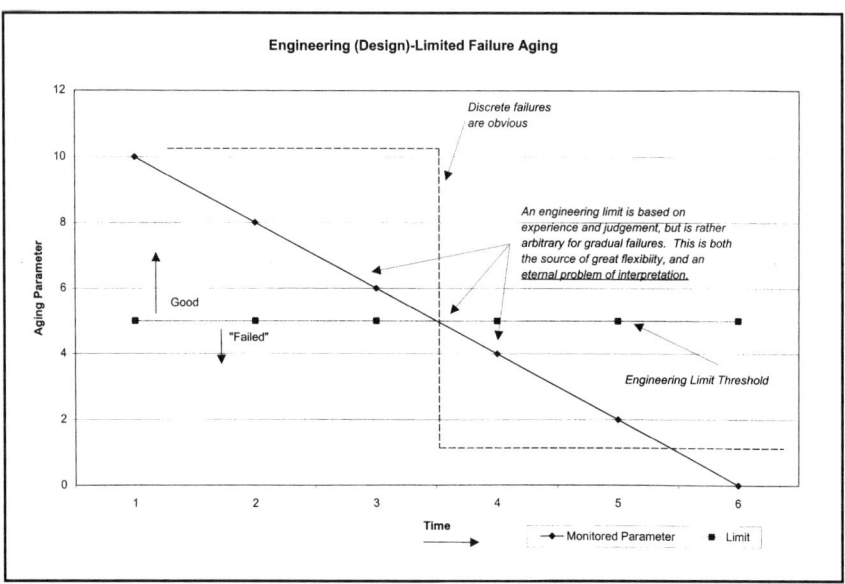

Figures 2-10: Engineering (Design)-Limited Failure Aging

ditions trigger a response (Fig. 2-9).

CDM in the proactive organization is structured around monitoring, rounds, and PM tasks—traditionally, TBM. These tasks provide earlier warning of failure and allow more planning time. CDM differs from CM maintenance (CNMM) by explicit versus implicit failure definition parameters, and limits that act as hard-triggers to initiate corrections. Maintenance performance requirements are specific and clear.

The Achilles heel of electric power generation has been the inability of operators to exercise discipline to establish operating limits. An organization practicing RCM knows those limits (Fig. 2-10). PM is at the heart of such a program.

Preventive Maintenance (PM)

PM is an "add-on" in traditional maintenance programs. "Fix things first—PM can follow." Managers and staff talk about PM importance, but many PM programs are low on the resource scale. This condition reflects the fact that life-cycle maintenance (LCM) lacks the simplicity and appeal of CNM. The LCM approach uses all maintenance performance information to schedule inspections and replacements, during which CDM is discovered and performed. The primary difference between CNM and LCM is the level of focus, organization, and effectiveness. LCM requires a higher degree of equipment knowledge, monitoring, and scheduling than a simple CNM program.

As companies' stated objective is to continue to operate aging facilities, useful equipment life will eventually be exceeded by operation. Knowing how equipment wears out, and the techniques to restore it, has become highly valuable. In many cases, projected facility lifetimes have elapsed. Major components—turbines and boilers—require major maintenance on shorter intervals than original facility life spans. "Maintain-design" performance objectives—heat rate, cooling, and generating capacity-deteriorate on a relatively short time scale-months to years. Outage performance—simply PM on a larger, less frequent scale—is a major aspect of any large generating unit's operations.

Traditional PM programs are based around three basic unit operating modes—on-line, restricted load, and off-line. Many organizations segregate scheduled outages from their overall PM program; they view "outage maintenance" using a traditional CM paradigm. Practically, though, only deferrable work can be scheduled for a planned outage. Scheduled outage work, accumulated in reserve and worked to completion in priority during an outage, involves restoring fault-tolerant equipment, instrumentation, and reserve capabilities. Some of it is time-based restoration. Scheduled outages are essentially comprised of CDM and TBM. In short, scheduled outages are all PM work!

Startup crews develop routine, online PMs during plant start-up and rise-to-power testing phase. Skilled mechanics, planners, and schedulers review vendor literature, combine this with their own experience, and develop PM activities to perform using a CMMS scheduler. When complete, they have exhaustive lists that faithfully recreate vendor recommendations. Yet, most of these "vendor-based" programs are only fractionally worked. Why? Because the scope of the PM program is so large, and credibility of the PMs—especially the performance intervals—is questionable; once staff realizes they can extend intervals with low risk, rigorous performance drops. Most vendor-identified, TBM intervals are grossly conservative anyway, because vendors cover every application with one recommendation, unaware of actual service conditions for the equipment they supply. Service conditions largely determine the appropriate intervals for performance. They include the operating environment—temperature, cleanliness, moisture, intervals, loading—and compliance with operating guidance.

Vendors do their best to identify necessary equipment maintenance. Some PMs can be stretched with little risk, some must be done religiously. Maybe not on the vendor's exact recommended interval, but on some service interval.

This illustrates a great opportunity for many plants. By reviewing and eliminating low value or no-value PMs (those that offer low or even negative benefit-to-cost ratios), the value of average PMs can be substantially raised. This adds to PM credibility and supports performance, because as PM credibility is achieved, performance barriers drop.

PM models

A PM is any scheduled preventative task intended to reduce the probability of failure. Key ideas are:

- scheduled
- intended failure prevention
- effectiveness

A PM can be scheduled by a computer, a repetitive round, the human memory, or another method, but it has a time stamp that triggers PM performance. Whether the PM is effective or not is another matter. Organizations have differing levels of PM feedback measurement. Some systems have no PM performance measurements at all, in other cases, when measured and presented, the results diverge from those intended.

Some PMs are conservative to a great degree. They have no real value, rarely get done and have no operations impact. Other PMs have impact but don't get done consistently. Some PMs are actually detrimental to equipment conditioning but get done anyway! Not surprisingly, programs with a high incidence of ineffective PM tasks have ineffective feedback. Since few companies have a R engineering specialty, such deficiencies are not surprising.

PMs can be done on "calendar," "clock" or many other bases. For example, time limits for reactor refueling and boiler outages are regulatory and fuel depletion time-based. These outages occur on nominal 18 month intervals. They were extended from a 12 month interval that was common more than 10 years ago. Equipment run-time provides another common time measure. This is suitable for continuously run motors. Processing facilities often use production age indicators. Tonnage (coal belts, dumpers, crushers, feeders, mills), total air moved (compressors), or integrated flow (pumps) can provide suitable age measures. "Demand" equipment, like medium voltage breakers, see most of their aging during operation, and so operation cycles are a suitable measure. Manufacturers typically set the parameter(s) most suitable for equipment age measurement, and they often provide suitable age-measuring instrumentation.

Some measures require an integrator. Coal tonnage through dumpers, feeders, or mills, resin regenerations, and breaker trips are examples.

Aging parameters are so important, when known they should be explicitly identified. Engineers may imply an aging parameter but the operator, mechanic, and engineer in the field need explicit aging parameters identified for all equipment. For PM timing, aging parameters must be explicit (Fig. 2-11).

Time Parameters

Clock Time	Run Time	Starts	Cycles	Refueling Cycles
Weekly	Integrated Use	Breaker Counter	Seasons	MWhr
Quarter	Revolutions			
Annual	Product			
	Mileage			
	Tonnage			

The only requirement for any parameter is correlation with an aging (time) mechanism

---

Figure 2-11: Time Parameters

Sometimes regulations establish inspection standards with the force of law. Such inspections must be performed as prescribed. For simplicity and convenience, they usually follow a calendar interval. Fire inspections, when mandated by laws or insurance agreements (which may be endorsed by law) are done annually. Many inspections have been grouped with implicit or explicit activity (such as reactor refueling or boiler outages) to assure the activity is performed.

In an ideal world, all time-based PMs would be specified to "catch" equipment at significant aging marks. Regulated, time-based inspections can lead to exceptionally conservative PMs when viewed using the equipment's natural aging parameter. For example, many in-service inspections for nuclear power plant valves prescribe a quarterly test to assure function. For many of the valves in these programs, the test will be the sole operation of the valve during the quarter. If the test is to detect in-service aging, such a calendar PM is simply too frequent. Most of these valves could be tested successfully on much longer intervals.

Time bases fall into several natural categories based on interval (short, intermediate, and long) enabling us to select the appropriate method to perform the PM activity (Table 2-1).

PM perspectives

Organizations view PM performance differently. A PM activity issued may be considered as good as complete at some facilities. Others treat "work complete" more formally, allowing equipment interpretation based upon their last performance.

Other organizations are PM intense, performing every vendor-recommended task. This approach initiates effective monitoring and time-based PMs processes but can also break down if operators or the craft discover they can skip task intervals with little or no failure consequence.

When equipment doesn't fail, people tend to continue with an existing program, even though it is over-conservative and performed too often. The only way to find out what the equipment can support (in terms of lifetime and PM replacement intervals), is to perform "age exploration."

And craft workers don't uniformly perform all PMs to completion.

Short (hours to weeks)
Performance checks
Area checks
Equipment checks
Alarm checks
Operator rounds

Medium (weeks to months): On-line, PdM
Lubrication level checks
Cleaning
Standby equipment tests
Performance tests
Alarm checks
Filter replacements

Long (years): Intrusive, PdM
Planned outages
Large equipment overhauls
Main control cals
Trip tests
Lubrication replacements
Relay cals

Table 2-1: Example of Time Base Intervals

If essential PMs slip and failures occur, then program credibility is cast into doubt and everyone's effectiveness is diminished. The systematic "clean-up" of casual PM programs is a significant first step on the path towards effective RCM implementation.

In reality, programs based on manufacturing recommendations can typically extend task intervals greatly with little risk of failure consequences because of conservative vendor recommendations.

So, what's best? Craft and operators need to critique and adjust recommended PM task intervals. Craft feedback on intervals is required for the dual purpose of finding the best intervals and maintaining commitment to actual task-monitoring performance. Craft worker feedback is also essential to the CMMS (and to engineering) concerning how well parts perform, in service, to continually manage and reduce parts costs. (In ARCM, this is a formal, continuous process.) Fostering close operations-maintenance ties, whether by intent or accident, yields more effective PM programs.

Operators, in fact, provide first-level monitoring in plant PM systems. Like maintenance PMs, operators' rounds (routinely scheduled checks that monitor broad areas and systems) should be based on value. Traditional rounds put operators into the plant on a non-specific, "just-in-case" basis, but rounds can be based on the frequency and risk of failures. Operators may extend certain rounds with no consequence but they bear the responsibility to support their decision. For equipment that requires no action until an alarm goes off, a monitoring and maintenance strategy must be based upon that. Actively recruiting operators to develop, review, and "turn rounds" is a continuous, high value process.

Support the craft workers doing what they know needs to be done by means of a task list. Once the PM task list is developed, work processes determine how much gets done. Some organizations have a catch-as-catch-can approach. Others have a "work-all" approach. Some leave work scope to the discretion of the workers. Others try to work equipment that is available. Few systematically measure the degree to which they adhere to and complete their plan.

In the absence of a measurable plan, there's reason to question maintenance effectiveness. Good programs are carried forward by knowledgeable and committed craft workers. Workers still lack information that points in the direction of improvement—unaware of the degree to which they are dependent upon the collective memory of the workforce to accomplish PM work. In an environment with turnover their success is diminished.

Power plant outages are planned and executed to restore production capability and to prepare for the next sustained operating period. How a plant identifies and performs outage work tells us something about their general work perspective:

- Some plants will work all outage work to completion
- Most have very aggressive targets but partial completion
- Many traditional plants don't recognize how an outage is equivalent to any other PM period
- Most plants are much more aggressive with their outage workscopes and management than they are with their ongoing PM programs
- Many fail to see opportunities as they occur. With proper work planning and coordination, upwards of 40% of outage hours can be worked online

When the plant is down, restoring production is important. But preventing the plant from going down in the first place is achievable, through highly effective levels of work, understanding, skills, coordination, and implementation. When PM is obscured by more visible activity, it can lead to a "PM is not real work" mentality. The craft often avoid PMs because they aren't organizationally focused, and prefer overhauls that give them more opportunity to perform disassemble/reassemble work management may even let them select.

PM delivery is taken for granted by many organizations even though failure records indicate that existing PM programs are not being followed. Documented performance, with periodic checking, is an effective tool to assure PM performance is real. Peer self-checks and periodic manager or operations checks are other ways to monitor PM completions. An outside audit can help establish the delivery credibility of a PM program.

When failures occur in a well-designed program, investigation is in order. Companies that review and assess their monitoring programs historically experience few random, surprise failures. When PM is treated with discipline it will add confidence to the program that stands behind the PM.

Identify the essential features of any PM program as a productive first step. In fact, if everyone can agree on the essential elements, it will save a lot of wasted effort. After years of consideration, I've identified some essential PM program features:

- Activity list: a formally maintained list of PM activity the organization is committed to perform
- Scheduling tool: a method that delivers routine PM performance and assures priority for PM even when crises occur
- Selection and issuing methods: ways to establish the scope and routine issues that PMs must address for peak performance
- Completion reporting: a feedback system to report completion
- Measurement of completion rate: a system for management to measure PM program health
- Assessment: periodic review of the program for effectiveness
- Standards: there are guidelines for performance, deferrals where appropriate, grace periods, and so forth

I characterize these elements as memory, execution, and discipline. Basic craft skill is necessary to perform the work—that's assumed. But to establish an efficient, basic program, you must have:

- focus
- group work organization and delivery capacity
- effective tasks

Organizations that have only one or two of these elements suffer fundamental flaws in their PM programs. Programs with major gaps are not viable PM programs.

Before developing tasks, you should determine whether or not you have a viable PM program. A little effort to create a viable program before too much work is put on the list can help the organization grow into PM more easily.

It is common to lack a PM "delivery" process. (*i.e.*, High value PMs are identified, but there are no means to assure that work is performed consistently). This may be part of a larger problem, such as overall work prioritization and scheduling. In any event, PMs are deferred to the lowest priority level and get worked as fill-in, or catch-as-catch-can. High-value PMs can truly be painful to track as premature failures reoccur. These are avoidable with a credible, consistently performed PM program.

How are typical PM programs operated? A PM sheet gets kicked out of a computer CMMS system and goes to the shop, where a maintenance foreman personally prioritizes and schedules work based upon his personal experience, workload, pre-assigned priorities, processes, and organization. Program reviews indicate that many PMs get completed randomly under this system—they're performed with high variance.

Another weakness of traditional PM programs is found in typical completion performance reporting and trending. Any program with 15% performance—and acceptable results—is in need of some tuning! Whether such numbers indicate low interest on the part of managers, or a complete lack of understanding about the program's value, PM is a complex subject that requires sophisticated understanding of equipment, failure, statistics, people, and processes. Its complex scheduling starts where project management ends.

PM triggers. PM, like any a repetitive activity, must have an initiator—a time trigger, condition identifier, and condition-based action. It could be as simple as a clock, or as smart as a specific condition. It could be something in between. (Fig. 2-12) Seasons, weather, age on a component, or a "sixth sense" that some mechanics seem to have all provide clocks and initiating events. An individual often must work for years to learn the subtle clues the operating plant offers about its condition and needs. The more concern individuals have for reading these clues, and the more plant knowledge and experience they have, the more success they enjoy. Training improves anyone's ability to read equipment.

There are seasonal, conditional, and equipment operating cycle triggers. Many do not reside in the plant CMMS, either because they won't fit or because time parameters preclude their use. Other times, triggers are just habit. Understanding what causes actions to take place, is important if you want to change a maintenance system. If, for example, people work off personal "scratch sheets" rather than the organization's

PM Initiation Reminders
(Triggers)
Paper
Memory
Habit
Outage
CMMS
Other

Figure 2-12: PM Performance "Triggers"

CMMS printouts, the value of the CMMS as a tool is diminished.

Current CMMS installations provide the capability to schedule PMs from flexible time parameters. Most CMMSs handle common time equivalents—process tonnage, machine hours, or operations cycles. Some can read plant DCS systems automatically. The principle—the time trigger—is the same with each. A known parameter, correlated by experience with an aging characteristic, initiates a work trigger based on elapsed "time;" monitoring with specific limits identifies an out-of-spec condition and triggers work. The more work that's identified on the CMMS, and the more time that's spent to assess it, the more effective a maintenance program will be.

Scheduled outages can pose a PM dilemma of sorts. Outages comprise large amounts of deferrable PM work on degraded, degrading, or failing equipment. Performance loss can be tolerated to operate up to the scheduled outage. However, at some point the benefit of deferral is overruled by unscheduled outage risk. Also: Outages made up of major, planned work (such as turbine overhauls, boiler cleanings, and reactor refueling) that involve very large workscopes and personnel resources. Not formulating an exact schedule raises the cost to carry these people and resources. Finally, major work such as turbine overhauls provides "the tent"—an outage under which we can "park" a great deal of other necessary work.

Organizations committed to optimum maintenance performance can use outages to introduce PM techniques to the plant. Scheduled outages are the most fundamental PM interval and activity. They represent opportunity.

The challenge of "life-cycle vision"

There's a fine line between maintenance and CM, and it has to do with how people are dispatched to perform their day-to-day work. Scheduled work—in which most of the time is scheduled—minimizes reactive responses. Such structured, planned work improves maintenance effectiveness and reduces costs. LCM (in which the work is structured around available equipment windows) means that equipment is more nearly in a state of continuous monitoring. This supports routine, continuous operations. Organizations adopting LCM must understand that a key aspect is the implementation of an effective maintenance measurement plan.

Informality and chance. Planning, implementation, and effectiveness measures that are vague characterize an informal maintenance program. Informality in a large, complex facility will not support a cohesive plan. An LCM approach to maintenance adds consistency and structure to all maintenance performance indicators in a plant. Routine, habitual work practices lead to consistent maintenance. An organization with millions or billions invested in plant assets cannot afford random maintenance—particularly when there are tools that can greatly increase effective task selection and maintenance performance.

Traditionally, plant PMs came from vendor manuals; others were initiated in response to failures that weren't adequately understood. It's also common to see PMs initiated, suddenly, where equipment had been operated for 20 years or more with no PM with reasonable success. Such PMs cannot pass effectiveness actionable tests. "Failure-prevented" applicability reviews and expert cost/benefit assessments are missing—assessments that need to be supplied by maintenance strategy specialists. There's no obvious connection between work performed and failure prevented in such cases. Workers and plant staff whose commitment to PM programs is lukewarm, take risks when they defer or miss PMs, betting that they can get away with no activity. And because frequently they can, the program loses credibility.

Credible PMs—and workforce PM program support and compliance—arise from explicit "failure-prevented" associations, the more explicit the better. PM priority then naturally falls on an equal basis with other work. The appropriate plant personnel must perform PMs. Nonspecific tasks and checks should be performed by operations. Intrusive, technical, or direct work needs maintenance performers and their support.

Vague PM scopes are of low value. Because they're indefinite, they're time intensive; workers fumble through what operators could do better on their rounds. Non-specific scopes cannot be assembled into credible work packages. One way to identify ineffective programs is their lack of distinct, actionable PM tasks assembled into performance packages. Many PMs involve quick, easily performed tasks repeated many times. When they're developed, planned, and blocked into work tasks, they optimize work time and increase organizational PM focus.

In some organizations, individuals select their work on a daily basis, even when there's a work backlog. Effectiveness depends upon the priority that they assign to the work. When they pick work they like to do, where does this leave organizational priorities? A significant part of PM performance involves the discipline to select, issue, document, and report. Scheduling discipline comes from aligning people and work. Cleanliness, operations professionalism, and cost-management are other indicators of discipline. Disciplined organizations have focus. Their goals are clear and they work to them.

PM's functional elements

Time-based: clocks. CMMS issue WOs based on computer clock time, or events such as an outage or even memory. Scheduling software does no work—people in an organization do work. Once a PM WO is issued, how does work follow?

Intrusive work ranges from low skill to high skill. The more intrusive the work, typically, the more skill required, and the more specialized the personnel. Sometimes only an expert can interpret data. It may take a mechanical engineer to evaluate a yielding failure or a civil engineer to examine a railroad loop fill. For non-intrusive tasks, we assume that people have skills to perform simple tasks, or they can learn them quickly and effectively on their own. In each case, training introduces consistency into task performance. Consistency in performance leads to consistent O&M.

VM, oil analysis, and ultrasonic diagnostics all require training and skill to enable workers to interpret equipment conditions. However, most people in the plant are capable of learning about reading gauge glasses, "oil slingers," and oil contamination; recognizing smells (such as burning coal); finding steam and water leaks; sensing vibration through foundations, and picking up on normal or abnormal sounds.

Working in a production environment, one must call equipment failure causes like an umpire calling an out at first base—and then get back into production quickly. Learning should be a program attribute. Change must be encouraged. Continuing on simple paths—changing perfectly clean instrument air filters, rebuilding valves, replacing like-new parts—instead of adjusting programs to reflect real-world equipment and plant needs indicate static programs. When a production environment accepts it, it's not structured for change. The systematic extension of replacement intervals to discover how long equipment and materials last is a high value attribute of a living program that can be missing in traditional facilities.

Operational based: surveillance. One reason that fossil units experience higher equipment failure rates than nuclear plants is the absence of "surveillance programs" (SP) that are structured into nuclear. Nuclear plants have extensive specifications imposed by their license and SP demonstrate compliance. There are more preliminary technical identifications of deterioration before the equipment proceeds to functional failure. With few fossil equivalents, periodic alarms or protective devices and standby systems run higher risks of failure. This in turn increases equipment failures.

For any plant, a SP is defense plan against random failure that captures the operators' value. The inherent design of fossil plants is more nearly optimized, as well. Originated in the first quarter of the twentieth century, their design has been evolving for nearly 100 years. The more a design advances:

- the more it supports NSM performance
- the less its overall function is impacted by component failures

Fossil units are relatively fault-tolerant—and more equipment is taken to failure limits. They come down when real parts fail—not abstract specifications. On top of regulatory constraints, nuclear plants are "younger" than fossils—both of which mean that it's difficult to make significant design improvements. Nuclear units must operate within specifications, and while failed "specs" are less interesting than broken components, they cost less to correct.

Operate-to-failure (OTF), preplanned failure, and "no scheduled maintenance (NSM)." OTF is widely misunderstood. NSM is a more apt term. Nolan and Heap never use OTF in R-centered maintenance. Rather, they use "NSM."

Refered to as "Operate-to-failure," OTF poses a barrier to operator maintenance cooperation. Most workers don't want equipment to fail—but they like to work maintenance. They need to be backed off inherently reliable equipment that can be trusted to self-identify failure during monitoring with low risk. Where a failure has no functional impact and maintenance can be readily scheduled after failure, OTF strategy is cost-effective. This is what RCM is all about-taking advantage of the fact that most equipment doesn't benefit from scheduled maintenance.

Where redundancies and risk management are absent, however, OTF isn't effective. The organization's risk tolerance—the period during which the redundant train, alarms, or other equipment can be impaired prior to work performed—influences OTF effectiveness. If operations can't wait for a few days to schedule work on backup equipment that's down, then savings will not result. OTF does not work for "nervous Nellies."

Design process evolution calls for equipment failure to become less important. Advanced designs tend towards fail-safe by automatically removing equipment from service or initiating self-shutdown rather than creating severe events. Advanced designs demand abuse to force failures. Most users give up well before the equipment becomes a hazard. Advanced designs include equipment in production for more than 20 years that has been through three design iterations. After 50 years, they approach their design limits. Few survive in their fundamental form for more than 100.

Measurement. CMMSs lack the capacity to differentiate a functional failure from other types of failure. Functional failures, as a practical matter, show up in operator's logs as shutdown initiating events—fires, accidents, and other facility compromises. They influence long term plant availability.

A nuclear plant's failure interpretation by law is much more conservative than its fossil counterpart. As the definition of functional failures becomes more common, improved work-order measurement can be expected and we can develop better ways to compare apples with apples.

Availability is a telling measure. According to NERC submittals, fossil units are actually more available than their nuclear counterparts. (Shorter scheduled outage durations appear to be the cause.)

Overhaul. Overhauls are grouped activities in which intensive, intrusive work must be performed, such as disassembly of a turbine to clean blading. In RCM lingo, an overhaul as a PM doesn't exist. Rather, it's a group of small tasks, such as:

- blade cleaning
- root tip crack inspections
- rotor inspections
- blade erosion inspection
- gasket replacement
- rotor re-balancing
- bore inspections
- lube oil purification
- cooler inspections
- instrumentation bypass line inspections
- casing bypass flow erosion inspection
- generator winding examination
- balancing
- stop valves inspection
- stem blush removable
- weld repair

Many of these activities are either time-based or established failures that require periodic checks and restoration. Risk is managed by a combination of instrumented detection (like VM) supported by periodic internal inspection.

Because internal examination work is so expensive, it's usually cost-effective to "do it all" once a machine is apart. Many PM tasks are blocked around one intrusive activity, so that the overhaul-to-overhaul life of large machines is prolonged. (Fig. 2-2)

In RCM, we restore components that are aging. The philosophy is very useful for examining overhauls, since it may lead to eliminating activities whose failure is determined to be low-risk or never possible. Large activity PMs should be reduced to individual tasks so that each added-value task can be separately assessed. A healthy, "question everything" attitude must be encouraged from the top down.

What measurement level is appropriate? Workers often don't see value in documenting work. They rarely see anyone use documented results. Accountants are the only ones who routinely assess completed work. Traditionally, maintenance organizations "do maintenance" rather than avoid or reduce it, so they don't recognize measurement value to reduce work and they're insensitive to cost.

Traditional maintenance organizations with "in-house" horizons haven't benefited from out-of-company maintenance observation. Self-training marks the traditional maintenance organization. They can even view maintenance support staff—planners, schedulers, and engineers—as outside the "maintenance process" and may harbor skepticism of new technologies and analysis. They're eager to work but planning, preparing, and training suffer from the preference to "do work." However, "doing work" requires tools-measurement, documentation, and analysis are also valuable tools.

Maintenance is a cost-plus proposition. Traditional maintenance rewards are proportional to the volume of "maintenance" produced—not value. Financial incentives to increase maintenance productivity don't make sense to those who suffer consequent income loss.

Maintenance measures have historically been cost-oriented, but they're measured predominantly at the high end—the unit, station, and corporation. Major process areas—procurement, outages, stock, man-hours—often lack performance measures. System level measurements can be lacking, as many older CMMS systems cannot support system level cost allocations. Yet, detailed system measures would allow diagnostic trend reviews of the maintenance processes for high value improvements.

Traditional plant managers focus on expenses. They never directly received benefits of increased generation revenues in the regulated environment. Completing additional maintenance work could bust budgets, even though increased revenues could result. Even today, many managers lack cost-management skills or CMMS systems to be able to track such fine points to hold the line on costs.

Measures are valuable when useful information results and improvement areas are identified. CMMS offers greater measurement capabilities but information is only as good as data received. Achieving consistent data input—hour reporting, costs, problem descriptions and work done—is the essential first step. Several alternatives may be available to provide immediately accessible benefits.

"E" MWRs. Emergency maintenance WOs ("E" MWRs) correlate to equipment failure rate. Since true functional failure often generates an emergency—the "E"—work-order, tracking by "E" MWRs is a simple, functional failure rate measure.

Overtime. Like "E" MWRs, overtime results from functional failure events. Though less definitive than "E" MWRs, overtime provides another useful way to measure functional failure rates.

"Failures." Most utilities that write a "home-grown" CMMS begin the measurement process by assessing the types and quality of information collected. Once measurement is established, organizations usually find that:

- there are many measurement capabilities they have that they don't need (the trivial many)
- there are a few measurements they need and don't have (the critical few)
- data quality or accuracy may be low (they need training or simpler methods)
- they need measures of both process simplicity or quality, or both

An organization that recognizes the need to measure has made a giant step towards performance improvement. At the time it recognizes that its measurement system needs improvement, it's just a matter of time before broader measurement needs and improvements are considered.

Maintenance rule. The NRC's "maintenance rule" [Title 10 Code of Federal Regulations (CFR) 50.56] can be meaningfully interpreted in RCM.

Three fundamental attributes of maintenance—fixed time maintenance, CNM, and operate-to-failure—must be followed for "systems, structures, and components" that influence safety and are not inherently reliable. The rule requires "monitoring" for things that don't meet organizational goals for availability and maintenance-preventable functional failures (MPFFs).

The rule is workable, if ungainly. Areas of confusion involve definitions of "run-to-failure" and "have no PM program." If we don't have a traditional PM maintenance program, are we running things to failure? And, if we do, must it be entirely and completely documented, and with documented performance?

An example: A large, single-unit boiling water reactor (BWR) has between 50,000 and 100,000 coded component identification codes (CICs), for which there are between 5,000 and 10,000 PMs. Though many PMs address more than one CIC tag number, do we have a PM for everything? No. Do we come close? Hardly! Well, then do we have a PM for every essential component? No. All environmentally qualified (EQ) components? No. Many items lack a formal PM program; are we running them to failure? Emphatically, no! First, remember that a failure must be a functional failure. With so much redundant equipment installed, and with many minor and inconsequential failure modes, the vast majority of reported failures are proximate, not functional. If there are 100,000 CICs with 3 failure modes each, we should have up to 300,000 PM tasks—but there aren't. Why not?

Operator monitoring covers about 80% of the work. Operator rounds, surveillance testing, and other non-specific monitoring, such as area checks, will identify the vast majority of failures. If we exceed target levels for unavailability or MPFFs, the maintenance rule says we must initiate monitoring and corrective action—what's known as "setting goals." If we're already monitoring, we merely need to document and in other ways demonstrate compliance. This means that corrective action may entail the same monitoring program we had before—if we had an excellent program—to ensure that we never experience either an

MPFF or availability below our target.

The system level challenge has been how to measure unavailability. In the absence of simple "unavailability measures," clearance tag-outs or log entries are used to declare equipment to be inoperable and so track unavailability. While these processes are simple and conservative, they overstate unavailability, thereby artificially forcing equipment into an "a (1)" category. Yet, probably the single greatest maintenance-rule benefit has been the requirement to track system level performance. Prior to the maintenance rule, few plants did this. System level availability and failure analysis would greatly benefit many non-nuclear units.

The Hawthorne principal (based on manufacturing performance studies done at Western Electric's Hawthorne plant in the 1930s) states that management's visible interest in performance improves it. This is exactly what happens after performance measures are instituted and tracked at a production facility. It's a powerful case for measurement.

Costs

Operation and maintenance practices based on reactive (not proactive) philosophies are very costly. Direct costs include rework, increased scheduling, and increased risk. Risk ultimately translates into "operating events" that impact equipment and employees. High risk organizations mean higher cost operations, just as speeding drivers mean higher costs for insurers.

Analyzing failures and costs confirms this intuitive knowledge— maintenance performance correlates with insurance claim losses. Insurers periodically inspect client facilities to assess their insurance risk and help clients better manage that risk.

There are wide variations in electricity cost, in part because some producers are more expensive, based on their plant outage profiles, while for others it's routine maintenance practices. If, in fact, competition re-invigorates the generating industry, it will happen because companies will be forced to re-evaluate and improve processes that have been slow to change.

Case Examples

SBAC

A 7000-acfm soot-blowing air compressor (SBAC) experienced premature filter pluggage, filter element tear-out, and prolonged operation with unfiltered intake air. The rotary compressor uses five high speed compressor stages, each driven off a common bull gear. The compressor required overhaul two years after the previous performance—about two years short of its early projected overhaul date but much shorter than the previous stage overhauls. Installed as one out of three compressors—two in continuous service—this unit achieved more than five years' service until its first overhaul. The nominal life for the compressors was placed at four years, aside from the "new operating period" when all units ran almost eight years. On these staged compressors, the high-velocity fifth stage ordinarily wore out first, establishing the overhaul need.

The diminished compressor life (between overhauls) was two years. At an overhaul cost of between $250,000 and $300,000, the shorter life cost an additional $150,000 (around $75,000 annualized). The missed PMs cost three hours every quarter—or $1,000—annualized, including filters. Cost benefit is at least 75-to-1 based on maintenance costs. Such a PM cannot be missed without adding substantial maintenance costs.

While "down," boiler convection passage plugging increased. Because operating staff had to be pulled aside for the compressor overhaul—a two-man, one-month duration job (with contracted help), normal schedules were interrupted. Because overtime was required in this union shop, the whole plant was authorized overtime, further driving up costs.

During operations, big-ticket failures mean major unscheduled events. Non-routine, non-turbine or boiler costs can be tracked by frequency and cost category. Major unpredictable equipment failures can cost up to hundreds of thousands of dollars. Such failures are an obvious target for reduction! They can be identified, counted, costed (annualized), understood-by-cause(s), and corrected. Taking on unplanned but statistically predictable "big ticket" events in a systematic manner results in gradual improvement in equipment online cost performance.

Ultimately, all equipment wears between service intervals. Achieving maximum predictable service intervals is a goal of a PM program. Big-compressor "soot-bacs" are monitored over time. Performance monitoring, combined with in-service PM between overhauls, should identify the projected "overhaul period" well in advance of actual deterioration.

Compressor performance drops as the stage blading wears until a combination of low flows at load with increased surging identifies the need for overhaul. The overhaul consists primarily of removal of the bull wheel (actually, the pinion), stage inspections for dimensional bearing wear and visual balding wear, discharge plenum cleaning and inspection, and reassembly. Normally, no bearings or gears are replaced. Only the fifth, finishing-stage compressor blading is reworked due to high wear from moisture and particulate erosion. Normal overhaul focus is on the fifth stage and normal part reorders anticipate replacement of these parts.

Premature turbine blade failure

A forced turbine outage due to blade deposits occurred in a unit for which nominal turbine overhaul interval is five years. Overhaul at four years was required due to derating from the lowered stage efficiency and literal valve-wide-open generation loss due to plateout deposits. To schedule an early overhaul, system dispatchers bought replacement power, because the base loaded unit's outage was in the near-term unit outage forecast and other unit outages couldn't be rescheduled to cover the outage period.

The turbine overhaul took around two months—annualized, 12 days on a 5-year interval, or 15 days at four. The cost of replacement power was the difference between the marginal production cost at that date and the scheduled replacement power cost (around $35), coming in at (the ballpark) price of $15-20/MWhe; 350 MWe (rounded) for two months at a price differential of $20 (the difference between $35 and$15), is around $10 million. This outage would have occurred anyway—just two years later, as a scheduled outage. At that time, the replacement generation would have been provided internally or with lower-cost, long term contract scheduled power. This wholesale power rate averages lower, thereby reducing its cost. The differential cost of replacement power, planned and scheduled, would have probably been in the $5-10 range rather than $20/Mwhe. Factoring internal generation, it could have been as low as $2-3.

Considering present value (PV) calculations, from an accounting perspective, the benefit of meeting scheduled outage periods includes the PV of the deferrable expense. A $5 million expense, deferred two years, costs $3.5 million (a finance charge of $1.5 million). In a financially competitive environment, this is real money. While this sort of calculation is new to traditionalists, it's a day-to-day calculation for IPPs and co-generators. It's something the rest of us need to learn.

Generator retaining ring

Consider a generator retaining ring failure on a 350 MWe unit, that cost around $10 million in direct repair expense and another $25 million in revenue and replacement power expense. The event had precursors. Two separate vibration amplitude step increases had occurred in an otherwise stable trend. The first preceded the event by about 10 days. The second occurred the morning of the final event. The second offset—ranging upwards of 10 mils—so concerned operations that they called in a VM crew to check the instrumentation that morning. Checking the generator bearings-using hand held instruments—the ring final failure developed.

The event was complex but had precursors and warnings that were ambiguous and/or missed. The cost of crack detection and repair is so high there's no real effective deterrent, except to run-to-failure. The key to operations success in this area is never to initiate stress cracking. One way to do that is to exclude moisture—the most common controllable cause of stress cracking. This is the generator manufacturer's strategy. OOS, with the generator disassembled, moisture is difficult to control, but preventive measures must be taken. Tented heaters and/or dehumidifiers can maintain non-condensing environments.

This particular event occurred in the West, and so humidity was an improbable cause for the retaining ring failure. Hydrogen coolers, on the other hand, are known to leak because during shutdowns, the hydrogen cooling water in service with low hydrogen pressure created potential for leakage into the hydrogen system. This, in turn, would provide the opportunity for internal condensation and stress-corrosion cracking in the generator. By use of periodic hydrogen moisture sampling, original equipment manufacturers (OEM) intend that generator internals—including rotor and retaining rings—are kept above the dew point. When hydrogen is monitored for moisture, leaks into the hydrogen system can be detected. Bleeding and feeding dry hydrogen can avoid condensation even when cooler leaks are present.

In addition, liquid detectors identify and alarm for liquid that accumulates in drain areas at the bottom of the generator frame—areas with isolation, vent, and drain capability. Any water or condensation accumulation can be monitored, trended, blown down, and inspected. The float-operated switches and alarms can be checked and replaced.

In this event, a third party performed root-cause investigations because a large insurance settlement was at stake. It was rumored that the generator in question (as well as its twin) had had cooler leaks. Whether or not the plant had consistently performed recommended monitoring was a matter of speculation, based on its overall PM completion rate. Taken as a whole, its PM program was weak, with overall completion rates of scheduled PM work in the 5-10% range.

The presence of moisture in generators has a known, deleterious effect on many components. Besides stress-corrosion cracking, insulation windings can fail to ground. Generators have been in production for more than 100 years and designs are highly refined. To remove great amounts of heat internally generated, manufacturers have evolved hydrogen gas cooling and water inter-coolers. Water coolers always present the potential to introduce moisture to a gas, usually during transient or shutdown operations.

Vendor manuals warn operators to monitor moisture internal to the generator at several levels:

- H_2 leakage monitors (offline)
- drain alarm switches (alarms)
- drain monitoring and blowdown
- H_2 moisture monitors online (optional)

This illustrates a typical, integrated equipment failure defense strategy—direct operator monitoring supported by condition-based maintenance (CBM) upon failure detection. Essential monitoring equipment—instrumentation and alarms—is maintained by an I&C group. The failure-avoidance strategy relies upon monitoring and alarm equipment availability, backed up by direct operator draining of blow-down lines that require only periodic performance checks.

But moisture potential is everywhere. Twenty years of plant experience tells me to look for water where you don't want it and least expect it because that's where it's certain to turn up. This applies to instrument air systems, air purge systems, or any kind of cooler with water on one side—sensing lines, damper lines, Helium lines (gas reactors only), and other gas lines. Every water cooler is certain to leak so make sure water quality is high and then monitor appropriately for leakage. Fix any identified leaks as early as possible. Avoid direct dependence upon operator/human intervention to directly control failure events. Designs should support operator activity by minimizing direct intervention requirements.

In many cases, like this one, maintenance didn't happen and operators learned to tolerate the deteriorated condition. Over time, an expensive repair would ultimately occur as secondary problems developed. Addressing degradation before it becomes excessive is key to a solid maintenance program—and also the weak link in too many of them. Operators typically do a credible job of identifying deteriorated and unserviceable equipment but follow-up is where success or failure is determined.

The problem is rarely that the installed instrumentation is inadequate. The problem is usually the absence of a maintenance strategy to successfully operate and maintain essential equipment. It's a problem when an operating philosophy includes reactive maintenance—not fixing things until forced to do so. This is a problem that instruments can't address and may even make worse.

This example illustrates the importance of critical instruments and of CNM. I use the term essential instruments—in essence, the instruments that provide identification and protection from unacceptable failures that must be avoided. Such instruments and their associated monitoring programs must be maintained.

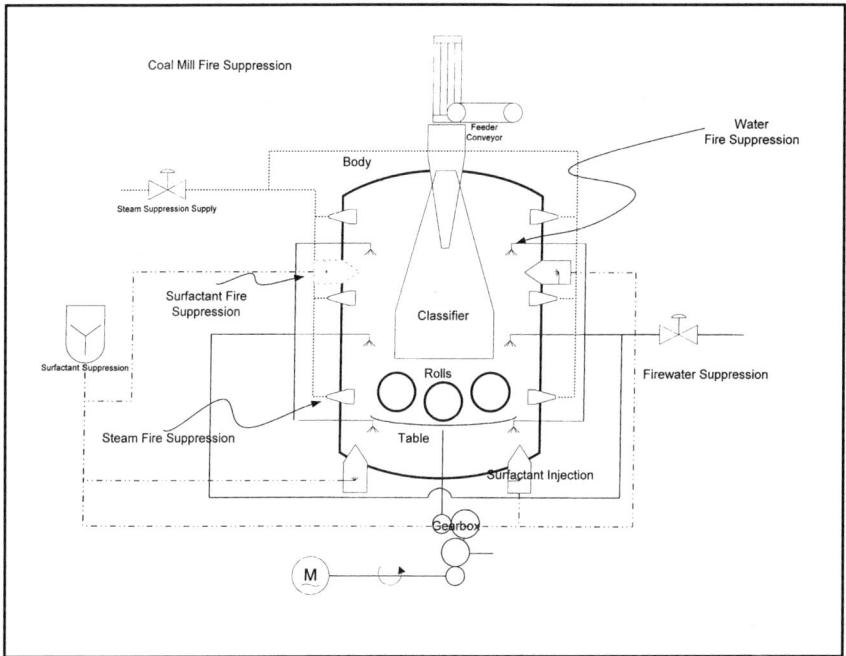

Figure 2-13: Coal Mill Fire Supression

Coal belt fire

A coal gallery fire occurred at a coal facility's inclined yard belt. Coal spillage, and a chronic station history of coal fires, combined with a defective fire protection system causing a major loss. Fire alarms and the deluge system failed. Insurer loss estimates (and rates) were based on fire alarms and equipment availability. Inclined coal belts and the right environmental conditions (fire doors left open, for example), generate an excellent draft. In this case, within an hour the belt gallery had burned and collapsed.

Fire, casualty, and industrial insurers decided that coverage was the thorny issue. Loss-risk assessments are based on actuarial assumptions, including systems and alarms. Events in which devices are OOS fit another loss category—perhaps one that can't be insured.

For both companies and insurers, the best defense against catastrophic losses is predictable, consistent operations. A fundamentally sound CNM program, supported by an effective CBM program, delivers such consistency. A great deal of fossil generation equipment has fire protection alarms and instrumentation. These provide direct means to manage fire risk, and should be treated with caution. (Fig. 2-13). These fire detection alarm circuts need a planned maintenance program. Coal mill CO monitors have great value—if they work!

Chapter 3
RCM Performance

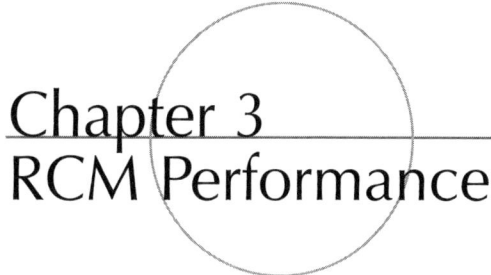

Left to themselves, things will go from bad to worse.

-Corollary, Murphy's Law

TRCM analysis is time-consuming, and companies must balance time against value. Without implementation, there is no value but many organizations have not benefited from RCM although they have done complete, system-by-system RCM analysis. Perhaps they're too interested in quick results. Perhaps it's a lack of patience—ultimately, everyone suffers from that, too. However, there is no value in academic RCM exercises or dismissing RCM altogether because you fail to take all the steps.

As a practical matter, I've rarely (probably never) had the luxury of time to do a "complete" analysis. In fact, the "complete RCM" analysts I've met have never claimed that their complete analysis was final. (No one would care [or dare] to do so, except when pressure to implement RCM comes in major part from regulatory agencies.) My focus instead has been implementation, as opposed to theoretical basis development. It's most effective to pass along a maintenance optimization philosophy

Figure 3-1: Modern Blending Coal-fired Power Plant: Apparently simple, looks are deceiving. This zero discharge plant ranks with the last nuclear units for complexity. The plant is running at full load.

in the context of solving operational and maintenance problems.

Nuclear plants have extensive formal maintenance plans because the NRC drives the need. Fossil plants have little to nothing. R rules are the same. Nuclear plants simply have greater documentation, justification, and "traceability" requirements. Each environment has needs, for which the maintenance optimization answer is the same, as it should be. Studies by EPRI on competing methods indicate PMO outcomes are largely the same and depend less on specific process than upon skill and knowledge of reviewers. Rules and regulations aside, the operating goals of each plant type are generally the same.

Lower equipment failure rates at nuclear plants (based upon failure data at the plant level) are not due to inadequate documentation—there really are far fewer equipment functional failures. This makes failure studies at nuclear units more difficult because the statistical population is smaller. While fossil failure experiences at the component level don't transfer directly to nuclear plants, they do offer insights. Industry-wide databases can fill in for statistical data for any single unit (Fig. 3-1).

Selection Based on Importance and Cost

Two factors determine whether a piece of equipment deserves TBM—importance and risk.

Equipment that impacts production (due to failure) must be managed. If failure has random components, there must be installed redundancy to provide operational flexibility. This philosophy supports OTF—"wait until it requires maintenance, then do it."

If it was worth buying and installing in the plant, every piece of equipment deserves some maintenance. "Some" maintenance, in most cases, means "NSM," based upon function and CNM. An on demand/OTF/no-scheduled-maintenance strategy is appropriate for most installed plant equipment because one tenet of RCM is to do no maintenance on equipment that doesn't inherently require it. A default monitoring program captures most inherently reliable, randomly failing equipment through non-specific operator monitoring, "area rounds," and periodic checks made in the course of doing other things. Combined with alarms, redundancy, and trained operating staff, there is a great reservoir of operating depth to draw from.

Every manager goes after high value opportunities, and this means availability loss and the systems contributing to them. Generation costs—including lost production value—represent value lost when an unplanned trip occurs. The RCM approach takes the unit apart, system by system, according to how each system supports overall plant generation objectives. Systems affecting generation are high value systems, particularly during forced outages, when steam supply and turbine systems determine their duration. Understanding how all work contributes to outage duration and prevents failure is another availability benefit of RCM analysis, however, because careful RCM analysis can trim outage workscopes.

Consider the number of systems in any large generating plant—on the low end, perhaps 40-at complex nuclear plants, it may be 100 or more. Does this mean analyzing each and every system to get a useful answer? This is the implied requirement with TRCM (especially at nuclear plants). If we don't have to do everything, do we pick and choose the systems and supporting parts we evaluate? The answer is that we can select systems:

- provided we retain the results in common retrievable format
- commit to return regularly to continue analysis on an ongoing basis

If you elect to go this route, you must consider integrating the resulting programs. The death knell of many TRCM programs was the failure to integrate the work. This requires performing LCM integration. Without this dimension, PM optimization is risky business.

Consider the "mission" of each unit in an owner's portfolio. In the past, we assumed generation was base loaded; today more varied missions are applied. Some plants conserve corporate assets while others, provide peaking generation or support transmission services (voltage support). Transmission support roles must be defined by facility owners and be identified to plant operating staffs. Some units serve in dual roles, others provide generation reserves. Just as phone industry deregulation has opened up new services and markets, the generating and transmission industries have new markets, products—and missions. A maintenance plan developed around an obsolete or wrong mission will not effectively use resources.

Plant system units

There are several major classes of generating units. We can broadly group these into five types:

- gas-fired boilers
- coal-fired boilers
- CTs
- hydro
- nuclear

Size, vintage, fuel, cooling, architect-engineers (A-E), boiler, and turbine OEMs differentiate the fossil units. Fossil coal plants with zero discharge permits are the most complex. Many perform fuel receipt, fuel processing, fuel storage, fuel movement, combustion, waste processing, emissions, and water treatment for plant equipment. Fossil units can further be grouped by investment, production costs, capability, and staffing. There are representative equipment and systems (often from one to several major suppliers) in any category. Each unit is somehow unique and can be treated as an individual exception.

Fossil CTs have gained market share spanning the last 20 years as their missions have evolved-from providing fast-start backup generation to base load combined-cycle facilities. They fill many niches today, from base load to peaking. Complete facilities may support heat recovery steam generators (HRSGs), which in turn supply a steam-driven turbine. Such combined-cycle facilities are similar to large fossil generators in their complexity, number of available models, and economic value. Standardization has established common configurations. These units often burn gas, have been brought into production quickly, and can be modified over time to incrementally increase load capacity.

As a class, hydro spans a greater range of equipment and more initial service dates than any other category. Some of the earliest electric generators ever sold are hydros—and many are still in production after 100 years! Age variation makes hydro unique. There are principally just two major design applications-high and low head (or Pelton and Francis wheel turbines, respectively). Many hydros are "seasonal peakers," although some provide pumped storage units. In the Northwest, many see base load service. In the West, a lack of water limits hydro use to spring runoff or peak load periods. Under these conditions, nominal load ratings can be very misleading as many units run only a few weeks per year.

There are two basic nuclear classes-BWRs and pressurized water reactors (PWR). Major equipment suppliers further differentiate PWRs. Each has unique demands and requirements and all share a common regulatory environment. Although sizes vary, later-design base loaded plants are industry standards.

Each plant category has many standard systems in common-design features, aging characteristics, operational features, and other "personality" traits. Some are suitable for a single type of service-others fill multiple roles. "Benchmarked" plants can be identified through NERC, Institute of Nuclear Plant Operations (INPO), NRC, supplier, and industry data-suppliers and can provide reference points for performance self-assessment.

Plant system functions

RCM maintenance analysis centers on retaining functionality, and so focuses on system functions at a high level.

Systems are the building blocks of power plants. In a power plant (or any large facility), they break down complex conduit, cable, and piping into discrete, understandable units. Classified by type, systems share many similarities from application to application. (There are many types of turbines, but turbines share many common subsystems and functional requirements.) Experienced workers implicitly know all primary functions of the major common plant systems. It's the unique, uncommon, plant-specific functions that occasionally surprise even a seasoned engineer. Engineers continually strive to enlarge the number of functions a given operating design can perform to stretch economic benefits.

A-E design/owner documents identifying system functions include process and instrumentation drawings (P&IDs), system descriptions, training materials (where available), major modifications, procurements, and engineering files. Additional documents include licenses and virtually any maintenance or cost information that can provide functional insight. Maintenance optimization requires that this information be assembled into a current snapshot of system and equipment status. For nuclear plants, training materials often provide a current, complete overview of current systems status. For other plants, an RCM- or PMO-type analysis may be the first effort to piece together a "big-picture" systems-operating overview since startup.

RCM system analysis also documents operator insights that often identify implicit functionality or hidden assumptions—particularly in vintage plants, where design documents are out of date. Capturing operator insights into system operating goals, problems, equipment, and integrity-items gleaned from years of experience-is a valuable exercise with many maintenance plan benefits for companies seeking to extend facility operations.

TRCM identifies system functions, boundaries, and interfaces in great detail. In non-nuclear applications, the detail can usually wait until a need presents itself. Streamlined ARCM identifies systems and functions implicitly. Nuclear facility functions are identified in detail due to the maintenance rule and so receive little benefit from recreation of system design documents. Identification of functional failures at the system level, which in turn shapes performance monitoring and testing, is key. Further definition can be added as needed when problems arise during equipment grouping.

Functional equipment grouping requires selecting system boundaries and interfaces, which can be done implicitly. Traditional analysis stops short of the kinds of grouping, scheduling, and other LCM features. Deferring such analysis into the working elements of performing effective maintenance, and identifying cost information and failure statistics, are yet other distinctions of ARCM.

Analysis often leads reviewers to discover simplified assumptions and methods that reflect the designer's intentions. Modifications made to older plants sometimes change original A-E systems but this can be captured by ARCM's basic, simplified documentation. This supports additional detailed assessments that speed analysis.

To establish system importance, it's important to explicitly identify functions that support:

- safety
- environment
- production
- license technical specification or other formal "commitments" agreed upon as conditions for
  - operations
  - practical support requirements
  - major equipment trains and redundancies
  - essential instrumentation

This review is best performed in conjunction with a full maintenance strategy review and PM development plan that examines performance, cost, component R and failures, operations equipment use, and monitoring. Extant PM programs are evidence that once someone viewed that equipment as important. This consideration should be car-

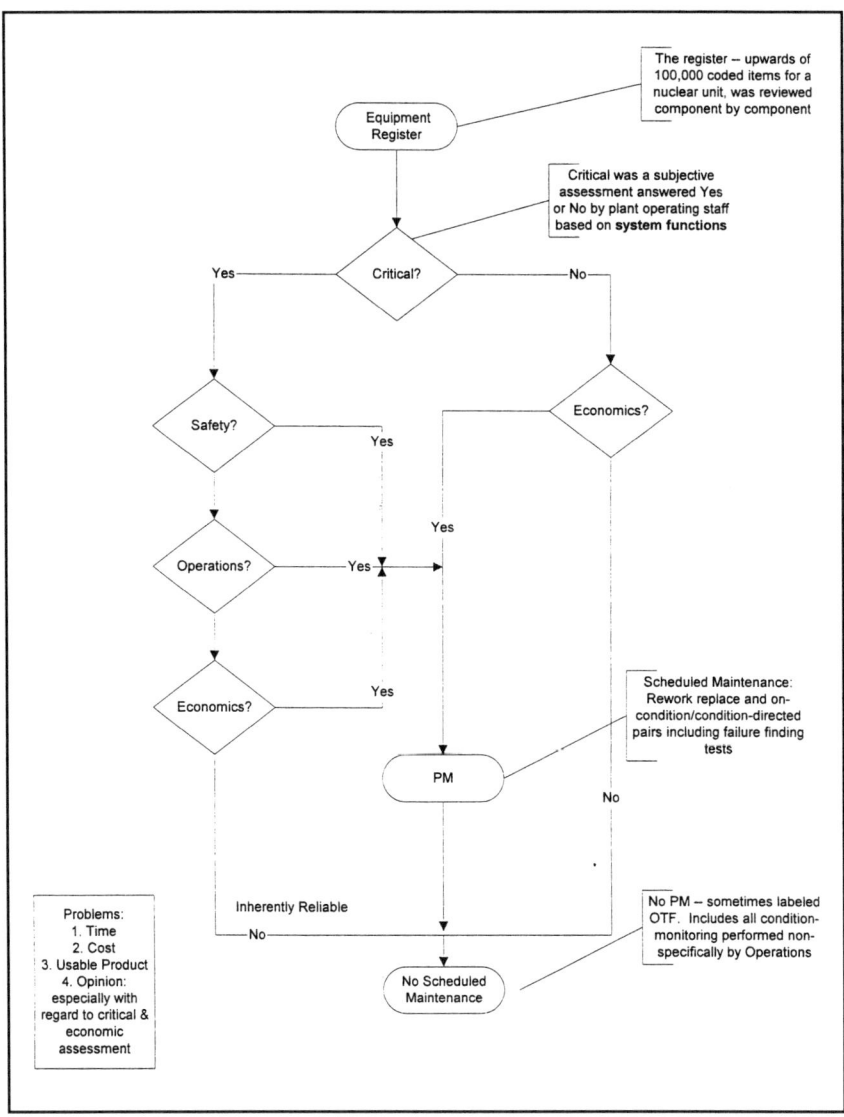

Figure 3-2: "Critical" Streamlined RCM/PMO Approach

ried forward, until more information identifies otherwise.

The analyst keeps these major functions in mind while reviewing equipment. The plant equipment master register should serve as a checklist for those reviewing systems and equipment. Consider equipment importance, costs, maintenance demand (PM and CM), and incidental requirements. The list should include:

- large equipment
- equipment covered by existing PM, calibration, and test programs
- equipment of regulatory, insurance or cost concern
- major redundant equipment

As a practical matter, organizations with PM programs in place should determine how it was developed and how much remains current to simplify the review process. For new facilities, determining equipment size and cost, production impact, and then benchmarking to standards, guides, and other facilities will identify important equipment. If a similar facility has already been examined, use its equipment as a reference. Chances are the analysis will proceed along the same path.

Efficiently performing ARCM requires standards, standard methods, and processes. Large, visible, high-value equipment with maintenance programs in place is often accessible only at outages and so receives high organization maintenance focus. Such important programs cannot be altered until staff "cut their teeth" on more basic equipment and simpler maintenance strategy evaluations. Developing and applying standard templates at each facility, based upon site-unique equipment and experience, greatly focuses attention on equipment that is important in an overall sense, and requires routine resources to maintain (Fig. 3-3). "Important equipment" is that which warrants standard maintenance plan development and includes any equipment that has production impact, is used repetitively, and demands a significant amount of maintenance resources.

Although the critical streamlining approach for RCM/PMO is used and useful to many, its arbitrary splitting of equipment into "critical" and "non-critical" classes has led me to search for other methods. By reviewing and summarizing my actual method for performing PM optimization applying RCM, I developed an inversion flow process that I feel is more intuitive to follow (Fig. 3-2). The reverse logic, if you will, is that we evaluate looking up the importance hierarchy. It is easier to see why an item does qualify for scheduled maintenance, and add that than wrestle over why it does not, and develop a supporting justification.

Plant Maintenance Standard		Coal Belts Standard				Component: Belting	Required:
						Code Type: BLT	(1) Performance
REV.	DESCRIPTION		DATE	BY		Manufacturer	Assessment
REV. 1	Issued		9/97	August		Goodyear,	(2) Age Exploration
						United Rubber	Belts & Splices

Failure Modes	MTBF Est (years)	Task	Interval (months)	Strategy	Performer	Criteria
Aging:						
checking	10	inspect/replace	6	CDM	Engineer	Checks < 1 cm between cracks
fraying	4	inspect	1/6	CM/CDM	Operator/Mtce Mech.	Frayed > 2 inch/side or 5%
wear	6	inspect	6	CDM	Mtce Mech.	Visible carcass
splice wear	3	inspect	1	CM	Operator/Engineer	Uneven surface or > 10% length
tracking	2	re-track	12	FTM	Mtce Mech.	Off-center > 1 inch
Age-Random:						
splice tear-out	2	inspect	1	CM	Operator	> 1 inch damage mechanical bulge,
Random:						delamination
punctures	2	inspect	Preop/1	CM	Operator	> 1 inch web damage or penetration
tracking	2	inspect	1	CM	Operator	< 0.5 inch side clearance

Alignment:
(1) Detailed inspections performed monthly by operations; pre-operational inspection on each startup.
(2) Maintenance inspection/lubrication six months
(3) Engineering belt inspection two years

Scheduling:
(1) operator inspection by round
(2) Plant PPMIS Scheduler PM sheet
(3) Plant PPMIS Scheduler PM sheet

History:
(1) Notes here
(2) Permanent MIS/CMMS
(3) Belt Life - Plant Engineering files

Figure 3-3: Equipment Maintenance Standard

Organizing standards around major classes of equipment has the added benefit of focusing plant staff around the spectrum of equipment at their plant, its varying needs, and the options and strategies necessary to maintain it (Table 3-1).

Standards focus us on what time-based and condition-based PM to perform and steer us away from performing low value maintenance. For example, it's common to find PMs on manual isolation valves though with rare exception, they have virtually no value as TBM. A more appropriate approach is NSM. This doesn't imply the valve should be allowed to fail or that once it has failed, to ignore it. It means "Do no planned scheduled maintenance and monitor valves until a problem is noted."

Because we cannot predict when valve maintenance is needed, it's

Electrical	Mechanical	General
Motors	Pumps	I&C
Breakers	Valves:	Lubrication
Switchgear	Air operated	Predictive
Contacts	Motor operated	
Relays	Solenoid operated	
Batteries	Hydraulic operated	
Meters	Manual	
	Pressure reducing	
	Safety	
	Relief	
	Filters	
	Compressors	
	Blowers	
	Traps	
	Dampers	

Table 3-1: Equipment for "Standard Templates"

more effective to use operator non-specific monitoring to manage the failure. For those rare valves with significant safety or other non-maintenance functions redundancy or TBM replacement may be effective.

Equipment

Hierarchy level

Equipment fits into the system hierarchy between "systems" and "components." Equipment is integrated into system support functions. It is often redundant, or supports a redundancy feature. Instrumentation also fills this role. Equipment is often identified in trains (Fig. 3-4). Equipment can be alternately viewed as susystems.

That being said, the ways in which we identify equipment is often arbitrary. It is convenient to view equipment as a combination of com-

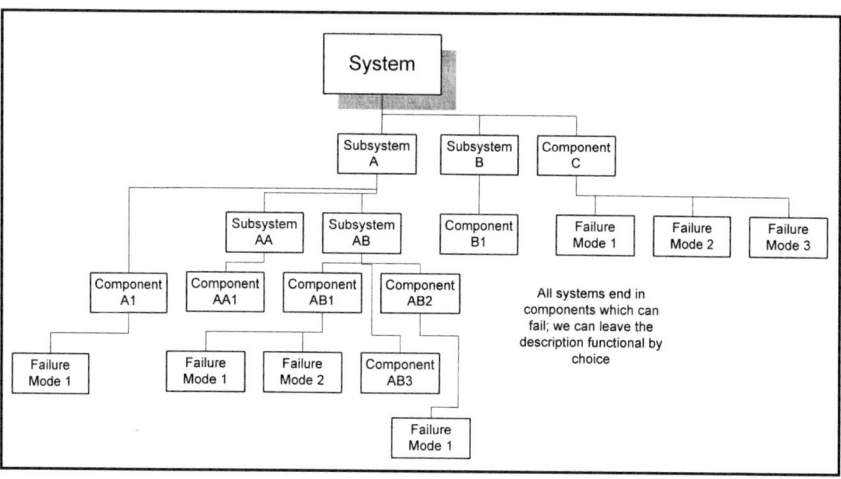

Figures 3-4: System Component Hierarchy

ponents that come out of a box or that is made up of other subsystems. The primary reason to differentiate equipment to the component level is components fail and are repaired. In doing analyses, many items "fit" as components while others work better as subsystems. The distinction is perspective. Can you (or do you want to) further differentiate it as a component? Do you consider the item functionally or discretely?

How you answer these questions influences whether you take a performance monitoring and testing PM perspective—appropriate in a general sense—or a proximate failure detection/correction perspective that is specific to a given component and failure mechanism. Once established, you can't easily restructure an equipment hierarchy—particularly when using a software product to establish and maintain it. The best structure-one that produces useful analysis-is one that provides the desired end product—common or relevant failure mechanisms (modes and causes), applicable and effective work tasks, and related essential information (who, what, when, where, why, and how).

Functional classification structures can be very effective at monitoring large groupings of equipment through the use of performance tests. This cost effectively simplifies monitoring (though, ultimately, components fail and must be maintained).

Components are further grouped through the use of codes. Plants code to different levels on a component register. Nuclear unit component coding is controlled (as a design document) so that it is best used. For fossil units, coding is largely the way equipment and components are entered into the station's CMMS for work order retrieval. Hydros (in my experience) code very little. Analysts must find an intermediate ground if there's a choice. Since the purpose in coding is performing maintenance, following the station's CMMS equipment coding methods is usually an excellent choice. Analysts who structure their work to follow CMMS equipment codes find benefits include:

- recognizable structures (for station)
- consistent equipment records
- direct applicability to work order tasks (PM and maintenance work order [MWO])
- simplicity of PM coding

Where existing CMMS coding isn't unique or has no logical structure, simplification may be needed. Elimination of repetitive equipment identifiers is a prerequisite to program consistency.

Equipment analysis can be developed by hand, spreadsheet, or a variety of software products. Although software products have limitations, they are a great asset to achieving consistency. Software must be applied consistently with standardized products supported by quality analysis. Software provides the ability to create and document on a program basis, which can be carried forward over time to support a "living" maintenance program. Software can provide essential standardization and grouping tools and also be used as a training tool.

Failure descriptions

System failures are described functionally. These are general and non-specific with regard to component and performance. System failures are ultimately caused by discrete component failure but can be identified much more easily at the system level than at a discrete component level.

For example, a system that provides hydraulic valve-position control might functionally "fail to control." Any number of other things can cause slow control response-fluid contamination, actuator plate-out, leaky seals, worn hydraulic parts, low pressure. The beauty of the performance focus is that it identifies the effect of the problem on performance without requiring specific cause. When we know "effects" we can project likely system failure causes.

We can take specific action when we know component failures. We cannot when we find system functional failures. System problems require us to diagnose a proximate failure at the component level to resolve the system failure. This requires troubleshooting and other analytical skills. Component failures ultimately cause functional failures. (Component failures are also known as "engineering failures" and some engineering texts refer to them as "root-cause failures.") These failures are not necessarily "root-caused" however, but are grounded in the R definition of failure—and components are at the "root" of the physical fault tree in an analysis. In a broader context, component failures can in turn have root causes.

Failures and their root causes are addressed by FMECA. FMECA is abstract, but grounds a complete, TRCM analysis. The inherent simplicity of a PM program is realized when we take a statistical approach. We want to know-statistically-what things fail, with what failure modes, and with what statistical frequency. Key points to remember in ARCM are that:

- components fail
- actionable tasks must address component failure
- failure mechanism = failure mode and cause
- typically, there are fewer than three common failure mechanisms for a component type. Statistically, there's often one
- ARCM perspective is statistical, not absolute; we worry about the common modes overall
- if a specific application has a known specific failure or failure mode, we can address that
- root cause does not have to be addressed for PM to be effective
- the objective is to manage risk

Normally, the focus of failure analysis is to get to root cause. Knowing whether you've found a root cause is a prerequisite to preventing recurrence. PMs don't absolutely have to be perfect to be statistically effective. For example, it's common to find misapplied designs in a failure scenario. A root-cause approach would involve a design modification to correct the application. PMs sometimes provide an effective, though not permanent, fix. Many in-plant problems have been effectively addressed in this manner.

RCM provides three basic strategy options to address failures—four, including non-specific CNM for the non-specific maintenance. Any successful maintenance strategy starts by applying tasks that are suitable and effective, based on equipment function and dominant failure mechanisms. The choices are outlined in Table 3-2 and further discussed below.

Strategy Options	Time Based Maintenance	Condition Directed Maintenance	No Scheduled Maintenance	Condition Monitoring
Abbreviated	**TBM**	**CDM**	**NSM**	CNM(NSM)
Characteristic	"hard-time" rework and restore tasks	Specific limits Time-based condition checks	"OTF" in some books; the bulk of activity in any maintenance plan	Non Specific operator tasks incorporated into routine monitoring and rounds
Notes	"dumb" – works off a clock, ignoring condition	Two part in simplest form: TBM inspection followed by "on-condition" restoration	Non specific area checks, channel checks, and other operator monitoring	The same as NSM – broken out for emphasis
Former Term Equivalent	PM	PdM	None	None

Table 3-2: Strategy Options

Time-based maintenance is statistically the least frequent task, but when suitable, it's often highly cost-effective. It applies to elements with a fairly defined aging behavior. When looking for TBM PM candidates, always consider aging characteristics—what parts age, how you expect them to age, what the aging parameter is. For low-cost replacements— filters, strainers, etc.—TBM is often the most effective PM in the sense that it doesn't expend time on CDM. Once a failure pattern is strongly correlated to a time parameter, TBM is usually a good choice. It works very well for inexpensive rework/restore tasks, less so for large-cost tasks that could be extended even slightly by CDM.

Condition-directed maintenance is also known as "on-condition" maintenance and "CBM." It's a two-part task in its simplest form: A time-based monitoring or inspection task is conditionally followed by an "on-condition" rework or repair task. The item is reworked or repaired "on the condition" that it fails inspection or test criteria. As applied in the commercial airline industry, "on-condition" maintenance is a highly controlled inspection procedure based upon a known, proven onset of deterioration that is detectable prior to final failure. It always has a specific "go/no-go" failure criterion. A "failed" item receives the directed-maintenance rework/restore task associated with the inspection. The rework/restore task is also direct, known, and determinate.

Establishing engineering limits that determine "on-condition" maintenance requires intrinsic knowledge of both the equipment and its failure mechanisms. It's time-consuming at first and tough to do—and tough to "sell." Until engineers and operators accept that limits are definable, and management agrees to follow them, a CDM program has no teeth. As a result, at many plants with informal PM programs, equipment continues to run to failure. Establishing limits and the culture that can live by them is a fundamental paradigm shift for a traditional organization.

After CDM tasks are identified, implementation consists of building inspection tasks into appropriate vehicles for performance— rounds, tests, inspections, PMs-and assuring that rework/restore tasks are defined in the form of procedures, checklists, or other formats for quick, consistent performance. The full benefit of a planned maintenance program requires a plan that the organization knows it will perform repetitively. Herein lies another ARCM benefit-the ability to reduce a substantial amount of maintenance to a production performance basis.

NSM. The choice to not perform scheduled maintenance is a profound one. When there are no applicable, effective tasks, and no safety or environmental issues, the default is to select NSM. While this selection seems the obvious choice, the role of the "PM gatekeeper" is probably the toughest in maintenance. Ineffective or inapplicable PMs arise from many sources. Regulators, managers, and engineers all feel uniquely qualified to make maintenance decisions. Systems engineers at nuclear plants have this in their formal job descriptions. The perception is that PM is free—write up a repetitive work order, and it just happens.

Properly developing PMs is as tough as laying out a facility design—tougher, where the concept of "engineered" PM programs hasn't been "sold." Organizations without gatekeepers perform lengthy lists of PM activity—only a small percentage of which get done. Nuclear plants spend inordinate sums on PMs that drive up their costs yet do not benefit operations or safety.

Every organization needs a gatekeeper-type R engineer with authority on the same level as the "chief engineer" at an architect-engineering firm. Such an individual controls PM scopes and helps to achieve implementation on those PMs that matter.

Hidden failure. "Hidden failures" are those not evident to the operating crew under normal conditions. They usually result from instrumentation and/or control failures, where a component identifies a functional failure not otherwise evident. Some relate to failure of redundant and/or standby systems. For all "critical functions"—those involved with safety, that would not otherwise be evident to the operating crew—an instrument is typically provided to make the equipment failure "evident." These can further be hard-wired to arm pre-set trips for critical functions where the trip response time is essential for safety or economics.

Nuclear units have many more "hard" critical trips than fossil plants. In both cases, however, if the instrument, trip or alarm is the operator's sole line of defense against a critical safety failure, then the component that provides that function is also critical. If it fails, an event could occur without the operator aware of it or able to take action. Consequently, instruments provided for safety require maintenance to assure their "hidden" function is available.

Auto-start systems and standby trains in nuclear plants have the same roles. Emergency diesel generators, standby core spray, coolant injection, and other systems must be available in the event of an accident. Fire detection and control systems at all plants have similar roles, and strategies are similar. Those fossil plants equipped with huge ID fans have automatic high vibration alarms to protect against catastrophic failure due to imbalance.

Maintenance strategies for critical instruments, alarms, and trips should include periodic function tests, especially when the potential for random failure is high—*e.g.,* when alarm functions and subsystems are complex. Checks of overspeed trips, periodic "surveillance" and fire protection, and other routine, scheduled maintenance test these functions.

The "functional test" is a general default strategy for any hidden function failure prevention.

Whenever instruments and controls are involved, they should be considered for hidden functions and functional test condition-directed activity. Failure to alarm, trip, or otherwise perform the protective action initiates CDM.

Rework/restore. For either time-based or condition directed maintenance, a rework/restore task may be specified. Rework means to rebuild or otherwise bring a component back "into spec." Restore returns to an in-spec state by replacing parts with qualified spares or by performing repairs. "Repair" has a specific maintenance context in nuclear work. Reworking a cracked weld back to specification involves a design change. In RCM, it's only a restoration task.

From an ARCM-PM development perspective, the two tasks are the same-from a performance perspective they're extremely different. "Performing rework" is often classified as light maintenance; "performing restore work" that involves repair or welding is classified as heavy. A large, sophisticated, and capable facility will have more in-house maintenance performance skills than smaller ones but in all organizations, there are repetitive needs to assess, and the decision to be made, whether to rework it or contract services to achieve in-specification conditions. A combination of the two is needed at most facilities, and must be factored into the RCM program.

"Blocking" tasks

After applicable and effective tasks have been selected, they must be blocked for effective performance. (Fig. 2-2, page 26)

Blocking starts at the task level. For instance-achieving performance effectiveness for a large turbine overhaul requires selecting and performing between 20 and 50 major TBM and CDM rework/repair tasks. Many of these in turn will be performed hundreds of times. We incorporate these into the disassemble/reassemble schedule, as a project to assure task completion and coordination. This theory applies across the board, even at the instrument calibration level. (We would never send a technician out to calibrate just one instrument in a rack.) A good measure of the effectiveness of PM programs is the degree to which they achieve blocking to conserve performance trip time. Blocking also reduces equipment outage duration (Fig. 3-5)

Motor Operated Valves: EQ, Essential, and Non Essential in Normal & High Temperature Areas (Hotwell Steam Tunnel)

EQ Hotwell/ Steam Tunnels	18 month EQ MO lube			18 month EQ MO lube	
	18 month Electrical Check			18 month Electrical Check	
		18 month Mechanical Check			18 month Mechanical Check
Other EQ		36 month Mechanical Check		36 month Mechanical Check	
		36 month Electrical Check		36 month Electrical Check	
Hotwell/Steam Tunnels (E)		36 month Mechanical Check		36 month Mechanical Check	
		36 month Electrical Check		36 month Electrical Check	
Essential: Other		54 month Electrical Check			
		54 month Mechanical Check			
Non-Essential		72 month Electrical Check			
		72 month Mechanical Check			
Hotwell Steam Tunnel (N)		54 month Electrical Check			
		54 month Mechanical Check			

Figure 3-5: "Blocking" tasks reduces equipment outages duration.

Assembling PM activity into natural groupings depends on:

- skill requirement
- the task
- the interval

Once grouped, tasks indefinitely stay together for scheduled performance. There is a temptation in traditional PM programs to attach additional checks to scheduled maintenance packages that add no value. The gatekeeper's role is to keep these activities out. PM integrity hinges on absolute credibility. In a discretionary environment, the absence of credibility means such PMs would likely not be completed.

PM tasks and vendors

Maintenance addresses component failures and their causes. Identifying common failure modes is an intermediate step to selecting effective PM. TRCM formalizes this step and consciously evaluates maintenance options and their effectiveness.

Vendors have the ability to provide adequate PM programs if they write effective manuals. Many do not. PM is an art—a specialty. Vendors don't always know the best ways to present PM options for their clients and their unique needs. They understand common failure modes for their own equipment because product-development cycles address these before products are brought into production. It is relatively common to find the full spectrum of RCM options represented in vendor recommendations:

- do nothing until a problem manifests itself (NSM)
- monitor for specific problems at some intervals, then fix them (CM [CM]/CBM)
- do rework/replace tasks on intervals (time-based maintenance (TBM))
- inspect to specific criteria (CDM)

Functional and performance tests are also prescribed. These insights are excellent starting points but they don't guarantee an effective program. Taken literally, and in full force, they often assure excessive and redundant monitoring or overly intrusive maintenance with excessive parts replacement. Furthermore, vendors can't anticipate a user's exact application. The application determines which of the common failure modes become dominant. Fortunately, almost every vendor calls for adjusting program intervals and recommends performance based on experience—the universal out.

In most PM programs, vendor-prescribed tasks are adequate. From an analytical perspective—and when cost-effectiveness and managing risk are involved (*e.g.,* R engineering)—we need to go further. Ideally, our statistical frequency-of-failure information identifies dominant failures, their occurrence frequencies, and the risk they pose in each application, so that our overall strategy can be tuned to manage risk. One needs failures to do this from a R engineering perspective, as RCM is essentially a R engineering derivative.

Basis history

The link among failure mechanisms and tasks, and our selection criteria, is what's called the selected tasks' basis. Using it, over time, we can track and understand why a given maintenance program is in effect at any point. This is important from a regulatory perspective—the maintenance rule requires that a basis be carried   forward for all PM tasks at nuclear power plants. A basis is desirable to maintain a "living" program in any plant, however.

It's much easier to assess changes made to a program when you can trace its origins. The lion's share of changes are based on an assessment of the current state and attendant equipment needs but why a given program was in place is almost never specified—even in the nuclear plants. At best, change-histories justify an existing program and provide the basis for it but in a regulated or LCM maintenance program, a documented basis has value. A basis is an important step to developing an effective, living maintenance plan.

PM work packages

At the equipment level, we can package multiple PM tasks to facilitate efficient work performance. Work can be organized by the performer skill group-instrument techs, electricians, mechanics-by frequency (weekly, monthly, quarterly), or by other convenient grouping attributes.

A PM program should be designed for each equipment group, based on the degree to which PMs resemble and differ from one another.

Information sources

Many sources of information identify viable PM tasks and their associated failures. Vendors provide more information about PM activities than failures, but it's easy to infer associated failures from their recommended tasks by analogy, comparison, and experience. Manuals from OEMs provide maintenance, diagnostic, PM, and calibration guides. Performance interval information they provide is typically not directly applicable, so the user needs experience and judgement to support intervals—or, better yet, a diagnostic capability with an age exploration program. Users who presume continuous service, and lean toward literal internal applications, design overly conservative intervals.

In addition to vendor operations and maintenance (O&M) guidance, there are:

- standards (industry, user groups, professional societies)
- legal guidance (particularly in regulated fields)
- insurance standards
- shop practices
- basic failure study and analysis
- benchmark plant or equipment practices

Occasionally more sources are available. If we take boilers, for example, there are:

- OEM manual guidance
- state laws
- American Society of Mechanical Engineers (ASME) Boiler and Pressure Vessel Code, V1
- insurance agreements
- EPRI guidance
- plant practices

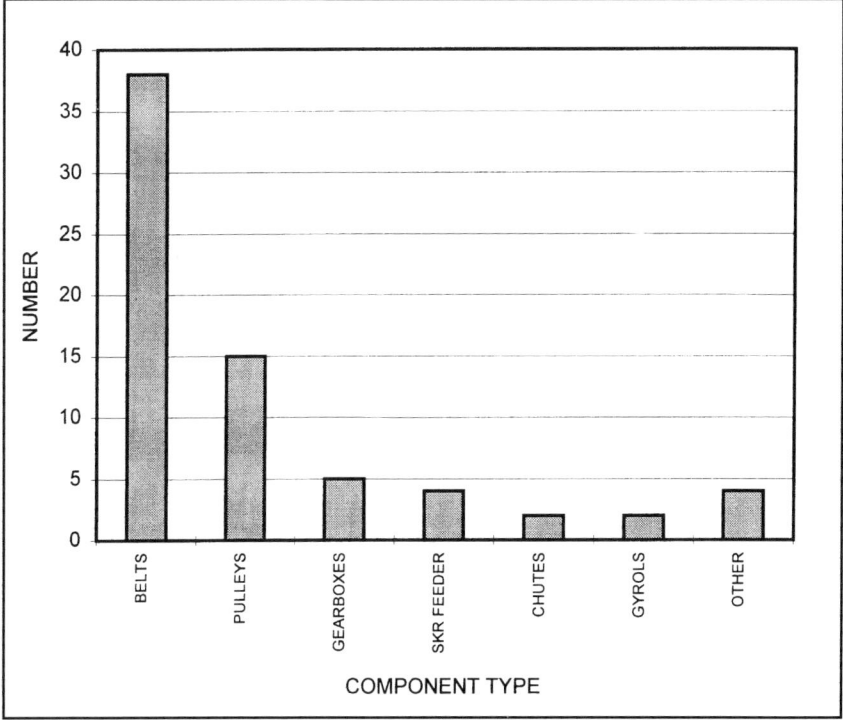

Figure 3-6: Coal Belt Assembly Functional Failures

- specific site-failure experience
- specific code interpretations
- failure experience (such as safety relief valves)
- industry event experience

Peer programs, supplier processes and literature, and published professional society papers provide a wealth of additional information. In some instances, company or plant licenses may identify additional specific requirements, particularly environmental or risk-management requirements.

The point is that to develop a complete equipment failure perspective, we must review all information sources and establish a relevant program based upon plant operating schedule, maintenance capability, and policies. The plan, whatever it is, must meet optimization goals, work simply, and be supported by workers. Worker commitment is crucial to PM success. Successful plans are those developed by worker teams, incorporating their ideas.

Effective, simple PM standards can address general classes of equipment (supported by generic industry performance information), site-specific failure experience, and unique worker insights. Developed by teams, they not only capture different perspectives, they gain buy-in. Even imperfect plans can rapidly optimize, under "age-exploration" constraints, to quickly correct for any initial absence of data. Selecting which classes of equipment to address can be based on value—deciding operational impact and cost. Once standards are developed, they simplify life for workers, schedulers, and planners.

PM and maintenance performance consistencies support R in many ways. Standards can reflect unit failure experience (Fig. 3-6).

Failure mode, failure mechanism, and cause

When components fail, the failure mode describes how they do so. Mode and cause together define a failure mechanism. A failure mode can be managed if you understand the failure mechanism. The goal of FMEA analysis is to identify—concisely—the failure modes and mechanisms of interest (Fig. 3-7). Successful plant operations depend upon achieving design failure modes and full component life.

The effects of failure types vary. Inconsequential modes and effects can be ignored while major, intolerable instances must be understood and managed carefully. Failure mode variations among manufacturers and within a given manufacturer's product lines can reveal radically different aging factors. For this reason, again, success with PM and ARCM

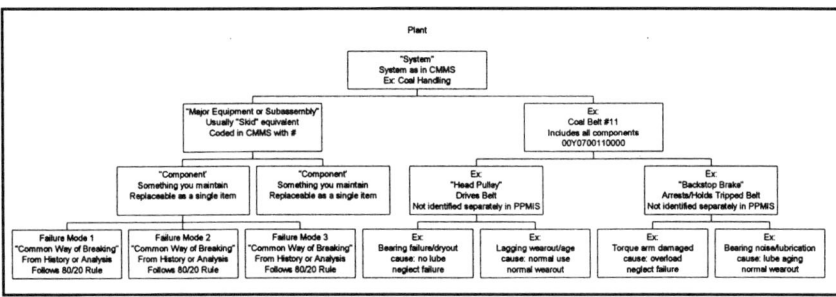

Figures 3-7: Equipment Failure Hierarchy FMEA

requires that you know the manufacturer's product lines and document your experience. Manufacturer representatives usually do a wonderful job helping to specify suitable products. These manufacturers cost more—but they very often warrant the extra costs. They provide valuable selection criteria service.

Criticality

"Risk" is mathematically defined as:

Probability x Consequence

where failure effects determine consequence. Analytical data collected from actual experience, industry data, failure libraries, vendor literature, similarity comparisons, or published failure studies quantify failure probabilities. Knowing failure effects completes a risk assessment.

However, risk itself has different measures. Overall equipment failure risk is meaningful, calculable, and measurable. FMECA (with "C" for critcality) and fault free analysis provides systems-perspective risk management tools. Overall failure "risk" for any major plant equipment, component, or integrated system could be assigned a numerical value at the design level. For example, the overall "mission success" goal for a system might be set at 0.995. From this, a FMECA of all system equipment—including supporting subassemblies and components—could then be developed and overall R calculated. If the calculated R didn't achieve the mission-established goal, then "criticality analysis" could identify R loss contributors to mission failure. These contributors could be re-evaluated and the largest risk re-apportioned as a design tool. Subassemblies or components in which failure-risk is high or which offer opportunities for risk reduction, are improved until the design-risk goal is achieved.

Risk-allocation looks at the desired final product and identifies an overall target failure risk. It is broken down to the system and component level, where FMECA identifies the main risk drivers. Design changes (substitutions or redesigns) can address and improve overall mission risk—systematically and on a budget.

The term "critical" has origins in R engineering and FMECA.

Unfortunately, the association with "R engineering" has been lost in common use. (FMECA ranks failure modes based on the contribution to failure. In this context, criticality provides a relative numerical rank.) Instead we have arbitrary delineation of equipment into critical and non-critical categories. The results are not useful. Once arbitrarily tagged "non-critical" it is generally assumed equipment can safely be ignored-even after clear-cut fault evidence develops!

This is not the same context used by Nolan & Heap. Their "critical" definition was that a critical failure is any failure that could have a direct effect on safety. The two differ on the basis of:

- basic purpose
- calculation ranking
- failure focus

Thus, "non-critical" has wrongly come to mean "low priority" and "ignore." In truth, failure modes, mechanisms, probability and consequences determine criticality risk—not equipment. This is exactly FMECA methodology. As an engineering FMECA points out, most failures aren't major mission risks, though a disproportionate few usually drive overall risk results. This is why risk allocation using FMECA works! Focusing on the wrong failures—exactly as if you blindly adopted the "critical equipment" approach—misdirects resources. This is the downside to a "critical equipment" mindset: Failure modes and mechanisms are important!

The importance assigned to equipment determines criticality-how high risk failure modes are converted into overall plant impact. Most of the day-to-day "failures" on "critical equipment" (e.g., work orders on important equipment) are not actually functional failures but are condition-directed work (from CM lists) that must be prioritized in context. An RCM mindset is an extremely powerful tool with which to prioritize work and ensure that operations work control is effective.

Practical Difficulties and Fishbones

Formal FMECAs are an engineering tool that can bog down with extraneous detail. There are literally thousands of ways that things can

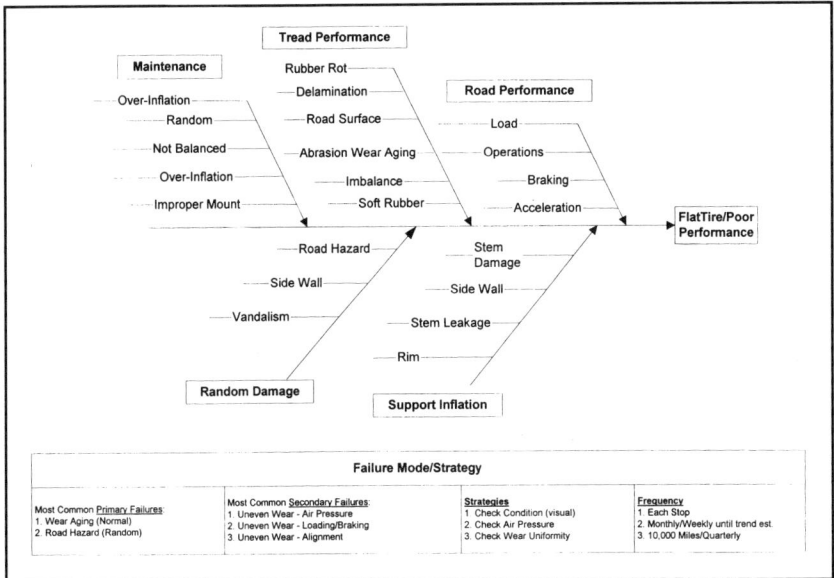

Figure 3-8: Ishikawa Fishbone Example

fail. What needs identification—at an appropriate level—are not the many ways in which things can fail, but the one or two ways that things do fail—predictably—in plants achieving their design capability. This means identifying "the critical few" failure modes—usually three or fewer—and assuring these are managed.

This "real data review" brings authenticity to FMECAs. Practically, equipment fails in a few, repetitive ways. To work smart, efficiently, and cost-effectively, we must take advantage of this fact. One of my complaints with nuclear licensing and regulatory bodies is their fundamental inabilities to come to grips with this statistical fact. Rather than focus on the way things do fail in this world, they theorize ad infinitum. Practically, this has no value in a mature industry.

Statistical Process Control (SPC) is a technique to focus work groups on high risk failure contributors in manufacturing processes. Ishikawa diagrams (named after the inventor)—or fishbones, based on their shapes—allow systematic group assessment and tracking of process failures (Fig. 3-8). Their effective display of statistically-based failure cause and relative frequencies, allow continuous process improvement to proceed systematically in a plant environment.

Fault tree analysis (FTA)

Once we understand an integrated system design, we can build a system block diagram and develop a mathematical logic R calculation. This is based upon the relationships among the assembled components and the probability of individual component failure. Such an exercise is an FTA.

Primarily a design tool, FTA relates overall system performance risk to the supporting component risk which quickly identifies where and how design can be improved to lessen risk. Once FTAs are built, however, plants can also use them for troubleshooting and sensitivity study. (Fig. 3-9, page 106)

Hierarchy and boundary

Plant structure proceeds in linear fashion from the site, with common support systems, to units, which have distinct and separate systems. Systems can be further expressed as subsystems, equipment skids, and components. Many components—assemblies replaced out of a box—can be further expressed as individually replaceable parts. This forms a hierarchy.

Functional failures can occur at any of these levels. They flow up the hierarchy depending on relational logic, redundancy, and design robustness. As components and parts fail, these failure incidents eventually generate functional failures. Functional failures occur where you define functions-the boundaries. Proximate failure occurs in components and parts. Making corrections to components and parts to restore function and performance is the ultimate focus of any maintenance strategy, whether it is TBM, CDM, CM/CBM, or OTF. Failure analysis is developed from the failure evidence. This will be found ultimately in failed components and parts, and the hierarchy and boundaries in which they lie.

A common hierarchy includes:

- unit
- system
- equipment/subsystem
- sub-tier subsystem(s) (if any)
- component
- part

- failure(s)
- causes(s)

The system/subsystem relationship can be replicated as needed. In practice, there are rarely more than two system levels, but exceptions can occur. Generally, analysts set the number of systems and levels of detail. They occasionally desire more levels than software allows. We must also remember that these are like an organizational chart—an abstraction. Real plant equipment can have dependencies and ties that don't show up in design hierarchies. Take any model with an appropriate grain of salt. Its utility is how well it provides useful insights.

TRCM establishes system boundaries, interfaces, and input/outputs; ARCM simplifies this process using existing, applied system definition that flows from plants, models, and CMMS structures. Not that extensive system design layouts aren't helpful. They are. But ACRM doesn't strive to translate existing design information into immediate forms. In addition to supporting the units' needs more exactly, RCM can be utilized to keep basic rules in mind:

- allow no component duplication
- seek closed systems, but work with open ones
- treat fluids like equipment—provide fluid failure modes
- anticipate equipment grouping for PM performance and rounds

Equipment groups provide one reason to understand system boundaries. Equipment boundaries simplify PM programs when selected carefully. They naturally define closed fluid systems and groups. When open-fluid processes are designated as systems, remember the open-system rule: Fluids (including energy) that flow into a system influence the system and can substantially influence system failure performance as they touch all pressure boundary components. Fluid performance tracking is simplified by treating fluids as another system component.

Functional Reviews

Engineers find new functions in familiar areas, and their design elements, major O&M systems, and equipment. These reveal functional intentions for plant owners and operators. Reviews of systems and equipment with which you are unfamiliar should be given extra considerations while old standbys, like feedwater, should be skimmed.

Design-to-design variations warrant quick review. Most A-E descriptions average 20 pages or fewer for a fossil system and are highly repetitive from plant to plant. They provide high-level schematics with important redundancies, which are useful for grouping. The degree to which the original plant design basis has been maintained is reflected in manuals, guidelines, and other available documents. Older fossil plants are often significantly different from their "as-built" configurations. The degree to which the existing plant deviates from documentation hints at the maintenance effort that will be required. TRCM reviews exhaustively create the system design basis, while ARCM captures the essential highlights for PM use.

Some OEM documentation and manuals are not available. Typically, in older plants, vendor manuals are missing or hard to locate. Where unavailable, an RCM-type review can reconstruct systems based upon similar systems already analyzed, station review, and previous experience. Supplementary documentation developed by walkdown may be needed. Personnel interviews provide valuable insights and the major historical events that have influenced the plants' overall production, safety, maintenance, and cost records. Incorporating personnel experience even when documentation exists, is a strategically valuable, cost-effective RCM review aspect.

How equipment is coded in plant CMMS systems determines the detail of failure review. It shapes failure, work, and cost-information sorting. Equipment review can be facilitated by the ability to download and sort CMMS information. Classifying equipment types for failure analysis by MWR and PM reviews is a useful and effective way to speed information processing.

History

Production plants develop a failure history quickly. Industry experience and history provide high value information for failure risk analysis. Nuclear plants list functions subject to significant risk and supporting equipment under the NRC's maintenance rule. In fossil, personnel—especially operators—know "risk significance" by experience. Significant risk factors for major equipment can be extracted from plant trip data submitted to the NERC at the unit level. Generic NERC data can also be consulted by unit class. In a new plant, it's essential to discuss projected risks with operators. For instance, fire risks differ radically with different fuels, and so an operator interview usually conveys such information quickly.

The importance attached to personnel interviews and comparison plant experience cannot be over-emphasized, both for gray beards who remember "that little valve up there" that blew its packing, initiated feedwater upset, flooded the drum, and tripped the unit.

In assessing equipment importance, obtain a representative history of between 3 and 10 years worth of failure data—usually as MWRs obtained from the CMMS by system and sorted by equipment tag number. Typically, for any given class of equipment or component, several hundred legitimate failures are necessary. This can be difficult to obtain in nuclear studies, where many work orders are hypothetical audit questions, not failed equipment. Depending on the plant's coding system, assemble information according to a given equipment type. Large-system reviews require methods to manage and display information that are fast and standardized. Generally, you need to collect:

- system equipment component lists with reference data
- P&IDs
- MWR lists for a statistically representative period
- vendor PM recommendations (from vendor manuals)
- existing PM plans and tasks (from CMMS)
- standard component programs (if any exist. They could be informal)
- management commitments
- federal agency information, whether EPA, OSHA, NRC, etc.
- state agency—departments of health, state boiler inspector, etc.
- insurance providers (fire, industrial, etc.)
- standards (ASME codes, National Fire Protection Association (NFPA), building codes, etc.)
- other local or regional standards
- operational reports—usually, summaries of operating logs and

production reports supplied to corporate managers, generation accounting groups, Federal Energy Regulatory Commission (FERC), and NERC
- the station operating license
- "unit trip" or other operating event reports

Operating reports provide insight into the station's major outage events to detail a plant functional failure experience.

R engineers review failure descriptions, item-by-item, from hard copy summaries, tallying up types of failures and operations impacts. They summarize dominant equipment failures, frequencies, and costs to provide raw statistical data for meaningful reviews of existing PM programs and what's needed to build PM templates. Once a failure history has been reviewed (and with a list of important equipment in hand) they review equipment importance and O&M recommendations. This prepares them to talk with operating and maintenance crews about plant strategies and experience.

Standards

In practice, there are two, possibly three broad categories of equipment and components:

1. High-impact items without redundancy that operate singly or in unspared-pairs and directly supports generation. Boiler feed pumps (BFPs), turbines, ID air fans, circulating water pumps, boilers, and reactors fall into this category. Large, expensive, and economically important, they also have safety implications due to rotating inertia, high temperatures and pressures of contained fluids, cooling functions, and plant trip potential. These components get maintenance in-depth analysis including equipment history review—MWRs, overhauls, PM program, emergency callouts, operating records, and historical practice

2. At the other extreme are replicated items found in significant numbers throughout the unit. A single-unit 830 MWe BWR has several hundred motor operated valves (MOVs) for instance. Smaller pumps, motors, valves, and other components are present in even greater quan-

tity. Some warrant maintenance but few warrant individual maintenance analysis based upon similarity of design and failure modes (unless identified by failure reviews). A nominal component model can be constructed and a standard maintenance plan developed to be applied in production as a plant standard

For example, an air-operated valve standard is developed, tailored by plant, based upon:

- operating air purity
- general plant cleanliness
- local environment—rural/industrial, etc.
- availability and quality of ventilation
- application

Standards must address substantial design differences only. When similar equipment types have similar programs dominant failure modes should be the same, or nearly so. Some of the equipment presented above is unique and needs a unique standard. On the other hand, many common motors, valves, and a host of other equipment are very similar—even when provided by different suppliers. These beg to be lumped together under a common standard.

3. Between the extremes, equipment maintenance plans can be broadly based upon standards tailored to an application. For example, multiple types of chain drive-style sootblowers have similar designs but different parts and unique characteristics. A standard can be developed for the general case and tailored to several special applications. Details in the standard can vary. Because blowers can be associated in different groups, a "high-soot group" could be maintained to one standard, a low-soot group to another. Unimportant blowers could be given NSM aside from routine lubrication

Development of standards is best done case-by-case, on a site-tailored basis, working with the maintenance craft and engineering staff. Special requirements from site to site will force standards to be tailored.

The environment, interfacing supporting system design, maintenance standards—all play roles. One site may have excellent, dry instrument air; another may have it comparatively wet. The former site may be in a dry, dusty climate; the latter, a moist, damp one. The former site might find that motor heater checks add no value at all but that air inlet filters are essential. The latter site may require motor heater checks to avoid damp and burned-out windings from motor startup.

Tailoring of vendor requirements to individual site conditions eliminates "canned" vendor-recommended PM maintenance and yields programs better suited to local conditions and practices. Skilled workers who understand equipment needs implicitly can perform off-track PM tasks and programs, year after year, without adjustment if their facilities start with OEM-based programs and never develop a PM "tuning" processes.

Standards also effectively address vendor requirements (especially since vendors go overboard on suggested maintenance at times).

Either that, or vendors advertise that, "Our equipment demands virtually no attention beyond periodic checks." This poses some dilemma for the system integrator who takes vendor promotional literature literally at face value!

A standard identifies those few things that must be checked on routine operator rounds, TBM PMs, and special CNM tasks. It provides an appropriate, site-specific interval to manage risk, as well as those things that need to go into a PM schedule on a longer interval. We identify the critical few by direct RCM analysis or comparison to a known analysis— and by using insights based upon other key information and lessons.

Comparison Analysis

TRCM always performs "comparison analysis" as a final key project step. These before-and-after snapshots hold limited value when compared to the value achieved with RCM-based PM reviews.

Comparison analysts require a high level of bookkeeping to maintain a spreadsheet documentation of project accomplishments. Documentation is suspect if plant staff delays PM changes, misses review meetings, forgets to prepare reviews for meetings and neglects rework analysis.

Comparison analysis is performed when plant owners want to know how well a PM optimization performed. Comparison analysis will continue to be performed for this reason. But numbers should be carefully considered to avoid a "paper" PM program that means little.

Summary

Detailed, TRCM analysis has great value primarily as an analytical learning tool. Standards—applied quickly and reapplied many times—speed PM assessment and implementation. People working to standards develop processes and production methods to perform repetitive tasks with consistency, speed, and simplicity to make the standardized application of formal RCM techniques effective.

However, dedicated equipment applications, failure consequences, and the ways in which we decide equipment importance necessitate adjustment that limits the depth of TRCM analysis. Fortunately, benchmark and composite references mean detailed RCM analysis isn't always needed. When we identify important components, develop appropriate programs and standards, then devise the appropriate PM task and measure the results, we can further adjust individual component PM programs to overall standards requirements, where needed. Simplify. Standardize. Implement the best ARCM-PM program.

Maintenance Process

Traditionally, work orders are based on noted problems. This is the CM maintenance model. A second model—scheduled maintenance—supplements and extends the fundamental model. But identification of problems, *a posteriori*, is how traditional maintenance works. Operators know the problems because they know the system, the equipment, capabilities, and what they need it to do. Response-based maintenance was the first improvement over disposable equipment due to its significant capacity to reduce cost.

Response-based maintenance is very cost-effective when compared to the alternative—nothing! It's the first basic step in any maintenance program. The next step is an intuitively harder one—scheduled maintenance.

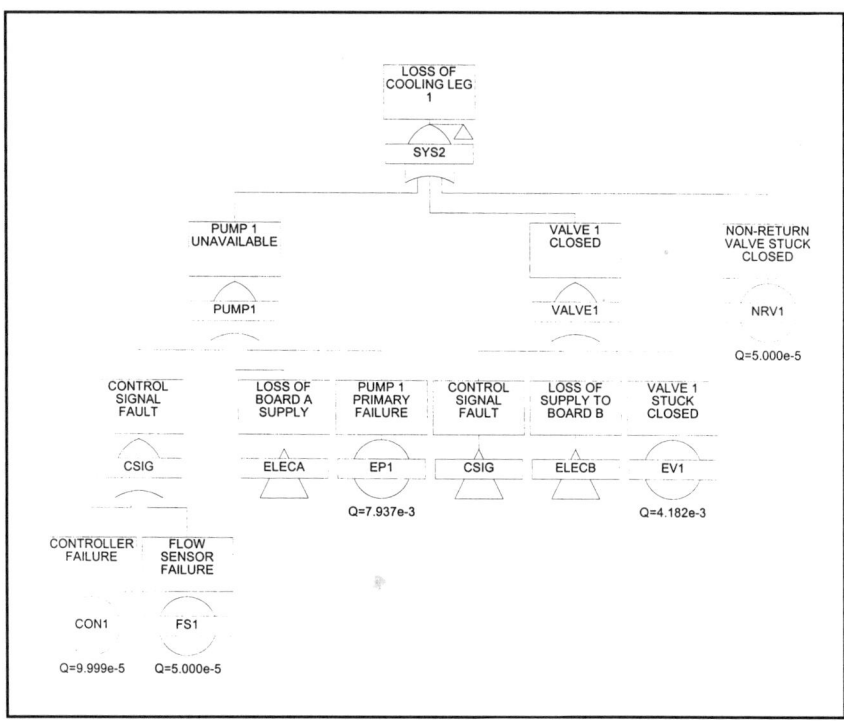

Figures 3-9: Fault Tree, courtesy Item Software, Anaheim, CA.

Scheduled (or preventative) maintenance is much harder to implement than failure-based maintenance, but many companies satisfy themselves with response-based maintenance implementation. To see what scheduled maintenance can do, consider first what it cannot (Fig. 3-10).

For any component, there is some residual failure rate tied to random failures inherent in the component (based on design and production processes). This "minimum failure floor" the best PM program can only approach, with diminishing returns. Between the response-driven (RD) and random limit (RL) floor is the area that scheduled maintenance can influence. This "RD-RL difference" can only be improved by fundamental design and manufacturing product process improvements.

The RD-RL difference varies with equipment. Factors establishing the floor include:

- design
- materials

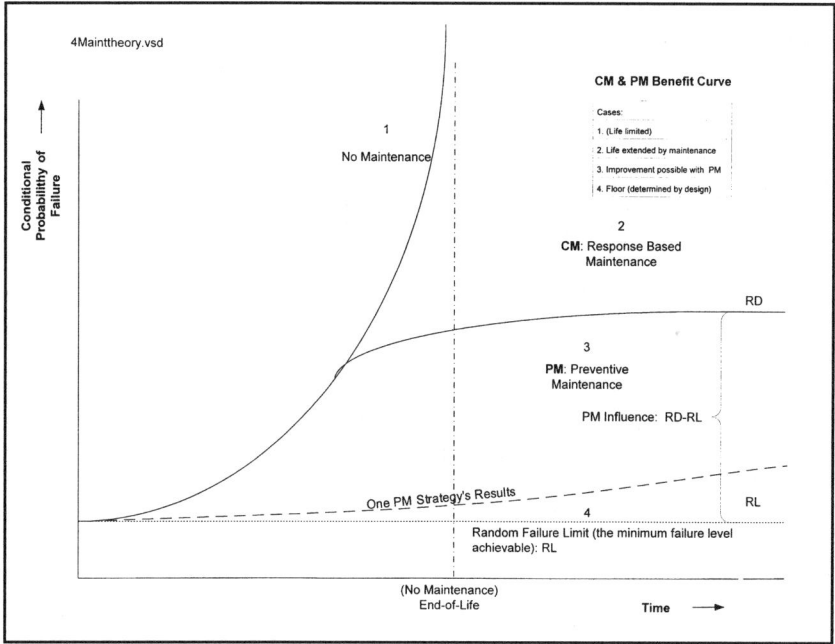

Figure 3-10: Random Limit

- construction
- environment
- operation

Failure results when stress exceeds capability. Design, materials, and fabrication provide equipment with capability. Operating stresses in a perfect, variation-free world would never exceed design limits but in the real world, they do. Equipment designers must anticipate field loads and conditions. Suitable materials, manufacturing, and dimensions assure products perform adequately with a factor of safety.

Designers build systems from components and equipment. They aren't exact. System designers work off experience. They stretch design envelopes with operating and environmental assumptions. Some application stresses exceed design expectations. A residual failure rate is often present in efficient economic design. A perfect maintenance program would achieve the residual inherent capacity of the design (Fig 3-11). Discovering this floor (with scheduled maintenance)

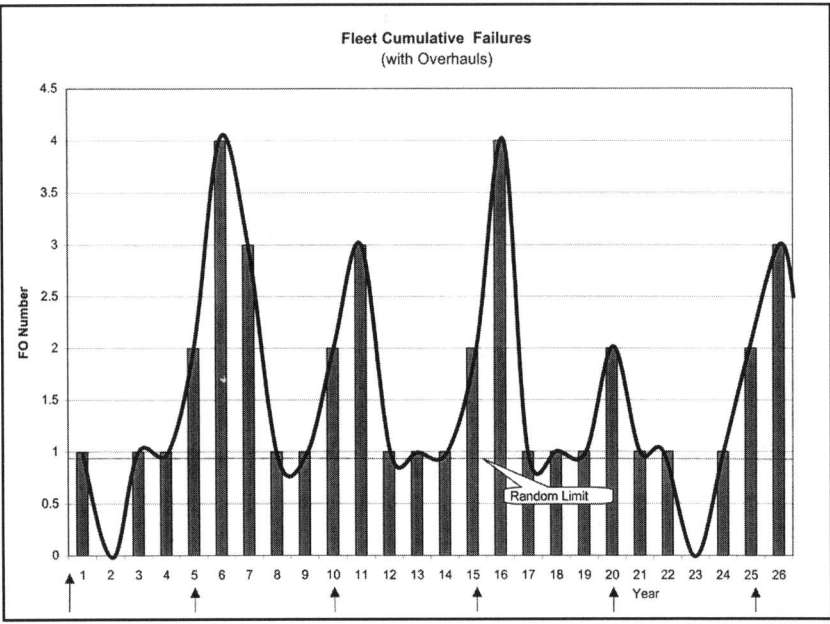

Figure 3-11: Fault Tree : Loss of Cooling

and extending it through design is the focus of ARCM. Ninety percent of component failure modes do, in fact, realize this inherent capacity with virtually no maintenance. This is the discovery of RCM and why we must use tools with great care!

CDM lies somewhere between "absolutely no maintenance" and "inherent R limits". Response-driven maintenance works well, as a first step-and this is where many organizations find themselves. Further strategies move closer to design-limited R.

Scheduled maintenance effectiveness has been validated by long term measurement of steam turbines. Here the failure rate curve looks like Figure 3-11. Load capability determines overhaul times.

Peaks are limited by suitable PM tasks that reduce failure rate to lower levels for a given period. "Performing maintenance" establishes an intermediate failure level R curve. Scheduled maintenance plans drive failure rates further towards the inherent R floor. To the degree reduction is cost effective, scheduled maintenance is effective.

For some failures, operation changes or re-designs are necessary. Failures caused by unanticipated external environmental factors require

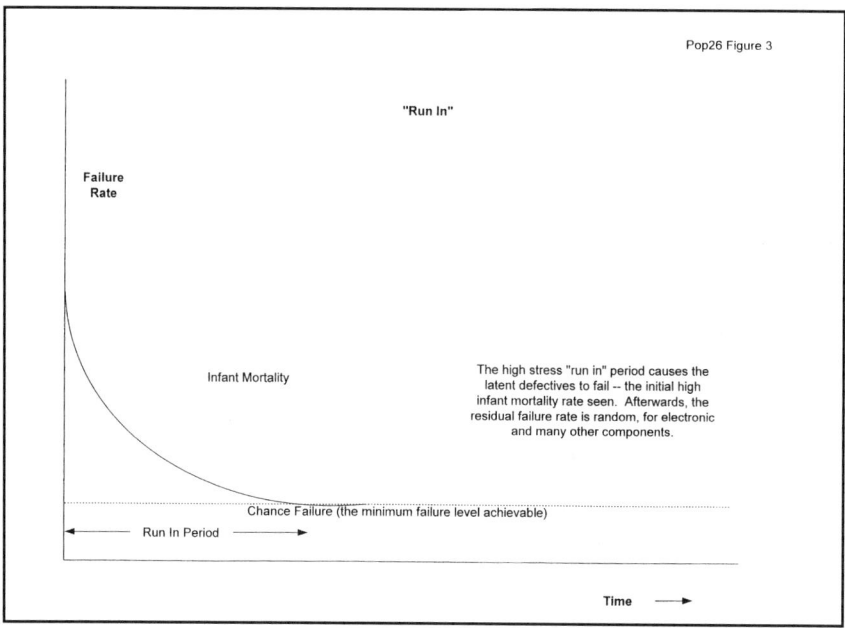

Figure 3-12: "Run In"

review of environmental control. Opportunities vary from one application to the next-some are minor, others are large. Other factors influencing task selection is the requirement that we add value. Infant mortality or quality can influence the "run-in" period failure rate. After some period a higher failure rate returns to an inherent baseline level (Figures 3-12, 3-13 and 3-14).

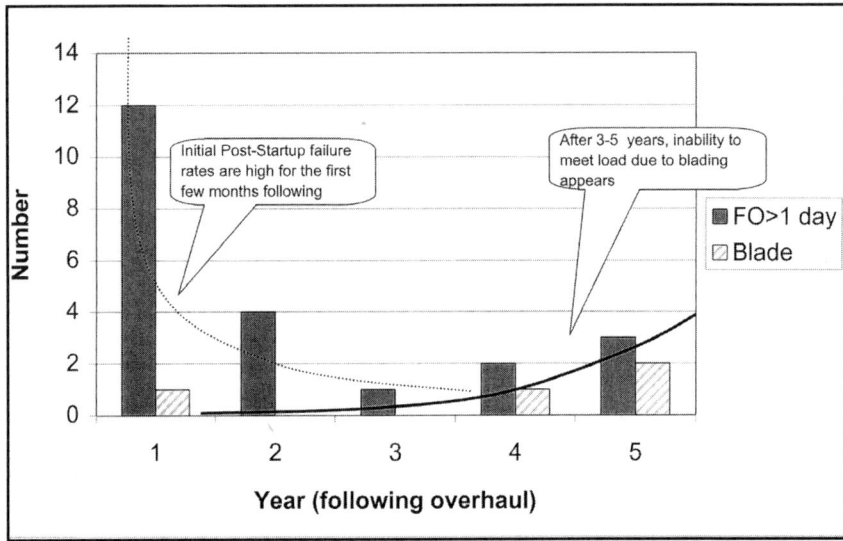

Figure 3-13: Turbine Failures. Although turbine overhaul failures following overhauls follow an infant mortality curve overall, the composition of individual failures is not so clear. Limited extension intervals suggest that extended lifetimes between overhauls are feasible for many turbines. Ultimate age-based turbine failures appear to be a composition of blade deposit and erosion failures for many machines. These cause stage efficiency to fall. Overhauls are then a question of economic production tradeoffs.

Assessing numerically small failure numbers means interpolating between few failure events using judgement. This is at best a risky proposition. The comparison of many machines with many failure modes at the other extreme is also fraught with risk. In the final analysis, an engineering inference supported by detailed parts examinations and performance tests is the most useful approach.

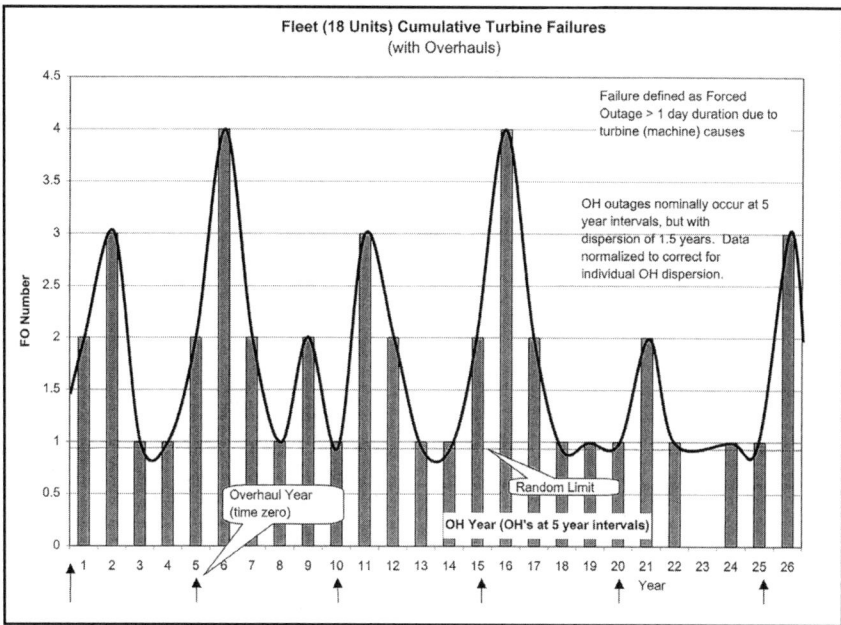

Figure 3-14: Cumulative Turbine Failures. With a fleet, turbine failure periodically approximates the overhaul interval. While this suggested wearout, closer examination showed most failures occurred following start-ups and reflected infant mortality problems. In fact, that best explains the timing! Data like this suggests that turbine overhaul intervals may be extended with minor risk. Until age exploration establishes a wearout interval with more exact failure experience, the predominant risk is under-utilizing the asset. OEMs complicate issues by providing traditional time-based overhaul interval recommendations.

Defining efficiency and load loss "failure" further complicates the problem. Some companies have vague standards for end-of-period performance that provide the basis for overall intervals. Without and exact efficiency standard, failure to achieve performance is a subjective call. Although the issues are complex, there are simple measures and solutions.

Lastly, the data support the idea of random limit. For this fleet, some failures persist throughout the overhaul cycle.

Chapter 4
Plant Needs

Operations is hours of boredom punctuated by moments of stark terror.

-Anon, Navy

Production and Delivery

Production processes have great variety—some are batch, some continuous; some three-shifted, some one. Most power plants are staffed "24 and 7"—but this is changing as operator functions are redefined from shift engineers (who can start up, shutdown, and reconfigure the plant) to utility workers performing minor maintenance and acting as diagnostic technicians.

What will not change, is that plant processes require monitoring. Even remote-operated site dispatchers assume monitoring status. When monitoring is separate from maintenance and other production support processes, there must be interfaces. "Operations" traditionally maintained plant configurations, production, monitoring, and control while others provided support services. Supporting processes greatly influence maintenance effectiveness. Operators identify CDM needs and initiate most plant maintenance. Various PMO implementations show that even with advanced diagnostic techniques, operators initiate at least 80% of non-routine maintenance MWRs. Operators form a key link in maintenance performance.

The planning, scheduling and other maintenance groups can provide effective support-to the degree operators initially monitor and identify equipment discrepancies—clearly, accurately, and expeditiously.

Operator value is even more pronounced when they provide flexible, immediate maintenance response—on the spot packing take-up to address a small leak, for example, or minor instrument adjustments. In many instances, operators can be trained to regularly perform these duties. IPPs have effectively trained operators for light maintenance roles.

Like all other staff, operators are a resource that must be conserved. Minimizing plant staff requires understanding exactly what needs to be done, when and then performing this (directly or remotely) with minimum wasted or redundant effort. This, of course, is in addition to the operators' fundamental job of managing and configuring the plant. Flexibility ensures that operators will always have plant roles. These roles may change or merge with other functions, but the basic operations role will remain until and unless plants become literally disposable.

How will operator roles change? With new information management systems, the control operator could easily be merged with a roving operator using virtual-reality technology—wearing a hard-hat with a "heads-up" display, like a fighter pilot's helmet. The operator would be free to control the plant, on the spot, in the course of making rounds. Functions such as plant monitoring, control, and startup/shutdown/reconfiguration capability are preserved in radically innovative ways.

Perhaps the correct way to say this is that operators who add value will always have roles. Companies must restructure processes and roles to add value for all personnel—or fade away. They need to adopt methods and processes that increase operator value. PMO and ARCM add operator value because they improve the ways in which operators are used for CNM. This improves R and maintenance performance by:

- identifying maintenance needs sooner
- initiating "condition-directed" maintenance earlier
- providing greater focus than a traditional corrective maintenance program

Delivery

The value of any plant improvement process is limited by the ability to deliver the benefits to the customer. When RCM methods drive PMO for equipment maintenance programs, LCM, and overall plant work scheduling and coordination, it's based upon an intrinsic belief that they represent better ways.

The next question is: how can a plant get there? Two basic processes are required. If absent, they must be developed. They are:

- a PM process
- an LCM scheduling process

There is no such thing as a completely effective PM process. An effective process must:

- consistently deliver a high PM-completion performance
- provide effective methods to set priorities and allow rescheduling to deal with contingencies (like outages) that compete for resources
- provide personnel opportunities to learn and to change processes, as needed

The LCM scheduling process must be able to deliver routine, consistent, non-outage plant activity. The scheduling window must be large enough that most work planning is feasible, and routines can be developed and learned. It needs to address management strategies for systems and conflicting equipment risk, "divisionalization" and equipment grouping and how it supports the adoption of schedules and routines. ("Today's the first Monday of the month. We always do fire protection." "During the third week of the quarter, we always do 4160 Bus 1B fault detection tests.") Such routines provide anchors. It allows groups to plan "exception work" around a system's or an equipment's "known work" basis. Equipment comes up repetitively and can be worked online. Risks are known, pre-approved, controlled, and accepted. Routine work has preset rules, checklists, conflict control routines and checks, and qualifications that avoid random performance, outcomes, and failures.

Management must commit to support process development and assure that maintenance orientation actually changes. Process changes must support work structures, software, and processes—*e.g.*, an LCM schedule that requires operations and scheduling, working with performance groups to lay out the systems they need to work on in some overall strategy.

Systems Approach to RCM

Development

To begin to apply RCM select a part of a plant. A large plant is complex—an integrated coal-fired or nuclear plant has upwards of 100 systems. To implement PMO and reap tangible benefit requires focus. You must identify, review, and evaluate high value systems in a project-oriented, standardized, cost-effective way. (Fig. 4-1)

It takes years to develop a "feel" for the operability and functional requirements of a specific plant. In-depth knowledge of 15 major systems—and working knowledge of the rest—is required. Specialists require more. Profound understanding of key generation-supporting system requirements, including R and cost, helps to ensure optimal plant operations.

Developing an in-depth understanding requires a cross-functional team effort, a process, and a place to hold the "collective consciousness" that results. An integrated LCM operating and maintenance plan provides the process and place.

Training, design basis, and needs awareness

At a facility, the high-voltage power supply demanded very little maintenance for many years. Then, in fairly rapid succession, a bus duct from the main power transformer grounded and a 4160 breaker failed.

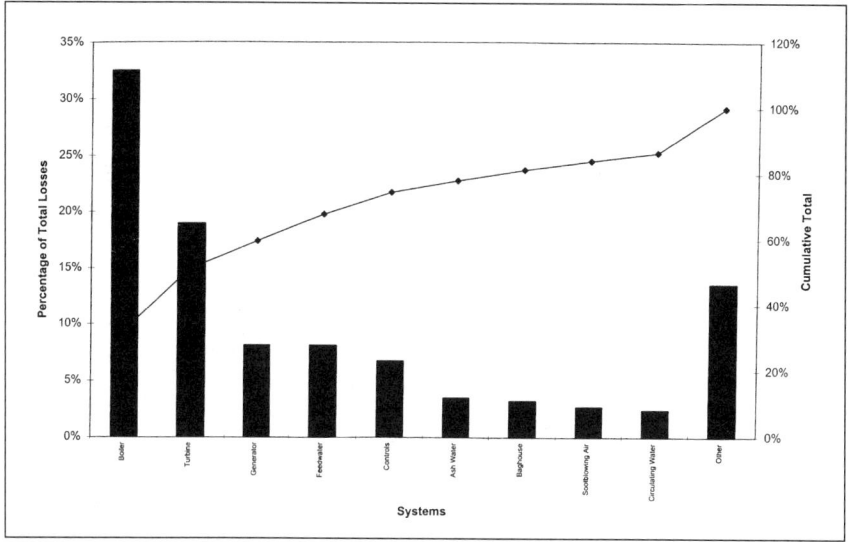

Figure 4-1: Pareto Chart of System Losses

At another unit in the same system, a station reserve auxiliary transformer blew up, along with a 4160 switchgear—all within several months. Reviews indicated that plant operators were unaware of modifications and requirements that fundamentally changed the systems. Had they not lost touch with their plant's design basis, they would have done things differently.

Plant personnel can lose touch with a design basis. In most cases, only significant events re-focus engineering on recovering them. Senior plant operating and support personnel often carry these requirements around in their heads but only rarely does this get captured in a safe environment. Nuclear plants fare better than fossil, because regulatory requirements maintain design bases; and refresher training. Even with that, there are problems in accessing, using, and modifying design basis information.

One consequence of a thorough RCM-based system review is that personnel "relearn" intended system functions, plant missions, equipment "criticality" and those weaknesses in design, operations, and maintenance inherent in an operating facility. The RCM process allows the team to recapture this information. Once recaptured, the RCM/PMO process can document it for easy retrieval and further use:

• Training
• Reassessment and updating of operator rounds for improved monitoring
• Identification and ranking of problems with redesign opportunities
• Extension of service intervals on equipment through a systematic application of age-exploration

RCM can be used to evaluate and rank design modification requests, separating "nice to do" from "must-do" modifications by objectively documenting risk. Even small modifications-like piping rerouting, ladder installation, or structural support modifications-demand substantial plant resources. Design engineering staff "cost" these innocent changes and charge for them accordingly, but plants often perform them under "maintenance." This could be to lower expenses, keep the "mods" off the capital budget, or to lower perceived expenses, but such seemingly minor mods take a disproportionate amount of the station resources. They often come in far above estimates and budget. One lengthy study of plant design modifications performed at a plant found that average design modification cost came in at more than 10 times estimated costs! This trend was sustained over many modifications and several years time. Modifications out of the shop are usually underestimated. RCM can often add the disciplined control that keeps the "nice-to-do's" on an even footing with the value-adders.

On a regular basis, RCM can screen two types of work that regularly vex plant management: capital budget requests and safety-justified work. Using RCM rigorously won't eliminate all ineffective capital spending, but I believe with strict implementation and enforcement, RCM encourages people to become more aware of options and possibilities before they commit to any one plan. It will help some utility staff to learn their jobs and will help control traditional committees plagued by little accountability, who use vague processes to spend money for projects that have no basis in actual need.

System definition

Operating and engineering personnel use the plant architect engineer's system structure to understand work and develop operating procedures. Maintenance personnel often lack such perspective, which can lead to missed teamwork opportunities. For example, when maintenance is repetitively willing to take systems down to perform corrective work, they add to operating costs. Operations' inability to recognize system-synergistic effects during troubleshooting leads to less-effective root cause analysis (RCA).

Understanding systems is a trait necessary for any competent operator. Operator training emphasizes systems and their requirements, but some plant training pales by comparison to simple U.S. Army boot camp gun assembly/disassembly drills! Usually it's because trainers forget that operating requirements stem from plant design but that operating practice deviates from design intentions, to some degree—sometimes substantially. Requirements not reflected in design documents or operating guides can be lost over time when designers' original intentions are forgotten, and increased costs are sometimes the result.

System performance measurement

The "system" metaphor is useful as a design and operating tool but it carries over into cost accounting or performance measurement unit operations areas. Only a few plants systematically follow system-based costs. Few old CMMSs provide the capability to systematically track costs below the unit level. Tracking capability is usually driven by FERC reporting requirements and codes, based upon specific components and FERC categories. FERC codes provide a place to start examining production cost contributors but they developed from an accounting/regulatory perspective. Better than nothing—less than ideal!

System performance measures break down losses into tangible, "graspable" opportunities. Fossil generation losses attributable to coal mill availability focus attention on mill operating strategies such as work performance (on/off-line), overhaul intervals (quarterly inspection schedule, tonnage based, etc.), CNM (motor amps trends, differential pressure trends, fineness trends), and even risk management approaches for startup mill trips—a common problem. Volatile coal fires suggest fire-risk measures are also important to select. System level performance standards and measures often provide warnings of problems to come.

Intuitively, I think we recognize that maintenance deferral on an asset like our house or car will result in higher costs later. This same level of concern is needed on systems in plants.

The NRC's maintenance rule mandates that nuclear plants measure the performance of "risk-significant" systems. When the rule was implemented, it elevated system management awareness in nuclear plants. Fossil units have equally complex systems-combustion, emissions, water and disposal systems all operate subject to complex, regulated requirements. Even without a maintenance rule, the value of system performance measures shouldn't be overlooked. In my fossil experience, system level measures identified many significant opportunities. Personnel at some fossil plants may look at their nuclear brethren—with the perceived luxury of system engineers—and claim it's not feasible to implement such a degree of effort, given fossil staffing levels. However, some IPPs and co-generators—some of the best competitors-have proven that they do have this capability, not just among staff engineers but on-shift operating crews, as well. Developing maintenance capabilities comes from an organization's attitudes and assumptions about maintenance-not resources. I believe a lack of "needs understanding" and training are as common among engineering support staffs as actual operating costs and problems.

System-forced outage contribution rates (e.g., chargeable losses) need tracking. Secondary effects from sootblowing, feedwater, circulating, and makeup cause boiler outages, for instance. The availability, performance, and R of important backup systems influences overall R.

Nuclear units declare "limiting conditions for operations" (LCO)—grace periods. Loss of fossil systems converts to a real outage! Nuclear units suffer LCO-driven shutdowns, but the loss of a single sootblowing air compressor at a Powder River Basin (PRB)-fired fossil boiler can cause convection passage plugging which, once started, typically ends with an outage, followed by blasting or lancing to free and remove slag and ash deposits.

System monitoring

Operations monitoring combines an operator's skills, knowledge, experience and senses (sight, smell, sound, taste, feel). It also requires the instrumentation needed to extend the senses to fully monitor all areas deemed important by equipment designers.

For operators to monitor the plant, the tools-whether traditional, discrete, physical devices or completely integrated DCS systems-must fit the maintenance strategy. A repetitive lesson of ARCM is the need to adequately define strategies for instrumentation and to tie these to the installed physical plant.

Existing work processes often:

- devalue critical and essential instruments (the "critical few" buried among the "trivial many")
- include extraneous setup and test equipment
- inadequately relate critical and essential instrument functions to operator actions

For operators to be effective with instruments-whether "working the boards," monitoring a cathode ray tube (CRT) "touch screen," or plying the traditional round-instrument roles need clear identification. Expectations for instrument response need clear, unambiguous guidance and instrumentation systems, again, must be part of an overall PM strategy.

While these rules are enforced in the aerospace and nuclear industries, fossil generation has lacked an integrated strategic perspective concerning the role and function of instrumentation. For many generators, the "low hanging fruit" in ARCM is the opportunity to organize the maintenance I&C sector. But ARCM is a powerful tool to add focus and integration to instrumentation maintenance.

Instrumentation is as complex as the plants it guides and guards. There's a lot of it and it's essential to operations. I&C specialists are among the highest paid and highest skilled personnel and the discipline represents a major portion of operating costs and capital requirements. Safety has strong ties to instrumentation as well.

What's the state of your plant's instrumentation? Do you have an overall strategy? Do you know, at a moment's notice, what any given instrument's function is? Its calibration status? Its annual costs? If you cannot answer these questions confidently, then the ARCM will help your operations staff integrate instrumentation into plant hierarchy.

System cost

Changes in system costs provide an early warning of items worth further investigation. They also provide key measures for benchmarking in competitive studies. How many generators know their air costs? How, then, would a fossil generator evaluate a proposal by an air compressor vendor to provide air at a unit volume price?

When system and service costs are measured, they provide serious numbers for thought. In a competitive environment, cost oversights raise unit cost. Loss of 1% generation in a year, and the associated generation R loss, is opportunity (and revenues) lost. Ask any IPP operator.

Assessing PM Programs

PM programs must pass the same muster as any other: They have to contribute to the bottom line.

This means measures have to be in place to assess PM costs and delivered benefits. Integrated effectiveness measures—the statistical and cost picture—are the key measures. PM activities must meet the bottom-line acid tests of technical and cost adequacy. Failing either test means the PM is probably an unnecessary expense. Like all expenses—time and otherwise—getting employees "tuned" to look for low value or non value adding expense material is a key to long term financial success.

Acid test #1: applicability

Applicability is a unique RCM attribute. ARCM uses applicability with fewer rigors than called for with formal statistical tests. It is still a rigorous test, however. A high level of assurance can also be attained by benchmarking tasks to others by similarity, using expert opinion and review, and performing statistical analysis of actual experience. In each case, simple tests assure that proposed TBM tasks are appropriate and incidences of ineffective PMs that we've examined are avoided. However, applicability is not easy to gauge. It shouldn't be presumed.

Applicability requires that a reviewer be thoroughly versed in the available technologies, their routine application, and results. Unskilled performance of suitable tasks can result in failure. This is an implementation issue. More basic is the application of correct technology in proven formats as predictive tests "on condition". The wrong test is not applicable. Using ultrasonic tests to search for early bearing failure could be fruitless if the test can't detect the fault in question.

Unfortunately, in the past some tasks have been based more on hope than proven fact, and it's not unusual to find inapplicable PM tasks. Occasionally their lack of value becomes a political football, rather than a learning exercise. Determining applicability calls for small teams of knowledgeable expert reviewers.

Many time-based CNM tasks include the specific measurement and assurance of chemistry limits, for instance. These activities fit the PM definition but have traditionally been included as a part of a station chemistry program. It's pointless to reclassify these as maintenance PMs if they're part of an effective chemistry program. On the other hand, critical chemistry alarms and instruments generally require time-based calibration or checks that are often not formally monitored under existing chemistry programs. Identifying, and finding a "home" for these checks can be a valuable aspect of an ARCM general review.

Until the station's entire scheduled work task list is reviewed and checked for applicability, there is an intrinsic barrier to implementation of any of them—and this general absence of value in *all* PM tasks destroys the integrity of the entire program. Specifically, every PM WO for an operationalized task must be reviewed and certified for effectiveness to assure specific benefit that is not repetitive or redundant to another PM check or activity. This detailed PM-by-PM credibility check builds a stronger foundation for the overall program.

It's also hard work that plant staff frequently lacks the skill to perform. Oftentimes, it has never been performed! However difficult, the benefits resulting from performing these reviews are substantial. Reductions and consolidations in PM activity almost become a by-product. "Cleaning up the books" has achieved up to a 60% reduction in system tasks, providing time for proven, applicable ARCM-based PMs. At the very least, a lot of extraneous work has been purged and the focus applied to the relevant tasks.

The applicability test is straightforward for the assessment of 80% of a typical program. In the absence of clear-cut evidence that predictive methods are applicable, the opposite should be assumed. Operational and economic-based failure prevention tasks should clearly pass applicability tests. Inability to confidently support a monitoring technical basis suggests it doesn't work!

Acid test #2: cost effectiveness

"Effectiveness," when used in a discussion or application of RCM, refers specifically to cost-effectiveness. Clearly, a non-applicable task will not be cost-effective and separating the two effectiveness criteria is unnecessary, for an obviously inappropriate task. Cost-effectiveness is, by far, the tougher to handle. Therefore, each should be given separate consideration to assure both tests are met and resources are applied well.

Timing influences effectiveness. Frequently performed activities can lose cost-effectiveness. Traditional airline RCM has rigorous applicability requirements for "on-condition" task monitoring that haven't been carried over to electric generation use thus far.

For example, all scheduled, on-condition operator tasks involve specific monitoring, incorporated as scheduled-rounds tasks. "Rounds tasks" have historically not been rigorously based and have included many non-applicable, non-effective tasks in addition to "NSM"-type, non-specific area checks. Yet, rounds offer the typical plant a major opportunity to reduce overall workload and scope while significantly improving monitoring.

RCM strives for cost effectiveness by assuring that all OCM tasks are performed at intervals that identify failures before the final failure phase begins. Because some discrete failures lack this terminal phase, often the best we can do is check for failed states. Since the design stage includes these elements with redundant equipment and channels, our tasks are channel, alarm, and status checks. (In some DCS systems, these are automated.) While ARCM strives to assure that rounds intervals are appropriate, it also recognizes that not all rounds will result in detection of final phase wearout—nor do they need to. The essential requirement for specific activity in operator rounds is to assure that tasks specifically address failures and that intervals are appropriate based upon failure rates. Traditional rounds failed to provide intervals for thorough monitoring. One had to be quick—and superficial—to complete a large plant's hourly round.

Sadly, the effectiveness test is shunned by engineering groups and maintenance support. Preparing cost estimates to support longer intervals and effectiveness hurdles is exciting only if you're an accountant. However, for these cost checks to be done correctly, a technical engineering support group must prepare them, develop effective, detailed cost benchmark cases, and then evaluate other comparable tasks. As much as I personally dislike this analysis, it sheds light on where the value is and where to place resources. Very often, the plea heard in plants, in defense of many PM tasks—usually distant from performers—is, "The cost is minor. Just do it!"

But once again, little things add up. This approach, carried forward over time, builds ungainly programs that can't be managed or maintain credibility. Typically these multitudes of things are only partially completed. PMs should be dropped if their cost benefit can't be determined with some certainty. In most instances these cases are straightforward. CNM efforts can often be trimmed based on cost. For big-ticket failures-when "what if" accountability could be an issue-responsible staff sometimes initiates a PM to wash their hands of further involvement despite the fact they're best qualified to manage the problem. This can't be allowed to happen. More than one PM program has failed based on over-scoped work. These cost assessments are the staple of any plant support engineer's work. If they can't create them, they probably don't have the basic skills to be effective in this gatekeeper role.

Process Improvement

Maintenance performance improvement must address two aspects:

- the establishment of a basic PM process
- high value, time-stamped tasks that really need to be performed

The two are parts of the same puzzle. Confusing the issue can be the problem of establishing a basic maintenance PM process when another system is already in place-even if it's not performing well. A pilot proj-

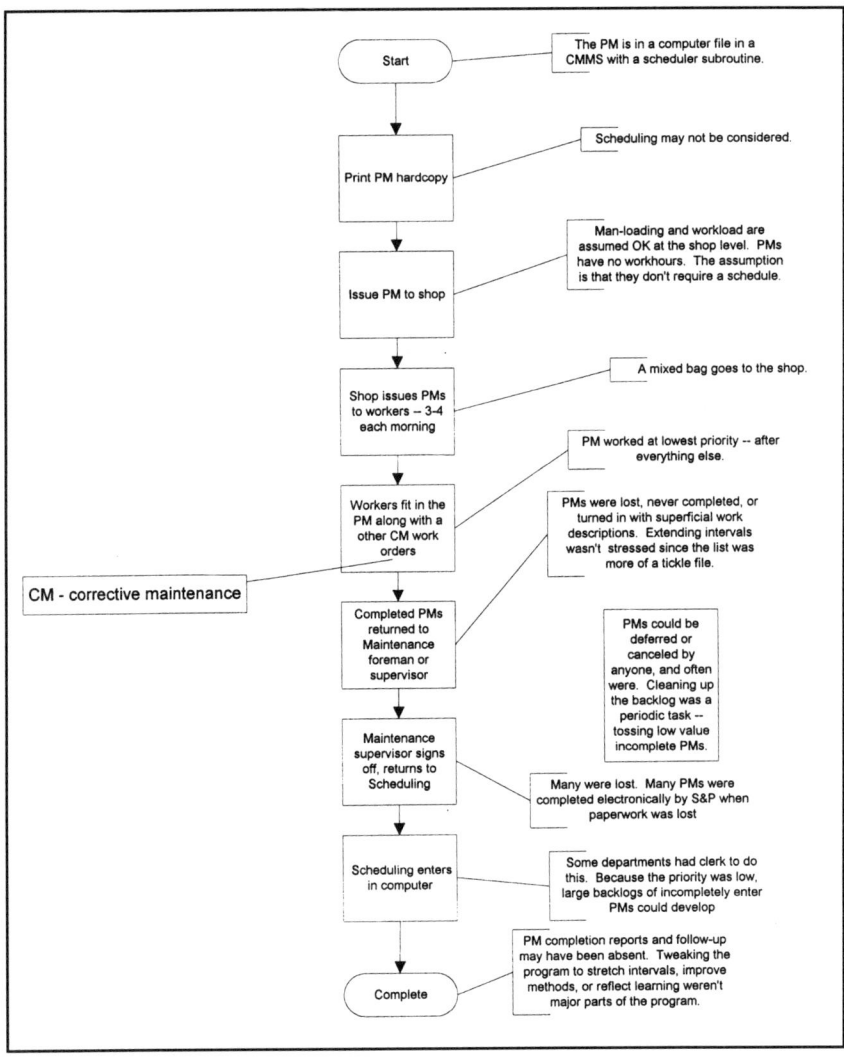

Figure 4-2: PM Performance Process

ect can be an effective way to initiate a workable process on a limited equipment scope when nothing's in place. You can work up from there. Because plants proceed on a "business-as-usual" basis, participant selection is one key to pilot program success. Voluntary participation by those with an interest in PMs and commitment to the long term well being of the plant is required.

Establishing a process

Since most large facilities have a corporate CMMS, it's necessary to build a CMMS-based PM system. Many companies treat their CMMS separately from their basic maintenance processes, yet it's the process by which they determine, plan, and carry out their work and the process they use to develop and maintain a maintenance strategy. Ideally, the CMMS is designed around a working PM process. (Fig. 4-2) Many legacy systems had PM added as an after thought. With so many facets of efficient PM performance and so much equipment in a typical, modern plant, concurrent PM process development and CMMS implementation is not feasible. Getting a basic PM process instituted around a small system or core group of equipment is a necessary first step to a comprehensive site-wide program. Tying CMMS support processes into the program follows. More PM process development follows additional CMMS "tuning".

CMMSs may lack someone to manage the program. In a crisis-oriented plant, the PM "portions" of the CMMS are implemented incompletely, so an effective PM process never develops. For these plants, attaining a fully-implemented PM process is an especially high value activity.

An effective PM process has several essential functions—perform ongoing PM tasks; rank and prioritize CNM results for time based, CDM work; incorporate improvements. These elements are based upon identified failure mechanisms, costs, availability, and other improvements to PM program processes. Plants that presume their PM program process is adequate often find, after performing an RCM effort, that essential elements are missing-PM elements, maintenance performance elements, support elements.

Among case histories of failed TRCM efforts are those which failed because of underlying assumptions. An ARCM focus creates the most essential PM elements-quickly-where they are missing.

Processes. Developing a "maintenance process model" sounds silly to those who have been "doing it" for years. On the other hand, why is it that some organizations do maintenance creatively and uniquely-as evidenced by their WOs, equipment, and other process aspects-and some do not? It's precisely because maintenance processes are so often taken for granted as a part of plant needs that occasionally we may need to confirm our model. (Table 4-1)

Like the maintenance process model, the basic PM process has many different interpretations. Maintenance outcomes are also influenced by organizations' different cultures and personalities. Some get many miles out of equipment, some get less, but as long as the organization extracts what it considers to be fair value from its assets, and it makes a profit, it makes no difference how quickly it's used up. In evolving industries, a facility's useful life may be five years. Typically, the high-tech and information-technology industries are radically restructured that quickly. Competitors adapt quickly or die. In microprocessor and memory electronics manufacturing facilities, plants are rebuilt or product lines replaced far more frequently than the generation industry is used to. In the electronics environment, extracting value from a facility in five years makes economic sense—it may be obsolete at the end of the period. Based on unit product cost, the least expensive alternative may be to entirely replace the facility with new when that happens.

Utilities and petrochemicals lie at the other end of the useful-life spectrum. Generators that cranked out MWs in 1910 are running

High Value Maintenance Areas			
Areas	**Sub-Areas**	**Definition/Discussion**	**Value Assignment**
Work processes			
	Prioritization	Consistently identifying and working high-value work	1
		Maintaining continuity of work elsewhere	
	"Age" exploration	Following parts performance. The systematic extension	3
		of maintenance intervals with engineering support	
	Flow	Keeping work, paper, etc. flowing	2
Planning			
	Outage	Developing and screening outage work for high-value areas.	1
		Establishing outage work priority.	
	High-volume work	Identifying and maintaining preplanned work plans for	2
		high-volume work	
	Daily	Appropriately selecting and planning daily work	1
CMMS			
	Cost management	Tracking costs by types and groups	2
	Failure analysis	Extracting, evaluating, and trending failures for opportunity	1
	Maintenance analysis	Evaluation and monitoring of maintenance work and strategies	1
	Reporting & measurement	Ongoing performance and cost management at the system and below	3
Procedures (for production work)			
	Simplicity	Keeping procedures available but simple and concise	2
	Consistency	Having standard procedures and work formats	2
	Completeness	Having complete work plans including appropriate limits and specifications	3
Parts			
	Selection	Identifying outage-preventing, unavailable parts as critical spares	2
	Stocking	Maintaining appropriate stock levels	3
	Return/use	Keeping spares appropriately available to avoid unneeded stock	2

Table 4-1: Value

today—and their product is little changed, right down to the cycles. Electricity is a commodity—or, rather, it has remained a commodity. What has changed are the differentiators and premium services packaged to make one supplier more attractive than another. Compare this to phone service. True technical innovation in 1983 was limited. Today (post deregulation) there are a host of exciting new phone services and options. What changed was that entrepreneurs discovered they could do a lot with phone service once it was opened to them. Expect the same innovation in electricity-creative players differentiating their products with exciting new services.

Petrochemical products add high value and profits to the value stream, driven by the demand for computers, consumer products, and synthetic fabrics. These markets are expanding and enjoying high profit margins. They are difficult to enter, however, requiring large capital outlays, complex production processes, and many different skills for effective production. Electricity, by contrast, is a single, highly refined commodity that can be produced with off-the-shelf equipment.

Companies' roles

More than ever, generation companies need units that operate within predictable costs. Since total costs include payments to co-generators and unplanned power purchases, in-house generation costs essentially control costs. Factors effecting random, forced outages vary from company to company, but for net power purchasers, the unplanned loss of a single unit means substantial costs. High electricity costs—whether generated by nuclear units or base load coal generation—is the factor driving large end-users to clamor for de-regulation.

Some companies are electing to get out of generation. Those choosing to remain find a tough environment. State public utility companies (PUCs) aren't granting rate increases for those who remain regulated and most are planning some form of deregulation. Companies whose benchmarks prove that they aren't competitive find it even more difficult to restructure for competition. In light of this new generating environment, the traditional arms-length relationship between plants and parent companies is likely to become more interested and concerned for plant performance—if it's not already here.

Despite stable fuel costs and new technologies, generating costs remain both high and uneven and this cost disparity is also driving deregulation. Some parts of the country are blessed with proximity to resources or "sweetheart" hydro deals that keep prices low. Others are on the wrong end of supply chains or committed to paying off late-built nuclear generation that keeps prices high. Users don't understand, nor do they care about price history—they see disparity in rates and want better deals.

The market will determine the competitive clearing rate for generation. Older plants will either be made competitive or shut down. Obsolete plants with high-heat rates or poor-performing facilities will be candidates for re-powering. A third option—performance enhancement— could be viable for marginal producers in which facility design and operations contribute to uneconomic performance. Companies will have several options for these plants:

- capital investment to improve performance
- people/process investments to improve performance
- shut down the asset
- re-power the asset
- sell the asset

Many traditional utilities have been slow to adapt to the changing environment. In a cash-bind situation, it's probable that they will elect to make a quick fix—asset sale or shutdown—to unload an unprofitable facility. Such facilities may re-emerge in the hands of companies that elect to become generators, provided that:

- improvement in performance can be gained with minor-to-moderate capital outlays
- human performance issues can be resolved with new contracts, incentive plans, training, and management

Companies with a "fixer-upper" will have to decide how best to realize a gain from each asset.

Nuclear generation

As an entity, the nuclear generation industry answers to regulatory masters at NRC. Despite outstanding operating records achieved by these plants, costs are high and hard to reduce. Nuclear processes-complex and slow to change-place burdens on plants that need innovative and cost-conscious improvements. Gradual attrition of high cost nuclear units will continue as competitive-pressure increases.

Although overseeing a mature technology, the NRC "generates" new regulations and findings, unabated. This maintains a regulatory focus—not a productivity improvement one. Nuclear plants face challenges to simplify their processes just as fossil plants do, but fear of NRC scrutiny gives rise to a conservatism that limits innovative jumps—and raises costs. Nuclear units need to be allowed to explore safe ways to manage costs and risk in the public interest.

PM Bases

Justification is the concept of a basis. In fossil work, the basis is the cost-benefit calculation. It should include safety, environment, codes, insurance, and other compliance and general concerns. It can be explicit, but more often it's implied—and never documented. In fossil generation the focus is "to do things." Nuclear has no such luxury.

Documented justifications are expected to support changes. Documenting a PM basis could be setting up changes to be blocked. This is particularly true where it's unclear why a task was even started in the first place! In nuclear generation, a PM change history usually provides such a basis. It merely needs to be collected and occasional gaps completed before it is grandfathered to the original PM program. Should something go awry, there's an opportunity to check the intent, results, and see how things got off track. Developing and retaining a basis—why a PM is needed, selected, and at what interval—is valuable information for history and review in either nuclear or fossil work (Table 4-2).

Nuclear generation, with great many prescribed PM requirements, requires the change-out of EQ components as specified in their aging design basis documents and compliance with all vendor-directed maintenance on essential components (unless a justification supporting an alternative is prepared and accepted). For a nuclear plant, a basis is a useful PM change tool.

Vendors usually provide excellent guidance, but occasionally specify activities that don't make sense. Occasionally, their equipment—some of it manufactured more than 20 years ago—can be maintained more effectively with other methods. Plant staff is often reluctant to question vendor recommendations, however, in any environment. In the fossil environment, vendor recommendations are often difficult to access. Even with overwhelming case histories suggesting certain programs, staff tenaciously adheres to vendor-based programs and recommendations. An age-exploration program with documentation is an asset, but most companies can't afford the expertise to develop a formal parts-aging program. Engineers are sometimes asked to make judgements on age exploration. As an occasional art, it's difficult to learn to practice with finesse. If you can't make an informed judgement on parts performance in-service, it's safer to stick with someone who has, like the vendor.

Yet, when applied to non-essential parts, it's relatively easy to make informed decisions on intervals based on experience and judgement. Life extensions based on service durations are also easy to justify. Occasionally, a part performs far beyond the vendor-specified capacity in service—even by accident. Once that's known, it's a simple extend-life decision. This is age exploration. Superior designs and equipment are insensitive to part aging. So, one solution to the aging dilemma is to specify high-quality equipment that includes ingenious methods of self-identifying aging performance, in service.

Most vendors specify replacement requirements that include latitude for experience-based adjustment. To do otherwise would be like a manufacturer directing you to replace tires at so many miles, ignoring your experience. (Nuclear EQ components are the notable exception to this rule because of nuclear regulations and potential fines.)

Obviously-especially in nuclear or other high risk applications-PM basis program development begins with vendor manual review. After so many reviews, and after working with common vendors, most R engineers memorize the dominant failure modes and applicable, effective

```
┌─────────────────────────────────────────────────────────────────────┐
│                              PM Basis                                 │
│                        (For individual CIC)                          │
│                                                                       │
│  Component Program and Change Recommendations                         │
│                                                                       │
│  PM Program                                                           │
│  The program [is/is not] consistent with the [approved CIC standard,  │
│  e.g., relief valve] standard for PM. If there is a standard,         │
│  discuss it.                                                          │
│                                                                       │
│  ["PM" activities: time-based condition monitoring or maintenance,    │
│  in the order below]                                                  │
│  SV Activity                                                          │
│          Frequency – RE1, RE3, WK, MN, ....                          │
│          Tasks – Replace, Check, Rework, Test, Calibrate [Reference   │
│          procedure, if any]                                          │
│          Basis (origin) – ASME Sec 11, Check Valve, Valve Repack,    │
│          Reg Guide 1.97, App J Pgms                                  │
│  EQ Activity                                                          │
│          ditto above                                                 │
│                                                                       │
│  PM Activity                                                          │
│          ditto above                                                 │
│                                                                       │
│  PM Change (Changes, PM only)                                         │
│  RE3 to RE6                                                           │
│  RE2 to RE1                                                           │
│  WK to MN                                                             │
│                                                                       │
│          Justification                                               │
│          Duplication of Operator Rounds                               │
│          Duplication with another PM/SR                               │
│          Site Experience                                             │
│                  No corrective maintenance                            │
│                  No significant corrective maintenance                │
│          Analysis                                                    │
│          General Experience                                          │
│          Comparison (to similar equipment in like environments)       │
│          Vendor Guidance                                             │
│          Risk Significance                                           │
│                                                                       │
│  CNS Historical Trends                                                │
│  [any failure trend noted is listed here]                            │
│  Significant adverse trends are identified using the CAP              │
│                                                                       │
│  Vendor Recommendations                                              │
│  None                                                                 │
│  No vender manual found                                              │
│  Vendor tasks not selected & justification                           │
│  [Selected vendor tasks, including modifications, are addressed under │
│  "PM Activity" above]                                                │
│                                                                       │
│  Commitments                                                         │
│  [if any] provide any industry experience-based PM tasks here        │
│                                                                       │
│  PMO/PMI Deleted PM                                                   │
│  [if any]  provide PM numbers for change sheet                       │
│  Change History                                                      │
│  [if any]  Change sheets have PM basis sheets attached. Change       │
│  history is tracked (in software).                                   │
└─────────────────────────────────────────────────────────────────────┘
```

Table 4-2: PM Basis Format

PM tasks! This also means that nuclear and fossil operators rarely need to develop cost/benefit bases for doing many PMs, but can apply templates that implicitly include cost bases.

The bottom line is that any PM that supports generation is a must-do. Countless calculations show that it's cheaper to do PM than lose generation. That's the reason for maintaining PM bases. The trick is to recognize the PMs that influence generation and those that don't. Marginal PMs cause problems. Such work virtually never has production impact in well-designed plants and shouldn't be done. Many studies have shown that PM is like playing lotto-given two work choices, we'd prefer the big payoff. Some PMs have big payoffs, more have small ones, and many are losers. Plants need to find the winners and shun the losers.

Conservatism

Nuclear and fossil plants are alike in that both are ultra-conservative in selecting PM task performance intervals. Both suffer limited access to expert analysis and support, and so depend heavily on vendors for analytical support. Part of what's driving this is conservatism.

Until one checks component performance, in service, first-hand, there's a tendency to grossly underestimate their capabilities. Combined with equipment's inherent fault tolerance, there are many unrealized opportunities to extend component life and service intervals. Use data that's available to you, if only to avoid severely penalizing a maintenance plan!

Conservatism offers "traditionalist" operators tools such as conditional overhauls and age exploration, both of which force them far outside their comfort zones. But this is where the significant savings are also.

Conditional overhaul is not more work. It is the directed rework of a component focused to restore original performance. While not intuitive, conditional overhauls have been demonstrated to be statistically effective for jet engine overhauls. Few generating companies have a formal repair policy of conditional overhaul, however.

Age exploration and PM interval extensions are a second opportunity. Virtually all companies that use age exploration extend intervals by minimums of 10-30%. Benefits from such minor extensions take a while to add up to real cost savings. A substantial lesson from ARCM-aggressive use of age exploration, can significantly extend PM intervals.

Where PMs support economic-based failures, extending intervals radically supports finding the appropriate time limits. Practically, most PMs avoid economic-based failure.

Over-conservatism

Task interval conservatism is a requirement for any PM maintenance program. That is: OCM intervals must be short enough to identify diminished failure resistance, but long enough to realize an item's useful life, so that "on-condition" maintenance tasks may be effective.

To identify appropriate intervals may require an actuarial analysis. Those charged with adjusting intervals are not trained actuaries—a skill that requires advanced training in mathematics and many years of special tests. A tendency in any program—including maintenance—is to utilize overly conservative requirements. Margin is hung on margins, until many basic intervals reach their constraining limit—the annual boiler or refueling outage. Of all PMs worked in power plants, the scheduled outage interval often has the greatest PM frequency. This represents the minimum interval that can be selected with no production interruption. In the absence of hard R and actuarial engineering analysis, these intervals have become accepted, implicit standards. They also must be challenged.

An interval that represents half the appropriate (or capable) design life of a piece of equipment puts severe restrictions on scheduled outage work and greatly increases expense. Statistically, annual or refueling-interval PM intervals disproportionately populate PM systems, and reviewing annual outage work is a highly profitable task. You may find that plant support staff selects outage replacement intervals in spite of performance, vendor guidelines, and other recommendations that support longer intervals. CNM performed with inadequate specifications identifies fault conditions early and overhauls prematurely. Overly conservative work can load up an outage with thousands of extra work hours. Properly selecting intervals represent an immediate opportunity for many plants.

There are other examples of over-conservatism. The asbestos inspection requirement is one year "in the absence of a monitoring program"—three years, otherwise—so a typical plant can inspect at three years! Yet, almost none do so. Others use annual replacements for parts faulted for a single failure. A nuclear plant automatically reduces by 25% its EQ service lifetime, just in case an EQ hard-time replacement PM gets missed. Conservatism adds up. Costs for replaced parts are high, as are infant mortality failure rates. Actuarial studies show that "overhaul" activities are ineffective at improving R, yet they remain a mainstay of the traditional generation industry.

Alternatives to regular and capricious applications of conservatism will address any number of oversights. Competent R engineering help, setting exact intervals, and age exploration standards represent excellent opportunities to advance along the same maintenance cost-management learning curve in generation that occurred in the commercial aviation industry.

If PM scheduling is a process problem go after the process—don't introduce common-cause conservatism. It can't correct the fundamental root-cause flaw an ineffective scheduling process presents. PM restrictions defeat the purpose. Process errors that occur because of complexity make it highly unlikely that more complexity in any process—just like our failure process itself—will reduce the error rate.

Based on documented parts performance, and provided the environment is maintained, quality parts usually exceed expectations. When environments must be maintained (*e.g.*, protected from water, excessive temperatures, caustic atmospheres, acid runoff, or excessive wetting/drying cycles), then use the best materials available, perform age exploration—and condition-monitor.

When fossil environmental control equipment (ventilation and cooling) is abandoned due to maintenance priorities or difficulty in using it, it's very much like abandoning chemistry specifications that are too difficult to maintain. There's no obvious immediate effect, but long term consequences can be serious. In a number of cases, restoring equipment to service was easily justifiable but hard to achieve.

Numbers can help define the objective-failure story-numbers that most traditional generation RCM analyses lack. Some RCM analysts go so far as to discount failure statistics and numbers. In my opinion, this is a serious oversight. Implicitly or not, we live by frequency and consequences. However, while numbers don't tell the whole story, those who

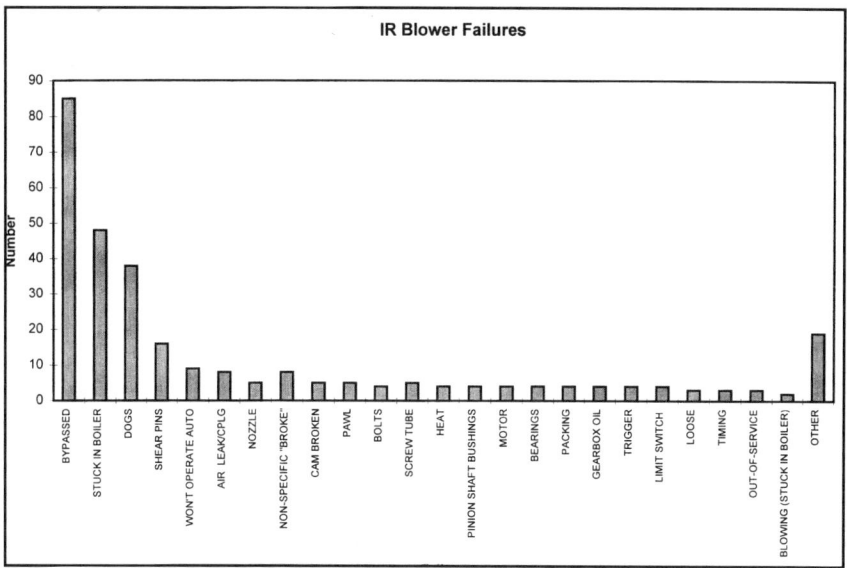

Figure 4-3: IR Sootblower Failures

ignore them often chase the "trivial few." This has given RCM a black eye. While some analysts bog down in endless pursuit of rare or imaginary events—things that don't happen in this world-my approach reflects interest in measurement. But I also review large quantities of data to identify failures, summarize statistics by failure categories, and make estimates (Fig. 4-3).

The numbers I work with aren't exact but they are in the right ballpark. I view them like dose rate estimates: they're order-of-magnitude significant and they identify sensitivity to costs. Costs need to be understood at a 10%, 100%, or 1000% payback during the period of interest. A 10% payback on a turbine overhaul may be worth chasing but probably not for a $20 filter replacement. A 500% savings on a $20 task clearly outweighs the same for a $2 task, so we want to structure our programs to capture that value. Practically, this means when it comes to a trade-off (and it will), we must give up the $2 tasks to make room for the $20s.

Ultimately, activities should reflect on-site statistical data and failure experience. Environments—including the work environment—are unique to each plant and influence what fails (and when). The cultural environment—levels of skill, knowledge, and other intangibles—can be inferred but is just as hard to measure, and also influences failures. Just as two randomly selected individuals will experience different success rates with the same auto (as measured in longevity and life cycle cost), two similar plants experience distinctly unique operational outcomes. These can only be explained in process and cultural terms (Fig. 4-4a and 4-4b).

Avoiding rare failure events—the root cause of most heavy production and financial losses—are the major "controllable benefit" from an ARCM plan. These events are worth understanding.

Those in my experience show common traits—inaccurate, failed, or unavailable instrumentation—often play a role. Second, general instrumentation status warns of structural process problems. Ineffective maintenance of critical instrumentation and failure to incorporate that into an overall operating plan, indicates a weak operating organization.

Just as there are strategies designed to reduce the risk of auto accidents—and we accept that a driver with a perfect record must be a good driver—so plants with high performance records practice risk minimization strategies. Conversely, plants with spotty records are those that fail to follow operating and maintenance practices that help to manage risk. In the absence of such strategies, they suffer more losses.

Before you conclude that the "better" plants must be higher-cost operations, note that insurance industry statistics and risk presentations indicate otherwise. Steady, consistent performers are not only low-cost performers but are safe, low-risk performers, as well.

After a rare failure event, managers need analytical support and the implementation of failure prevention strategies. The notion that a big-ticket failure—the equivalent of a meteor strike—is a random event that happens only once in a lifetime, is not true. In practice, they keep happening, over and over. Generator retaining ring failures, water chemistry upsets, bus-bar explosions, shorted buses, plant trips, fires...managers tire from probing questions or site trip reviews.

After many years spent examining major losses, I find that in most cases, a chain of events presents a history. The progression towards ultimate failure depends on systematic process weaknesses. Rare events occur more frequently in the absence of process awareness and controls.

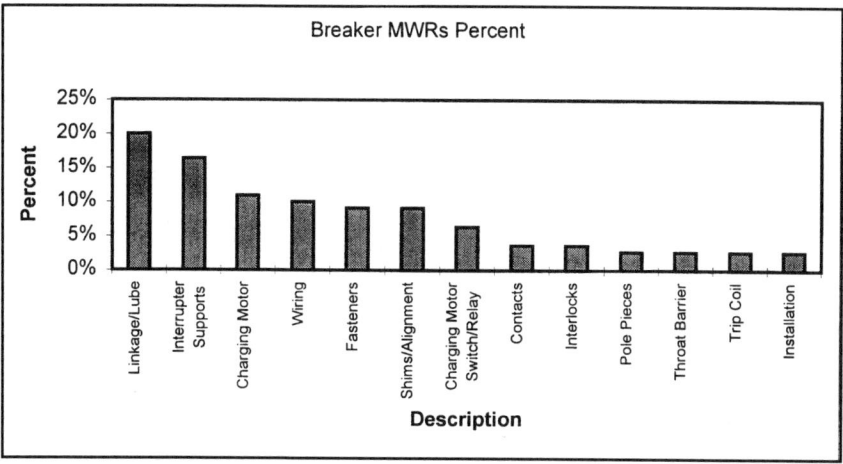

Figures 4-4a: Nuclear 4160V Breaker Failures

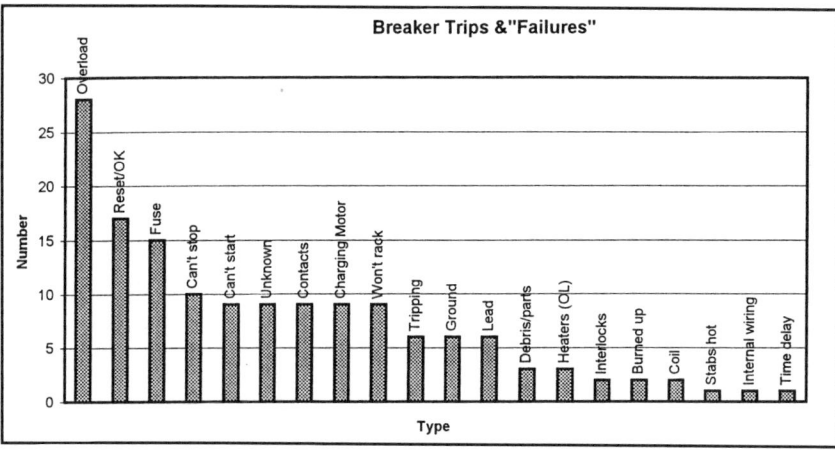

Figures 4-4b: Fossil 4160V Breaker Failures

They reflect random, repetitive occurrences that together convert to an operating event. The more and greater their frequency, the more probable they will do so. Ultimately, statistics tell the story. Rare events can be managed with conscientious, complete operation strategies. These rules are well known in theory and practiced by professional operating organizations. ACRM is another way to add clarity to an otherwise cluttered operating field.

Failure frequency

After leaving nuclear and working in fossil power plants for nearly a decade, I developed several strong impressions. For those who haven't worked both environments, there's a great deal of information sharing that is possible. Each has very focused strengths that are applicable to the other.

Nuclear is focused on identifying technical failures that add both clarity and certainly to help those working in unfamiliar terrain. Embracing "failure reality"—versus abstract considerations of imaginary problems—can greatly improve nuclear competitiveness at virtually no risk to the general public.

"Fossil focus" means using inherent design availability that is built into plants to perform work as needed. The advantage is the ability to perform real-time maintenance; the risk is potential functional failure because margins are expended. Fossil units' easier start-up and load cycle, for the most part, minimizes production losses incurred from a forced outage.

The ability to mobilize personnel and systems to get a job done is another fossil capability. Paperwork and organizational systems are compact, focused, and anchored in a vested and accountable individual or group. This focus supports the performance of CBM. However, because fossil maintenance is less formal, operating limits are occasionally stretched or overlooked and reactive failures or forced outages can result. Defining clear operating limits to trigger condition-directed maintenance is a fossil generation need. The opportunity for fossil (unlike nuclear) is the authority to make individual plant interpretations of risk and benefit when engineered limits are reached. This can provide great operating flexibility. There is absolutely no benefit when limits are blown over and failures result.

On many occasions, fossil plant staff clearly understands key operating limits from a technical perspective but organizationally, they fail to act in a timely manner. Expectations were not made clear, or managers failed to support operator decision-making. Again, the point of CNM is to identify and perform CDM prior to final failure. To do this well, those who perform monitoring must be expected and empowered to act.

Coded components. Nuclear plants have more coded components than a comparable fossil plant. By way of comparison: 500 to 1000 MW nuclear plants have 40,000 to 100,000 coded components while a coal plant of the same size will "code" as few as 500. Arguably, fossil plants have more complex equipment and systems, and plant coding effectively ends at the skid level. Beyond this, identification is by text description alone, which is adequate for equipment identification and failure analysis.

Fossil units are solely coded for maintenance and their CMMS equipment descriptions and codes are typically the only ones available. Design basis equipment tagging is usually absent or abandoned once CMMS equipment lists are prepared. Nuclear plants, in contrast, maintain regulatory design databases coded for configuration management, regulatory oversight, and equipment control. Fossil plants "tag" too little and inadequately. Tagging systems should uniquely identify coded equipment to the skid level using consistent descriptors. Perhaps a tagging standards committee is needed. The ultimate answer is that balance is required. Excessive detail introduces complexities and costs no one needs and presents a burden to use and maintain. Too little detail means that costs and failures can't be adequately traced. My experience at the nuclear plant was that about 5% of the equipment "tags" generated 95% of the MWRs. These items truly need unique identification. Based on this assessment, nuclear plants are over-tagged for practical operations.

As a practical matter, tagging uniquely identifies equipment for operations, maintenance, and modification. Detailed tagging is required if there are modifications that must be controlled. For normal O&M, skid-based tagging systems used by fossil plants are adequate.

Complexity. Technically, nuclear environments are little different than that of fossil or hydro. The risks are greater, but the complexity of the controls in a modern fossil plant is actually greater. The slowdown in nuclear plant construction has dated its technology. Fossil and CT plants, however, have had a full decade of DCS applications that nukes don't use. Yet, new technology introduces more inherently reliable designs.

Technical barriers to nuclear advances are countered by the (relatively) generous nuclear budgets. In a utility environment under regulation, cost pressure on nuclear plant management is slight. With large capital assets at risk, utilities have historically spent generously on nuclear projects-sometimes starving fossil cousins in the process. It's this "generosity" that has encouraged nuclear complexity in the regulatory sense; the NRC's insensitivity to costs has aggravated it and further damaged nuclear competitiveness. There can be no other way to interpret the tremendous growth of nuclear support infrastructure during the past 20 years, at a time when the technology has, if anything, matured and gotten simpler!

Organizational complexity introduces organizational errors—or organizational "failures," if you will. These "virtual" errors have become the focus of regulatory interest as much as anything real. The net effect has been an even greater complexity and more unreliable organizational systems. Management focuses more on covering its exposed regulatory backside than cost management while the NRC continues to operate as if there is no numerical threshold or objective measure of performance that objectively establishes "pass/fail."

The U.S. nuclear industry was conceived and built to emulate the philosophy of Admiral Rickover and the Navy in commercial form but has failed to learn several statistical and cost lessons. Statistical and numerical measures in the commercial nuclear industry are virtually absent and this has unfortunately led to inherently higher structural costs.

Complex failures

"Complex" failures in this definition include interdependent and logical-sequencing faults involving equipment and control interactions, multiple failures, intermittent failures, secondary failures, loss of redundancy, and drift. They're difficult to identify, troubleshoot, and correct. Analytical difficulty arises because many variable facets present themselves in concert. Each emulates the problems of a plant startup, in which defining and solving coincident problems takes a thorough test plan, expert assistance, and persistence. Teams and specialists are needed to ferret out complex failures.

Avoiding complex failures lowers costs and increases production. A thorough maintenance strategy helps identify failures before they generate secondary failures and propagate into complex failures. The converse is also true: When a plant gets behind its optimum maintenance curve, failures occur, failure complexity increases, and a plant loses production and increases expenses.

Troubleshooting skills are of great value in a plant suffering a high population of failed equipment and complex failures but such skills also raise costs. Timely, CDM performance and discipline reduces the number and complexity of failures, and ultimately allows a plant to be monitored at an overall average lower skill level.

Serious secondary failures with complex secondary failure effects—fires, systematic contamination, and deterioration of sophisticated plant fluid system chemistry, makeup water, and waste water equipment—are often identified but consistently downplayed. Over time, they blossom into serious problems that shorten the useful lives of equipment and ultimately, facilities.

Fires, flooding, and environmental changes can introduce pervasive, "common cause" failures-a failure that crosses assumed independence boundaries. Once they take hold, common cause consequences are difficult to correct. Because they cross boundaries they invalidate redundancy. Avoiding common failure modes has many paybacks, and implementing strategies to avoid them arises from R engineering and from understanding events that are influenced by O&M.

"Operationalizing." A professional society can develop standards but only an operating environment can "operationalize" activity—convert a subjective goal into a measurable performance objective. Once a goal is operationalized, subjectivity is gone. For example, you could operationalize a goal of "performing well during an academic career" as achieving an overall grade point average (GPA) of 3.4. Once the goal is set, you either graduate with a 3.4 (or higher) or you don't. This goal is very measurable. Operationalizing therefore supports goal setting. Governments operationalize tasks, as do companies and the workforce that labors for them. Developing operational guidance and implementing standards is a necessary step.

Many agencies implement standards that call for an effective PM program. But what is meant by an effective program? What are its attributes? What objective performance level is "effective?" How does this come about? To comply, companies must:

- determine what constitutes an effective program
- "sell it" to others in the organization
- set standards
- implement those standards
- assure themselves that they achieved implementation

This means there must be measures. If a program doesn't measure up (by all accepted standards) then:

- correction is necessary to remain within an effective program

When governments impose new standards, by law or regulation, utilities must determine how to operationalize them. For example, the Americans with Disabilities Act passed in 1992 and caused an outcry from industry, not so much over the breadth of the act, but over the need to develop an operational interpretation of the requirements. Now that operationalization has occurred—reasonable standards have been developed and implemented—there is less concern over the law.

Slow-changing environments can't operationalize as quickly as dynamic ones. As the utility industry becomes more competitive, those who respond quickly will be more viable than those who can't.

Implementation is how work standards are operationalized. Operationalizing tools include:

- goal-setting
- documentation
- procedures
- training
- measurement
- feedback

Common mode failures

Common mode failures denote failures that are common among

classes of equipment, equipment in common locations, or other common factors. Using an inferior grade of grease that separates and accelerates aging, failures increase. Perhaps we now survey a few of the failed values and find that the problem is really incompatibility of greases. I had assumed all the lubricated valves would fail individually; now, I have all valves loaded with grease and a common problem-a common cause failure (CCF).

CCFs violate the design assumption of redundancy. While they are independent failures, they are also programmatic failures, for the most part. It is the intention of all design and operating rules (and particularly regarding nuclear plants) to avoid CCFs. Common cause failure modes substantially change the overall probability of functional failure by negating design redundancy. The NRC worries a great deal about them-and rightfully so! Fossil environments also display CCFs where expected conditions changed, were never met, or were lost.

CCFs can show up as environmental problems, or as defective support system services. The recent year two thousand (Y2K) controversy reflected a common cause failure made for software. Another classic example involves the instrument air that operates many plant solenoid-operated pilot valves and air operators. Instrument air contamination has been a relatively frequent occurrence in industry, when moisture and dust—especially entrained rust—clogs service instruments, valves, and air-operated controls. At low temperatures, air line freezing has caused complete failure—the loss of instrument air and related services. Prevention is provided by drying the air, and air-dryness monitoring, using moisture monitors.

A second CCF involves exceeding design service temperatures-in a boiler enclosure. Ambient temperatures around one fossil plant boiler ran 30° higher than ambient in summer and colder in winter. In summer, "excessive temperature" failures—sootblower overload trips from misalignment (due to thermal expansion), ignitor and burner-flame scanner logic failures, and other control failures—predominated. In winter, ventilation damper instrument air lines froze and instrument drift was a problem. The common mode failure—due to the loss of environmental control—could be tolerated, but maintenance and operational expenses rose.

"Keep it simple, stupid! (KISS)": Maintenance is a complex process. Simplification is key. To borrow a phrase from the military, "When complexity beckons, think KISS." Maintenance reverses entropy—the progressive trend towards disorder over time. As thermodynamics students learn, a system's entropy can only be reduced with the injection of energy or control. An outside source creates order where disorder would otherwise prevail. Order is not the natural state of things and achieving it requires continuous, constant renewal. That injection of energy or control—or both—is necessary to be able to maintain. It doesn't just happen; it's hard work. It contrasts greatly with the traditional organizational maintenance view of the world!

It's not that we cannot achieve order. We can. But we cannot do so without:

- thought processes
- applied effort
- providing each in enough volume to offset the inherent disorder of the system

The KISS dictum says that by keeping systems simple, we greatly reduce the capacity for disorder and simplify the effort required to maintain order.

The very best maintenance performers often have very simple and effective implementation processes. ARCM says, in so many words, "There is value in our underlying maintenance processes."

The Japanese have a phrase, "poke yoke," which roughly translates to "make fool proof" or "make it impossible to get it backwards." So-called poke yoke devices have long been used by the Japanese to simplify production processes. Viewing American maintenance as a production process, I believe that it needs many more poke yoke devices. Intelligent engineering focuses on making every process a poke yoke process and the key to more poke yoke processing is engineering maintenance production teamwork. That requires communication.

Ambiguity. Plant information is often incomplete. Parts usage, equipment histories, WO entries, aging documentation—all lack completed CMMS fields or are never developed, leaving us to make decisions with the best information available.

Blame some incompletions on information systems-many main-frame-based systems disallow user "entry correction" so errors or incomplete information cannot be addressed. Worse, many information providers have never been shown how information adds product value. (Information must be used if it's to provide value.) Information fields that can be retained may lack standards for entries. Because system installations lack direct worker input, use, or records guidance, users stumble along, ignoring the CMMS unless a timecard is tagged to it. As a maintenance expectation, CMMS use, system training, expectations, and information retrieval have been tailored to engineers, schedulers, and managers. Requests for user-friendly systems have gone largely unheard and applications for common user problems missed.

PMO facility review projects can be intense—and rewarding—and if they actively involve workers in development of PM intervals they can include CMMS use. Quality and use can be improved with a few simple tools. One is basing PMs upon statistically representative, adequately large samples. Sample sizes can be extended by using multiple unit data, vendor data, and overall industry data. Consider industry surveys and vendor recommendations for new equipment when data is unavailable.

In the absence of unambiguous experience, it's wise to perform "similarity analysis" and benchmarking. If you can't find history for the specific model of pump in question use another, similar manufacturer's pump. Although it's not perfect, it's a fast way to "grow" an effective program. Skilled workers can suggest similar models, manufacturers, and environments with which they're familiar. Materials, like components, can be estimated for inservice life with benchmark comparison. Getting a similar, proven-life component from the environment in question provides an efficient way to identify service life.

Again, to limit ambiguities, there is a continuing need for standards based on equipment populations and composite environments that are as large as possible. Standards fill gaps in specific information. When information is incomplete, we must augment it and develop standards based on age exploration that can develop a wider, more complete experience base.

Culture

Maintenance delivery

It's my belief that American maintenance performance is disorganized. We cope with high rework rates, ignoring statistical (and other) tools to identify, measure and reduce, (or eliminate) rework. Substantial maintenance coordination and improvement opportunities need to include ongoing:

- continuous improvement
- innovative jumps

American culture supports innovative leaps and always has. Process improvement technology is an area where an active approach to performance improvement can pay off.

A single-unit nuclear generating facility with upwards of 10,000 scheduled "PM" maintenance activities sees perhaps 10,000 jobs worked annually. The sheer volume of items to track and coordinate is overwhelming, even before operational constraints and other complicating factors are considered. Complexity, though attributed to nuclear plants, is typical of any large production facility. A refinery, a fossil power plant, or a chemical process facility has comparable levels of complexity.

Plant complexity also leads to work complexity. An I&C technician troubleshooting a tank level controller requires the tank be "in service" to perform the task. A mechanic replacing a valve needs the tank drained first. Operations can't have the tank "back" in service until the work is closed and the paperwork is complete. Such a clearance "tag out" takes hours to prepare on the front end—and sometimes gets lost. "Right hand/left hand" stories abound.

Though the resolution to such conflicts is usually obvious in hindsight, it's difficult at the outset to remember that large, complex jobs require many different skills, schedules, awareness, and work conditions. Coordinating these difficult pieces requires the most skilled staffs and capable systems available.

The alternative to choreographed work is taking every day as it comes—embracing each work activity separately, distinctly, and singularly. This approach results in scattergun maintenance performance and repetitive equipment downtime. It's frustrating when workers find that conditions don't support the work, or have changed. Trenching a newly paved street to replace a sewer is urban-legend folklore in part because it's a common occurrence.

Yet, tools to improve worker focus and work coordination are more available all the time. Computerized maintenance planning promises better coordination. The widespread availability of PCs, Local Area Networks (LANs), and other electronic tools greatly enhance our ability to coordinate work. ARCM also promises better work identification—separating the wheat from the chaff, so to speak. There is so much equipment in a large plant, that "just doing work" doesn't cut it. Work must be structured and focused to high levels. Until now, work practices haven't changed in part because environments supported "business as usual." We must first understand work practices to be able to invoke meaningful maintenance changes and better serve plant needs.

Delivery of better services, once understood, is another issue. One great frustration milestone of my early RCM career was recognizing that, by itself, RCM offered limited benefits. Just knowing what to do, has limited value. Delivery is an equal partner to knowledge. Knowledge has to be packaged into delivery methods and systems that integrate with the organization to provide lasting benefits. Many organizations simply aren't ready for this commitment to change without extraordinary external forces working on them.

What are those forces? Competitive pressures—and a growing awareness that maintenance both directly and indirectly offers tremendous potential for cost improvements—have changed this perspective. In the last couple of decades, an awareness of RCM—applied in aerospace as a successful technology—has enabled the potential for benefits-transfer to other areas. Indeed, there have been considerable benefits developed from both the nuclear and fossil generation areas, as supported by EPRI. At the same time, there's a nagging feeling that RCM, like so many other programs, has "under-performed" thus far.

Maintenance performance

Maintenance normally occurs within an uncertain environment. Maintenance organizations often share information verbally, which has limitations. Workers cope with equipment problems with varying degrees of engineering support. There's little documented, easily retrievable information concerning equipment failure and many ways to approach it. PM programs are implicitly defined and rarely have a basis. Available PM information only implies the failures it addresses while prescribing monitoring or corrective tasks. Few vendors specify (or perhaps even know) exactly how to perform organizational maintenance or address appropriate maintenance intervals.

Times and characteristics of equipment failure mode attributes—actual or idealized—are very uncertain. Yet, it's from them that we obtain mean-life, conditional probability of failure curves, distributions of failure type, and mean life variation. Conservatism, built into maintenance task performance to compensate, could come from institutionalized monitoring frequencies that are too tight. There's also a lack of trust in supporting systems and processes, including the computer maintenance management/information systems. Many facilities compensate by over-performing maintenance.

Some conservatism arises from the very nature of large industrial maintenance and the craft's inherent desire to do good work. Part of it stems from the lack of effective CMMS PM systems. However, a huge part of the problem arises from the uncertainty of equipment lifetimes and use. Combined with a TBM model—the traditional PM model (assuming it was performed)—means huge amounts of conservatism have to be built in.

Maintenance doesn't need to be random, nor must there be so much of it. In disorganized facilities, factors working together to help control failures include conservatism, craft, design, and monitoring. The tendency to maintain wide margins for error (on the assumption that things will be missed) adds tremendous conservatism to part-lifetime calculations, randomness of failure assumptions, and other maintenance program features.

Craft workers in a stable working environment learn equipment needs and deal with equipment directly without organizational system intervention or support. In some cases, they fight disorganized control systems, succeeding with traditional techniques and skills that evoke the meaning of "craft." Insights shared among operators and maintenance crews are intuitive, perceptive—and valuable. Supporting systems that should benefit from this information, oftentimes do not. A PM program may fail to capture many insights known at the worker level.

Great conservatism is intentionally incorporated into equipment design, as well. To reduce the risk of equipment overload, use, or premature failures in uncontrollable environments, designers build in margins—excess capacity. Users become aware of these margins (particularly in fossil and hydro facilities) by pushing limits. Equipment tolerates these stresses or fails. Over time, practical operating limits are established based on experience.

If we don't have (or require) huge "conservatisms":

- costs drops
- production improves
- waste drops
- R improves

The case for improved R is therefore counter-intuitive: Less margin forces us to operate closer to design limits. This places additional stress on blading, tubes, casings, and other long-lived hardware. Design is based on more carefully specified conditions. Increasing operating limits without such insight is more likely to generate unacceptable performance. These days, owners increasingly foot the bill for lost performance.

Equipment groups

Work association (also known as work blocking) can speed and streamline maintenance performance. Associations can occur at the task, equipment, or systems levels among equipment, function, or boundary groups. Such opportunities arise when it's convenient or mandatory to work on elements within an equipment group.

The basic objective—similar to the objective throughout industrial manufacturing—is to minimize work performance time and labor. Maintenance time analysis shows that trip, part, and planning time represent significant amounts of total average maintenance work performance. A major maintenance performance cost factor in any large complex facility—power and chemical process plants, factories, or even transmission systems—is the trip that is needed every time work is performed.

Creating equipment groups (EGs) and work associations supports the systematic reduction of trip time and allows coordinated corrective and preventive work to be performed both within and between craft groups by appropriately linking tasks. Associations take advantage of what's been learned in previous work performances and support continuous learning. This can be lost in environments where the emphasis is on "doing work," not "making money."

For repetitive, planned work—especially PMs—it's especially convenient to "park" activities on the plant's schedule within a scheduled EG. Once work is associated, you avoid the need to schedule it on a detailed level within the group. This avoids PM realignment.

When performing a number of PMs, lining up intervals within and across groups speeds performance. If you make mistakes-if work gets misaligned and performance complicated-problems can be identified and worked around. Problems occur, day in and day out, in any scheduled maintenance program. Rescheduling misaligned PM work becomes quite a useful skill in a PM-oriented plant. Of course it's better to minimize the causes of misalignments so that few occur. Grouping, scheduling, and man-loading around a 12 week LCM schedule is crucial-and then working to the plan!

Routine work alignment can be based upon electrical divisions, for instance—the rotating 12-week quarterly schedule and the plant "surveillance test schedule." A routine "surveillance plan" is required at nuclear plants and at fossil plants with test requirements. However, every plant has an ASME code and insurance test requirements and a list of other required monitoring that is often specific and lengthy—even if it's environment is considerably simpler. One spin-off benefit of surveillance-level scheduling is the absolute need to develop and manage a plant's 12-week schedule to meet intensely monitored nuclear license requirements. This schedule can offer benefits to all plants. Where available, such a schedule provides a ready vehicle for a monitoring test program such as developed by an ARCM-based failure review. This benefits any large, complex facility operating and maintenance process. A CMMS can provide the schedule software "ticklers" that initiate the plan. The 12-week schedule offers a near-term window that fits with other scheduled plant activity to become a tool that allows the plant to perform more—and better—planned work.

CNM

Most CNM-initiated maintenance originates in operations. CNM-monitoring without specific failure criteria-can be hard to rank, prioritize, and perform due to its generality. In the absence of time-based and on-condition WO categories, an organization can measure its CNM-originated work, based on the work-fraction coming from operations. If an operation originates 70% of the WOs unrelated to operational tests, then about 70% of them are NSM. Scheduling and planning, and engineering, initiate most of the balance of the outage, PM, and modification WOs.

TBM comprises the planned maintenance that is traditional, and time-based rework/replace task work. If a plant can identify condition-based from time-based WOs, they can measure the RCM maintenance workload as as shown on Table 4-3.

A small fraction of CNM identifies functional failures. Measuring that fraction involves (1) reading WOs or (2) checking logs. Few CMMSs have fields to record "functional failures (FF)" and few operators discriminate functional from other failures. Logs typically record functional failures.

A quick way to re-align CMMSs to measure RCM-based work strategy is to relate CDM to on-condition WOs. You can also perform all "condition-directed" work as part of the original on-condition WO. This establishes three basic WO classes:

This approach provides a quick way to measure existing processes.

Now, what do the numbers mean?

Benchmark profiles are only now being developed in the power industry. Grouping numbers in this manner can show absolute WO numbers any way desired, but "hours worked" is a common benchmark comparison quantity. PM work hours are inherently low. Most non-out-

PM (time-based)	(1) TBM
	(2) OCM/CDM (including OCM FF/CDM)
CM (corrective)	(3) NSM/OTF (Failed)

Table 4-3: PM vs. CM and WO classes

age PM jobs are simple tasks.

Organizations generally need to increase their on-condition work fraction. This work involves an explicit failure resistance measure, which initiates condition-directed work. This requires explicit failure limits, work performance focus, and priority on work with a "failure limit" exceeded.

This combination is often seen in traditional instrumentation programs, where a significant amount of out-of-calibration and failed instrumentation work is identified. When restored, often at low cost, immediate operational R improvement results—a quick payback.

Two Perspectives on Failure

There are two failure perspective focuses—function and component. The functional perspective expresses what a component does, the component perspective, and how it deteriorates. Part names have functional roots—blower, pump, and breaker—often derived from a verb. The name suggests the primary function. Because the perspectives differ, remembering the context whenever "failure" arises can help avoid confusion.

When components, equipment, and systems fail, ultimately it's because a physical part deteriorated. This proximate part degeneration eventually causes a functional failure. (A bearing wears until a pump trips on high vibration, a motor winding resistance fails, the motor shorts, and trips the pump off, or an operator smells smoke and transfers pumps, shutting down the offending pump.) Functions affect work performance. Failures translate as lost functions. The operator's component is a "black box" that he or she may not understand. They only need to see the functional outputs or note their absence and act appropriately (Fig. 4-5).

We require functions while operating plants. When functions "break" (e.g., are lost) we diagnose failures, locate the source, and fix parts. Operators' perspective is inherently functional. But while identifying a functional problem is one step, tracing that back to its physical source is another matter. Success in managing failures depends on organizational diagnostic skills (Table 4-4).

Holding a functional perspective simplifies the operator's required equipment knowledge. Operators need only assure function availability-which involves the senses-and interpreting instruments. Facility instrumentation supports function monitoring, and the specific function, measurement requirements, and equipment redundancy determine the instrumentation needed.

The function-part failure dichotomy is important when we talk about failure and "operate to failure." There are few (if any) cases in which plants intentionally operate to system functional failure. This simply makes no sense. We provide robust equipment redundancy specifically to avoid it. In the hierarchy of systems, subsystems, and their functionality, however, redundant or incidental functionality is provided at subsystem (or lower) levels that can tolerate failure, to some degree. Risk accompanies function failures, but it can be managed.

NSM is meaningful on components where a failure will be evident, can be managed, or has no functional impact. Redundant instruments packages, inexpensive components, and even spare trains and equipment support this approach. If redundant equipment can be run to "failure" while maintaining system functions, the deciding factor is cost. Sophisticated microprocessors and sensors can identify and shutdown deteriorating equipment, limiting damage. The "cost" is loss of the equipment until maintenance is completed. This strategy is viable for wearout failures where there is installed redundancy (Fig. 4-6).

Consider a boiler feedpump in a 50% redundant train (three 50%

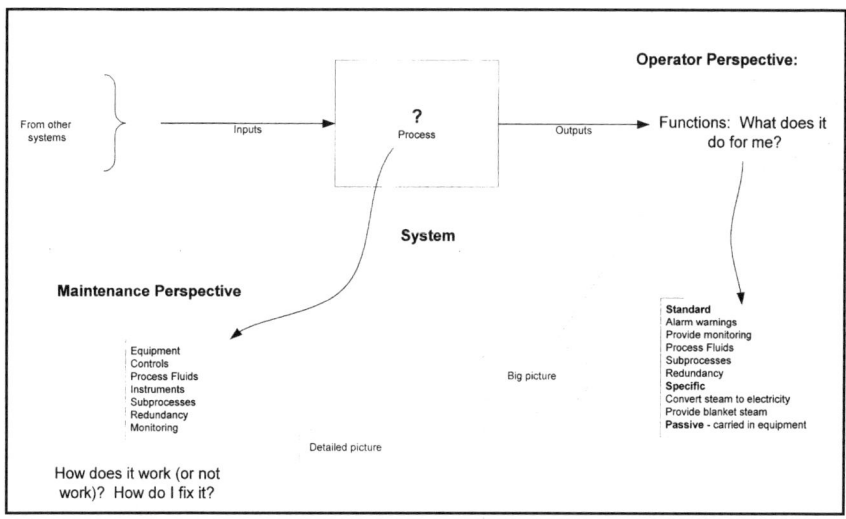

Figure 4-5: System "Black Box" Model

Component	Functional Failure	Part Failure
Pump	Won't pump	Seized bearings
Blower	Low volume at speed	Worn impeller blades
Breaker	Can't extinguish arc	Weak coils
Pump	Won't pump	Bad starter
Pump	Won't pump	Lost prime

Table 4-4: Component, Function, and Part Failure

pumps, any two of which provide 100% rated flow). (Fig. 4-7) This configuration is a standard plant feedwater design, and meets boiler head requirements for four to seven years of service. This approach is viable and effective, provided the standby-train pump can start and load reliably. Such assurance can be provided by periodic testing. When in-service feedpump failure is identified, capacity is shifted to standby. The worn-out, failing pump is removed from service and repaired. This could be online or off-line, during a scheduled outage. Although equipment must be restored, the systems functions are maintained (Fig. 4-8).

OTF as described here is a rational. We need to remember that the "failure" considered here is an abstract engineering proximate "function

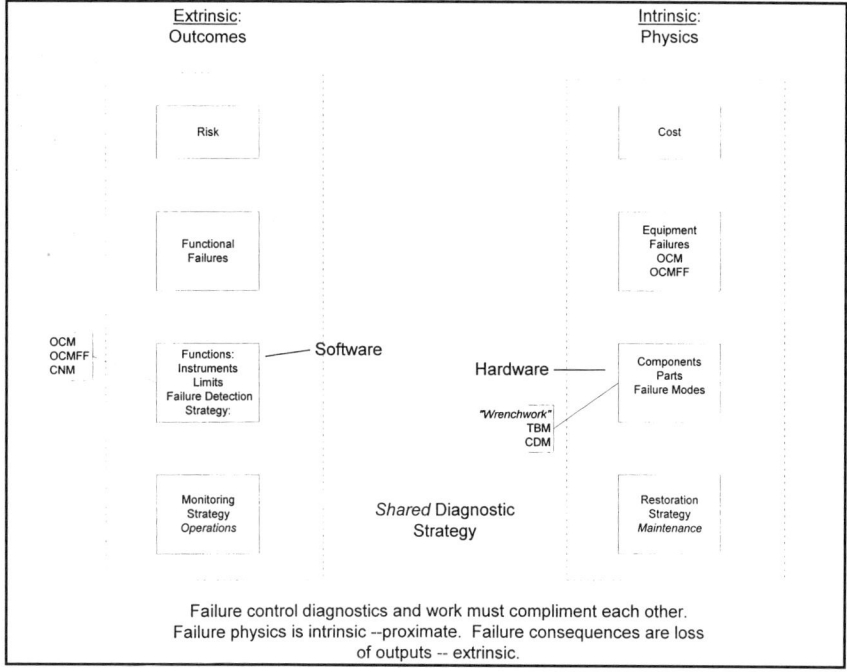

Figure 4-6: Two Sides of Failure Management: Risk and Cost

failure." (Fig 4-9) For many traditional engineers, this is not their perception of "failure". The function-equipment-failure seesaw makes OTF confusing for some. For an operator, maintenance is of no consequence so long as they always have necessary (or backup) equipment available. OTF means little as long as "black box" system functions work.

This approach may not set well for the mechanic, however. OTF must conserve equipment or economic consequences make it unreasonable. Catastrophic failure fears explain why many mechanics object. In fact, a great deal of equipment is designed to support an OTF strategy. Internal sensing devices initiate shutdown on fault conditions causing function loss. This limits equipment damage, but sacrifices functionality. Cases can arise where sacrificing equipment for extended functionality is preferred. Operators make the choice.

This function-to-physical failure mode relationship is summarized with Figure 4-10. Functional failures observed using a system perspective are the result of physical part deterioration. Functions can only be

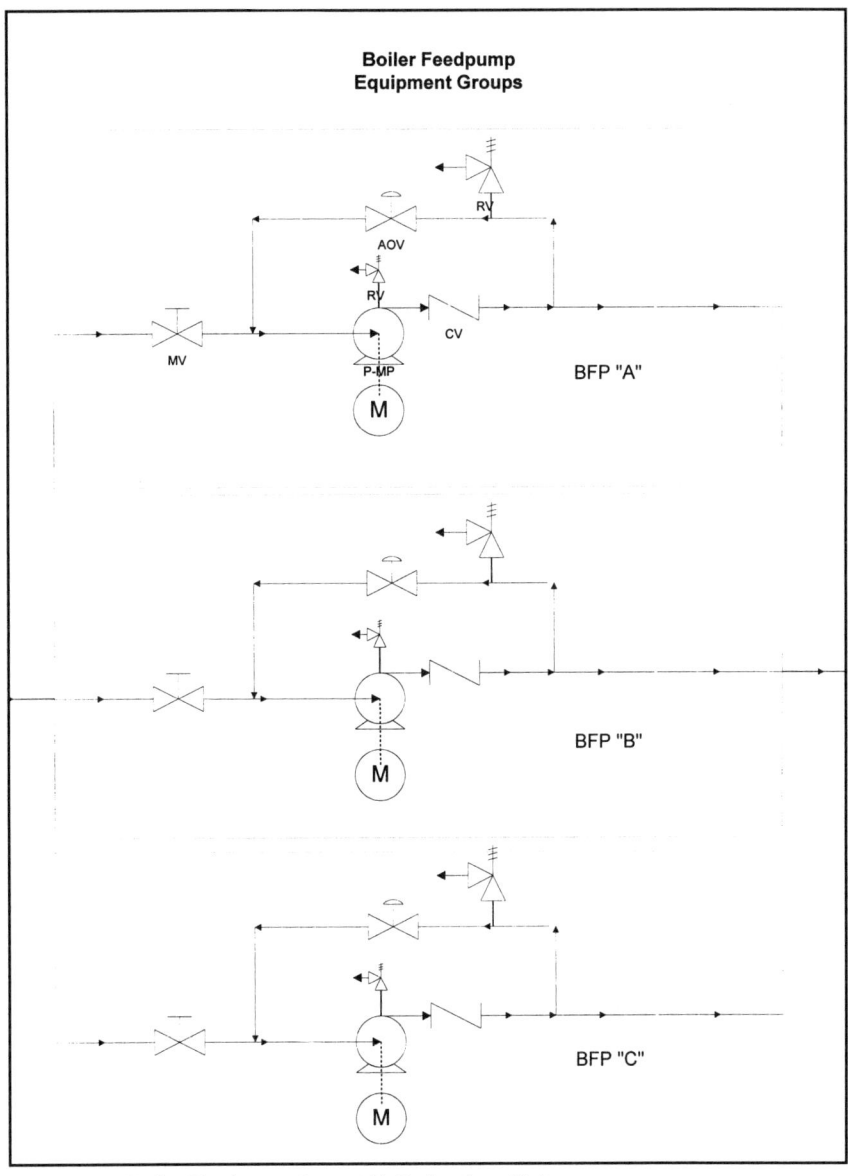

Figure 4-7: Boiler Feedpump in a 50% Redundant Train

restored by addressing the failure mode of the physical part level.

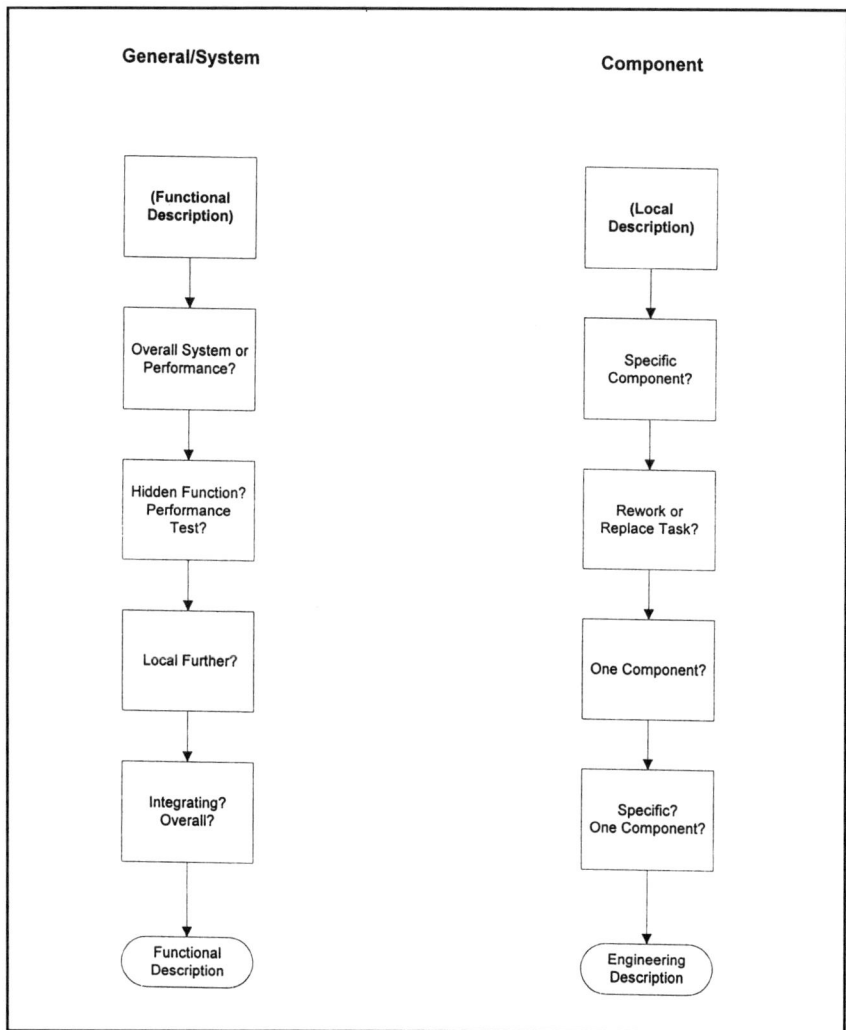

Figure 4-8: Failure Description

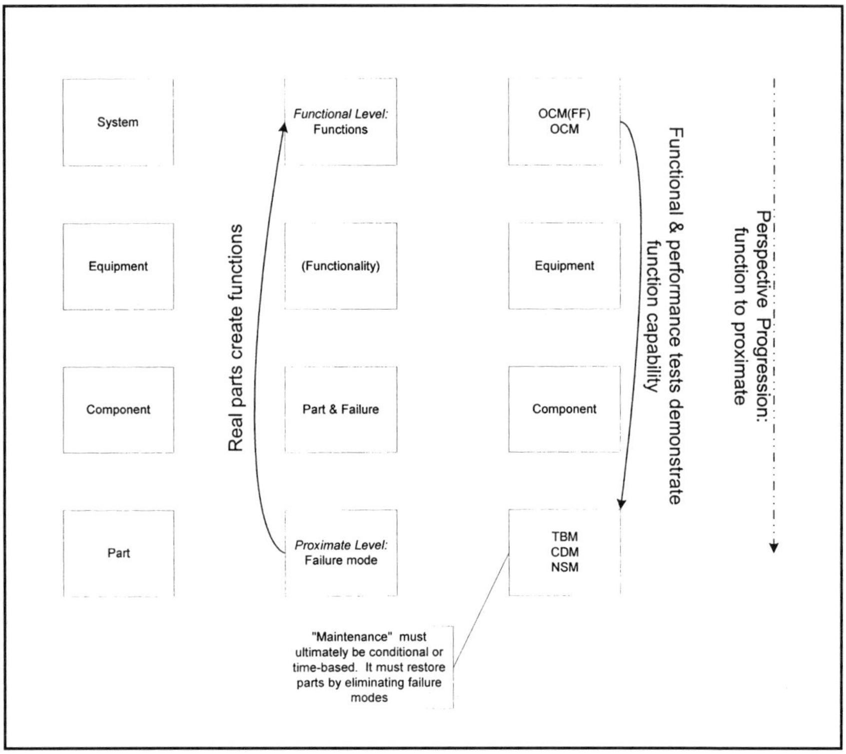

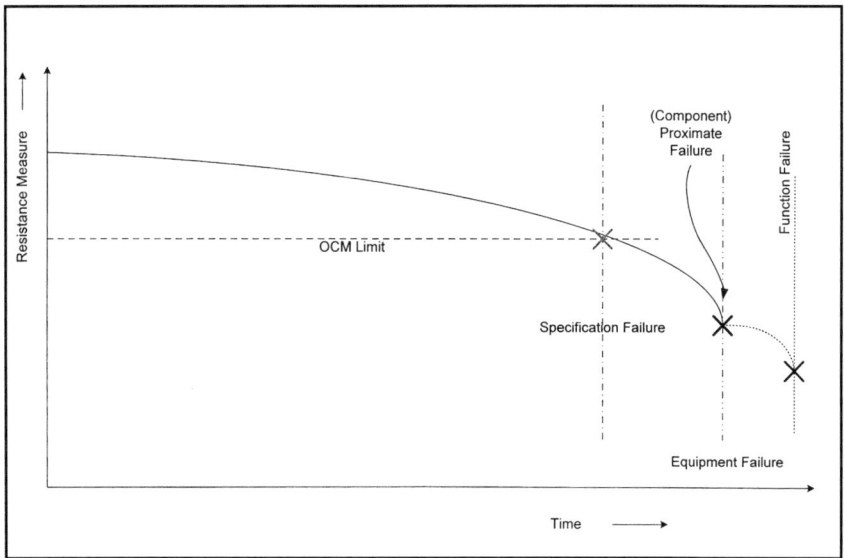

Figure 4-9: System Part Functional Relationships

Figure 4-10: Failure Progression

Chapter 5
Applications

Rule 1: Fly the airplane.

-Pilot saying

Overview

Operations

Generating plants employ upwards of hundreds, even thousands, of direct employees. Plants are complex, with complex needs. They're built with production capability and system-support roles in mind and with the outright goal of making a profit for their owners while keeping costs low.

This mission comes into play even before generating plants come into being. Decisions on siting, project management, and other issues depend on it. Once a certificate of "necessities and benefits" has been issued—quasi-governmental authority allowing cost and earnings recovery—and project construction proceeds, the utility traditionally engages an A-E who develops plans and specifications based on needs.

With utility guidance, the A-E refines plant objectives and performs initial scoping of required systems, their capabilities, and other plant requirements. A design takes shape as goals and objectives are reduced to paper specifications. Plant layout and supporting systems are based upon years of experience and proven practice. The A-E uses previous designs and experience for reference, but each new facility is a new design and guidance from the utility takes precedence over the A-E's previous work. Even when a plant is completed, new units are added, one at a time, as loads grow. The focus is on initial cost, so standardized plant designs are rare. Even common equipment such as sootblowers and boiler feedpumps differ on units adjacent to one another. Yet, for operating organizations, unique designs increase the complexity of operations. Why is it done this way?

Unit design supports high-level operating roles and goals. Plants don't just make electricity—they support company production goals, filling multiple roles in a complex generating pattern that includes seasonal load management, weather, system disturbances, long term purchases, and other unknowns that have to be taken into consideration. These roles are defined during the approval and design phase for any proposed plant.

Over the life of a facility, roles evolve, dramatically change, or even end. Some are identified that weren't originally anticipated. Conditions also change—business and political, as well as technical. Virtually all fossil-fired boilers have been retrofitted with emissions monitoring and control equipment to reflect environmental laws. After the accident at Three Mile Island, the nuclear industry radically changed as the NRC required major plant modifications.

When operating staff—the people who actually run the plants—understand current plant roles, they are better able to focus and manage competing needs. But operating staff see only part of the plant operations picture. Changes in mission, company production goals, regulatory intervention, load shifts, plant aging/obsolescence issues, fuel cost-many factors that change over time are beyond operators' direct influence. To maximize operating returns, operators must understand goals and mission or their performance effectiveness is undermined.

Operations personnel "operate" the plant—reconfiguring for the "mode" of operations (full-load, regulation, part-load, or shutdown). Reconfiguring the plant includes tagging out systems and equipment for service. But operations is also responsible for plant CNM—planned time-based equipment monitoring on rounds, as well as non-specific general area monitoring performed during plant operations and rounds. Operators initiate and prioritize much of condition-directed plant maintenance. Most originates from a functional text schedule. Lastly, operations is responsible for the material condition of the plant—cleanliness and general safety.

Overall, it's a huge task list. In a general way—and only in a general way!—plant training provides skills to accomplish these tasks. No group does it perfectly, but those that do it well receive a wealth of operating benefits—including predictable cost and operations.

Balancing department goals

Station operating departments that deal with plant performance, budgets, personnel, stock, services, and many other cost decisions on a day-to-day basis can easily sub-optimize operations. For example, a purchasing agent may want to minimize parts cost while a mechanic wants a quality part from a specific vendor. These objectives potentially conflict. If the agent is unaware of the quality of different manufacturer's parts, he may second-guess the mechanic and substitute an "equivalent."

O&M goals may also conflict over the performance and timing of maintenance, prioritization of work, work standards, and a host of other issues.

Value added. Everyone on staff either has a direct plant support role or an indirect service role. The former roles include operators and mechanics who keep the facility running, as well as onsite engineers, technicians, and others engaged in maintaining operations through direct equipment support roles. The latter category includes clerical staff, management, and off-site support staff. They enable those in primary roles to do their jobs. When an individual's contributions support either role, they add value.

Engineering support role

Engineering's role is to enable primary workers—operators and maintenance staff—to perform their work more effectively. Plant engineers support two major functions— ongoing maintenance and operation of the plant, and plant re-design for improved performance or cost. The latter role ties to original plant design and construction, but includes services such as redesigning parts for life-extension. Most engineers understand their traditional role as design/build. Few are specifically trained for supporting plant O&M roles, which must be understood to be effective. Traditional design-construct engineers struggle with this issue.

What type of engineering support is needed? How much intervention from plant operations is appropriate? Who plays Solomon when operations and engineering goals differ—for example, in plant operating envelope specifications? Competitive generating companies with strong engineering cultures are able to establish operating and maintenance standards that benefit both R and cost, reducing undesirable, unexpected events.

Plant engineering fills the operations-design interface gap. Effective system engineers require people skills, operating, and maintenance experience, and general engineering competence, supplemented by cost awareness and computer information management skills. Skilled system or plant component engineers favorably influence plant operations by reducing operating costs.

Instrumentation and control (I&C)

I&C groups calibrate, test, and maintain plant controls. Without controls, the plant doesn't run—particularly those with newer DCS systems. I&C also provides the operators' instrumentation "window" on the plant, whether by traditional analog instruments or more current digital or DCS display control screens.

I&C evolved out of maintenance, as a work specialization but I&C remains a very critical maintenance role. I&C technicians in a sense are "super operators" who know a plant functionally well enough to tweak its controls without "trips," yet fully understand the technical details of their trade.

Malfunctions in I&C can cause plant trips and other undesirable events but it's not a black art. Controlling the risk associated with I&C activity involves procedures, routine practices, and standardization of equipment that are planned, developed, and utilized. The real potential of I&C lies in improved availability. Cost reductions aren't an especially promising or even desirable goal (except perhaps at nuclear units). The tedious, time-slugging work of disassembly, rework, and reassembly of major equipment-the traditional mechanical maintenance role— is absent, because I&C hours could greatly increase or decrease with small impacts on overall costs. R comes from reliable instruments and controls and for plant R, I&C holds great value. Direct I&C influence on other areas is slight.

Understanding the factors that cause trips, and improving instrumentation until it plays no role is the major I&C goal. Instrumentation R and availability is a significant concern for operations. Operations and I&C must work closely.

Other players. Traditional mechanical, electrical, and I&C maintenance is supplemented by welders, insulators, and specialists such as non-destructive evaluation technicians, vibration analysts, direct-support engineers, and janitorial staff. As it fulfills its primary role of implementing time-based and condition-directed programs, maintenance must also coordinate with specialist and contract maintenance groups brought in for special jobs and outages.

Maintenance holds the greatest influence over costs, through planned and outage maintenance programs and budgets. Because of the time-intensity of any major disassemble/reassemble work, maintenance has tremendous leverage over operating O&M cost. In a forced outage, or a delayed return-to-power situation, traditional maintenance costs can increase with few questions because the value of lost generation is great.

Engineering

Operations: organizational relationships

Engineering, operations, and maintenance have historically been distant cousins. Engineering performs design-build roles. O&M run facilities. Their interactions were usually limited to day-to day operating issues. Engineering provides project management support for large modification projects but it routinely works alongside operations in plant support. Fossil plants may have two or three onsite engineers who provide minor project support, controls, problem analysis, and special plant needs but an organization in which engineering directly supported plant operations is more exception than practice.

But the absence of a working relationship can cause coordination and support problems. Special arrangements address outages, typically two or three engineers supporting a large, multi-unit plant. The obvious gap is that of ongoing, structured engineering support for maintenance and operations. To fill it, O&M call upon their own resources. Engineers from corporate engineering to plant engineering groups usually hold little operating background. Plant issues—CNM technologies, controls, failure analysis, and maintenance support—have to be learned informally along the way. Companies rarely have specific job descriptions, culture, or measures to identify plant support expected or measure how effective it is. Consulting engineers continue to fill out plant support engineering ranks.

The exception is in nuclear plants, where adequate engineering resources are mandated by law.

Yet, virtually every plant's operating life begins with design problems that cry out for engineering involvement. Stations develop strategies to cope with all sorts of high cost problems—non-functional equipment, failed instrumentation, analysis and introduction of new methods, materials, or equipment to reduce costs. At some point the size and scope of maintenance efforts increase to where facility re-powering is more attractive to manage long term costs. How do RCM, maintenance, and engineering combine to address these problems?

Traditionally, maintenance engages design engineering assistance during new facility startup. After that, engineering supports corporate-initiated changes such as new technology or replacement of large, existing equipment and facilities that are worn out, such as cooling water towers, circulating water tunnels, or large equipment foundations.

Maintenance problems given to design engineering staffs often lack problem definition. Design engineering traditionally accepts project requests regardless of projected payback. Cost justifications for many design changes simply aren't available before design engineering initiates fixes. Management-initiated design fixes are often made in response to regulatory, safety, or other cases without adequate research.

When engineers design fixes, or engage contractors, or incur other expenses, it's often without adequately understanding the problems, their causes, options, value added benefits, or costs. The combination of regulated environments and traditional engineering aversion to cost awareness has combined to allow this kind of project evolution.

This is where RCM can help design engineering groups-enabling them to add value by improving the design request prioritization process. Analysis of failure-mode statistics identifies those plant problems that have "design-only" solutions, the kind that need to go to competent plant engineers for resolution.

The flip side is that design-R features assist maintenance performance and support overall plant R.

As the industry continues to deregulate—with no additional capital to spend—the single biggest opportunity for generating companies and engineering staffs will be the preservation of assets. Unprofitable (or marginal) assets in the competitive environment will need assessments to identify their best options. Re-powering or topping cycles may improve basic production costs. Unreliability will present opportunities for improvement. Under-performing assets (based on competitive benchmarks) will benefit from R engineering in concert with ARCM O&M programs. They will be the quickest route to improved performance and lower costs.

Many facilities have been maintained with homegrown modifications over the years. Some of these plants will gain immediate improvement by having basic hardware unreliability issues identified and removed. In cases where new units were added with little consideration to R or common services, complex and hard-to-run systems resulted. Older, multi-unit plants—some hosting different models and vintages of equipment—added on equipment that compromised redundancy or added tag-out complexity issues. Resolving these problems can lead to quick improvements in performance at little capital and minimal operating expense.

Plant engineering support roles

When it first became clear that complex power plants need ongoing plant engineering support, the nuclear plants established "system engineers" to serve as system managers. They support operations with failure analysis, specific system design interpretations, modifications support, failure analysis, and many other useful functions. Ideally, fossil units have someone with the same capacity and training.

In practice, nuclear systems engineers focus on regulatory compliance but their ideal role is to improve system work, operating processes, system R, and lower system costs. An effective plant engineer, system or otherwise, is someone who masters subjects not taught in engineering curricula-failure analysis, cost engineering, maintenance, controls, R, and operations. Procedures, cals, test programs, and general industry requirements take time to learn, but an operator's role must be learned over time. Learning to provide real-time support to O&M doesn't come from sitting behind a desk pushing papers, but rather managing personnel at the plant level.

R engineering theory and RCM provide excellent guidance for plant engineers in this endeavor. Developing the skills and capabilities necessary to perform RCM, in streamlined format, can provide guidance for support engineering groups. RCM requires engineering, technical, and plant support competence. Plant support includes failure analysis, operating procedure analysis, "poke yoke" (human factors), engineering simplifications, maintenance support, process improvement, I&C understanding, measurement, cost and performance improvement awareness. It's a tall order in anyone's book. But it's so important—and those who do it well are so few—that companies need to develop new job descriptions and training programs to ensure it occurs. Capable plant engineers are required at any facility—no matter what type. This includes steel mills and food processors, not just generating stations, and includes whatever title by which they're currently known (plant engineer, production specialist, services engineer, maintenance engineer, project engineers, application engineer, and so forth).

This will not be easy, however. For utilities, the need is great, the position is new, but organizational inertia is a barrier to anyone seeking to fill this role. That inertia exists because utility generation has been organizationally static for 40 years. The last great change came from the nuclear units' special regulatory, safety, and operations support needs-the change that differentiated generation into fossil and nuclear. The latest changes—the proliferation of gas-fired CTs—may again split the industry. Deregulation and convergence with the gas industry will "fuel" additional changes organizationally. Fossil generators may also find battle lines drawn over issues of high costs created by re-regulation.

These battles can be best engaged through improved engineering functions, for engineering traditionally improves the product. Improving generation processes reduces costs and increases safety as it increases generation. In the 1950s and 60s, the generation industry— thanks to more efficient processes and facilities—enjoyed reduced costs, improved product, better customer value, and ultimately further industry growth. Today's CT technology is continuing along this path— the primary reason why new CT orders keep coming. Gas supply looks adequate to support much more electric conversion to gas. In the market-driven environment into which all analysts say we're headed, markets will decide these and other issues. New opportunities for plants and support personnel await those who can put engineering improvements to work.

Plant modification

Meaningful improvements will not come cheaply because they depend upon plant design modifications and such "mods" are expensive. Those initiated within the plant tend to cost much more than estimates suggest in my experience. Minor modifications managed on-site often have the lowest level of control and stand as the worst offenders. In concert they add up to a burden on operating budgets and available staff. When the final numbers are in, such projects can cost more than what's budgeted—about 10 times more, throughout my utility work experience, based upon final-cost figures for many "minor" design changes using a cost-accounting system that traced charge numbers to jobs. Given that original cost benefit, justifications (where utilized) were based on estimates that were a factor of 10 low, it stands to reason that there must be a significant volume of design work of marginal value— or, more likely, of no tangible value when the goal is reducing unit operating costs or increasing generation.

Improving the design change screening process will thus have great paybacks. ARCM can do just that. In RCM task selection logic, design modifications are the last choice when there is no effective PM that can be done, and failure can't be tolerated. In fact, these cases are rare.

"Effective PM" translates to technically effective, a case agreed to by experts addressing a failure mode. It points to fundamental design flaws uncommon in "production" components and equipment. More commonly in these cases, maintenance fundamentally misses the mark. *i.e.*, the task performed has no applicable—much less cost-effective—basis.

Until it's proven that a design change is required, redesign is a costly proposition. If a maintenance solution is at hand, however, savings and benefits will be substantial. To make this point, you must have done your homework, and there must always be analysis on which to base design changes and value. Formal RCM analysis provides the basis specifications for redesign.

Another common organizational weakness is the failure to pass design-developed equipment assumptions (and support requirements) to the facility operating and maintenance staffs in a manner useful to them. After problems arise and designs are reviewed, it often becomes apparent that plant management and engineering staffs never connected on procedures, training, drawings, or other key aspects of what was supposed to be a joint effort. From my experience, in about half the cases engineering did provide the product, but it got lost at the plant level because the plant lacked the infrastructure to use the material provided.

It's hard to recall faulty designs, and so developing thorough failure-based maintenance plans effectively identifies areas that can truly benefit from design. Such reviews ensure that operating and maintenance problems at the plant level get corrected at the plant level—with little or no engineering assistance-before going to the design engineer. In this manner, "step improvements" occur in O&M. Operating groups improve their understanding of plant design specifications—and limitations. Plant operators better grasp design and operation factors required for success. RCM considerations assure that design requests are those that design personnel and processes can and should legitimately address.

There is also value in having engineering staff work on product improvement. O&M staff can identify costly failure problems, develop alternatives (including the basic maintenance program strategies and tasks), and support age exploration. Design engineering can focus on "mysteries" that aren't understood or require re-design. One inevitable consequence is that engineering and maintenance draw closer-focusing on facts, on quantifying costs (and benefits), and on supporting equipment needs.

Engineering tools

A number of engineering tools provide R analysis for generating units.

Hand-calculated until just a few years ago, R analysis was generally not applied to complete designs. Instead, thumb rules, benchmarks, and standard solutions were applied. Today, personal computers (PCs) and specialty software offer greater capability to evaluate detailed designs for availability, R, risks, and other life cycle R aspects. Many analyses tie directly into plant operations. RCM evaluations support implementation of the unit planned maintenance program. Other products provide similar services. For instance, Markoff Analysis can be used to evaluate conditional probability of failure when important equipment is OOS for maintenance.

In a deregulated environment, with many new plant and equipment designs emerging, capital investments are put at greater and greater risk. This increases the need for R tools for these design assessments. Here are some of the best.

FMECA. A complete RCM analysis begins with a "failure modes and effects criticality analysis" (FMECA). ARCM limits analysis to the major "hitters" that can be identified and used, based upon experience. ARCM/RCM for an existing facility is an a posteriori assessment— experience limits the scope of the review and focuses on value. New facilities can be reviewed using *a priori* RCM, utilizing a variety of formal R engineering tools, including FMECA. Projections of likely problems, availability, and maintenance costs can be generated based solidly on analysis.

Analytical FMECAs have been used for years in aerospace applications to zero in on risk contributors and manage overall risk on a budget. This analysis logically fits capital-intensive, single-mission design applications that characterize the space program and many high-risk military missions. Government and general specification MIL-STD-1629A provides standards for FMECA preparation. Software to meet these standards is available commercially.

Fault trees. Fault trees, like FMECAs, focus designers on weaknesses in a developmental design. They allow detailed assessments of alternative configurations and any likely failures. Fault trees can be used not only to assess designs, but as a corrective tool to assess existing applications. As a side effect, they provide excellent risk-management tools for operator training. (Fig. 3-9, page 106)

Availability simulation. Although applied primarily for design consideration, availability simulation can project the impact of redundancy loss during major maintenance. Many times availability simulation tells a story better than mathematical analysis. If maintenance schedulers can see risk impacts, they can more wisely schedule maintenance. Often the design aspects of redundancy are not conveyed in ways useful to schedulers. Simulation results can fill that gap. Again, software products can perform this work.

Weibull analysis. Weibull analysis models failures into a Weibull distribution—the most generally available failure distribution (in the sense that both infant mortality and aging can be modeled). Processes, programs, and equipment can be tested for infant mortality and for random or lifetime aging failure behavior. Failure distribution can then suggest strategies for corrective measures, particularly as they relate to maintenance performance. More formal failure and design-out engineering can be included.

Weibull analysis is of particular interest to organizations evaluating suppliers and parts. Parts specifications can require a Weibull distribution evaluation and multiple parameters. Weibull analysis provides excellent information about the performance of parts in-service. This includes:

- infant mortality failure rate and period
- normal service period duration and residual random failure rate
- expected life
- dispersion of expected life

Weibull analysis can provide a competitive edge to parts suppliers. As a plant engineer, I would have much greater faith in a product that came with a Weibull specification than without.

Integrating "The Big Three"

Three basic functions are necessary to operate facilities over time-operations, maintenance, and engineering.

Operations is the first supporting leg in every plant's mission and implementation. Unless they're truly remote units, plants don't operate without operators. Even remotes have dispatchers!

Maintenance includes all those organizations generally engaged in maintaining the facility, including chemistry. Little-"m" maintenance includes direct service roles such as scheduling and I&C. (Cost accountants call them direct labor.) These people turn the wrenches and perform tangible work. Their direct support—planners, schedulers, and dedicated clerical staff—is also maintenance staff. Engineering "owns" plant design, design improvement, and design-cost reduction over time.

These three groups influence, to a large degree, overall plant performance and competitiveness within the cost constraints of supply and demand and corporate structure. They determine the degree of production success at any particular plant. This is where RCM improvements live!

Operations roles

Failure identification. An operations staff is primarily responsible for plant condition. However, operating staffs "own" the plants, to varying degrees, and so failure identification is a legitimate responsibility.

Recognizing failure requires knowledge, experience, skill, tools, and failure standards-a perfect fit with operators' plant-monitoring assignments. Failure identification, as a rule, is sometimes assumed, overlooked, or taken for granted. Again, operators have the abilities and the obligation.

Operations spends more time than anyone else in the plant—reading instruments, operating equipment, feeling vibration levels, smelling fluid leaks, hearing noises, and seeing how things do (or don't) perform.

They are naturally suited to recognize changes, identify faults, and initiate correction.

Successfully identifying failures depends on experience and skill. Some operators receive excellent training, either during career development or prior to hire, while what others receive is very limited. Effective operators in a competitive environment need higher-than-average skill levels. Turnover increases training requirements. Nuclear plants have excellent training programs because of license and industry standards while fossil plant training is more on-the-job, hands-on, learn-as-you-go. Both methods have their place. Training needs to be cost-effective—in fact, measurement for cost-effectiveness is a training need in itself.

About 80% of all failure-identification tasks originate with operators, based upon RCM failure analysis. That is: fully 80% of all RCM-based maintenance involves operator monitoring! In a CNM program, then, maintenance starts with operations. Because operations monitoring is so pervasive, failure recognition—a key feature of effective maintenance—begins with operator training.

Two primary operations tasks are CNM and functional testing.

CNM uses the senses and instrumentation to identify equipment failure and failure trends. Functional testing for hidden failure functions—alarms, trips, and other protective or standby devices—assure function is preserved.

Nuclear plants won't discover large, available benefits from increased functional ("surveillance") testing because they already have extensive surveillance plan requirements based upon their licenses, and they generally have excellent availability. Fossil plants, however, may find major gaps in their testing and equipment protection plans. Many fossil "surveillance" plants are informally controlled and miss critical and essential instruments and alarms. If implemented, these can assure design conditions are met.

The second aspect of the operations monitoring program is the testing program. Essential alarms and trips are typically tested on the largest equipment in both nuclear and fossil plants. These include turbine trips and vibration trips for large ID and FD fans. But other, lesser alarms don't get tested—"chemistry out of spec" condition alarms, for example. Some critical alarms occur in remote locations and may not go to the main control room. Calibration and testing programs for these alarms implies value and perceived importance. "Hard" (unit trip) calibration limits are frequently neglected. Operations and engineering personnel often interpret alarm values substantially differently, as if the two aren't reading the same set of guidelines.

Some utilities intentionally minimize "hard" trips when an equipment supplier provides a "hard-wired" trip or status alarm (on high vibration, perhaps) and the company uses a status alarm. The operator then acts on the alarm "appropriately." Such vibration status instrumentation is installed on virtually all the main turbines at one Midwest utility I know of. This approach undermines effective instrument maintenance. Their position was that, "We don't want any trips to occur due to sporadic alarms." They expected their operators to interpret ambiguous instrumentation from the same erratic instrumentation that no one wanted hard-wired for trips. There was a R problem with the trip instrumentation that the company was unwilling to address.

How an operating company addresses instrumentation indicates much about its operations philosophy. In the case of "critical" instrumentation, "critical" has two connotations—RCM and common usage. RCM is a direct safety consideration, common use is subjective, inexact intuition. Ambiguous instrumentation guidelines indicate unclear management philosophy. An RCM-based instrumentation review can help management select the instrumentation and limits for clear action.

OOS, uncalibrated, or otherwise unessential instrumentation abounds in a typical plant. A vast majority of instrumentation provides non-critical, non-essential status. Such instrumentation can readily have non-scheduled maintenance (run to failure or "self-identify") and be maintained as operators recognize their need for, or dependence on, its use. A few instruments provide early warnings of impending high cost failure—large-machine vibration monitors, for instance. These need attention.

Although concern that hardwired instrument trips will lower unit R is legitimate, there are more fundamental worries. Focus on essential instrumentation improves unit R and safety. Clear instrumentation maintenance standards supports safe operation. Usually, instrumentation and personnel protection for large equipment go hand-in-hand.

Concern that "hard requirements" get carried away (in the fossil world, anyhow) is driven by fear and culture, not careful analysis. The opportunity to establish operations-administered guidance can improve safety and performance.

Once identified equipment failures are entered into a plant's maintenance system, operators describe symptoms and provide other insights. "Getting the right maintenance" starts with identifying problems correctly and that means clear WO problem descriptions. Even someone with limited writing skills can quickly grasp WO specifications. The more defined the WO problem, the more diagnostics completed, the easier it is to troubleshoot, define scope, and perform work.

Operator monitoring. Operators monitor plant performance, remotely and locally-in the control room and on rounds. Automated DCS plants trend by CRTs or by automated-round logging devices. Monitoring via DCS CRT or control room panel requires a big picture perspective and the capacity to anticipate. DCS make monitoring the plant easier, simplifies work, and improves alarm response. DCS simplify "round" monitoring requirements because remotely monitored points can be trended and need not be replicated in rounds. Invariably there are instruments that aren't monitored, or that need a physical presence to visually review, or that can't be downloaded because appropriate "drops" aren't available. In these cases, a round is still necessary.

DCSs—like all other instrumentation—need oversight to control information going into the system and the alarms safeguarding it. Because DCS has the capacity to tie together large amounts of information, scope-of-monitoring is even more important. RCM helps prioritize and rank information value. Critical alarms can be emphasized and status alarms de-emphasized. On a DCS upgrade, an RCM filter can evaluate alarms and instruments for monitoring, and limit the scope of monitored equipment, hardware, and software. This substantially reduces the amount of instrumentation required, and saves money. Savings continue over the life of the facility because the scope of monitoring and maintenance has been limited.

Rounds optimization. Rounds consume the major portion of operators' time. Ideally, time not spent reconfiguring the plant is devoted to rounds and monitoring. Rounds are CNM tasks that incorporate failure-finding checks. Traditionally, rounds were a way to ensure operators were in the plant. An hourly round was a typical service standard but in systematically developing a CNM program, one finds that thorough rounds through a large facility take several hours. The one hour round is both not feasible and not desirable since it compromises monitoring quality.

Because RCM-based CNM guidelines identify expected failure modes for each item of equipment and equipment monitoring limits, they can be incorporated into logsheets and hand-held logging devices or remote DCS monitoring screens. RCM rounds data can be downloaded directly into equipment and CMMS files for detailed assessment, history, and trending.

By comparison, traditional operators' rounds were developed using vendor guidance and general experience. They offer substantial improvement opportunities including reduction in scope of numerically monitored points while increasing the "completeness" of failure monitoring. Combined with interval extensions (that random failure data suggests with benchmarked actuarial analysis), the simplification opportunity is substantial. The results include:

- reduction in rounds scope
- elimination of redundant and inconsequential logging
- addition of overlooked instrument and alarm checks
- development of a rounds strategy (grouping)
- consideration of regular and abbreviated rounds

During startups, it may be necessary to temporarily reduce "rounds-logging" to avoid adding staff. (The increased risk from reduced monitoring is presumably acceptable.) Automated-rounds data-logging requires that round tasks and routes be specifically determined. Previously, many facilities used informally developed and managed rounds. As a major strategy change, implementing "rounds-logging" is the ideal time to perform a "rounds review" and integrate rounds into an overall plant CM strategy. The assumptions and basis for rounds are well worth examining before to re-institutionalizing rounds. Implementing automated rounds-logging is a milestone to assure rounds are firmly grounded by ARCM.

Parts

Age exploration. Actuarial failure statistics from commercial aviation studies show most in-service parts (93%) never reach their design end-of-life. Components are replaced on a hard-time basis though only partly consumed (Fig. 5-1). This stems from conservatism and from untested assumptions about wear-out and overhaul. Commercial aviation experience and conclusions transfer directly to the generating industry, supported by appropriate information control and management. "Age-explorations"—service and wear monitoring on parts in service as they are replaced—provides information that can extend lifetimes.

To improve part utilization, the involvement of craft performing part in-service performance assessment is both necessary and logical: They remove, service, and replace virtually all parts and so their assessments of parts performance are essential for aging study. When skilled workers ask the question, "How much remaining serviceable life is there?" it orients them towards assessment of failure modes, criticality, and part service performance. CMMSs offer the ability to track failures and "replaced-parts performance information" with less effort. However, no information trail begins without a skilled craft assessment and data entry.

Evaluation of parts performance in-service is every plant person's job. The savings potential is simply too large for such work to be ignored, and many facets to parts service requires that all be involved. These facets range from warehousing lifetimes to nuclear environmental and usability issues. Sometimes savings come where least expected.

While most CMMSs have the ability to develop age exploration processes sensitizing the craft to age explorations as a routine practice is more challenging. A simple assessment of a part as it is replaced is more that adequate. Fancy material-failure analyses that are within the capabilities of some companies are, for the most part, not needed. Parts management subroutines in new CMMSs will enhance parts use and tracking-but even good guesses are helpful!

Component monitoring and age exploration have been practiced as

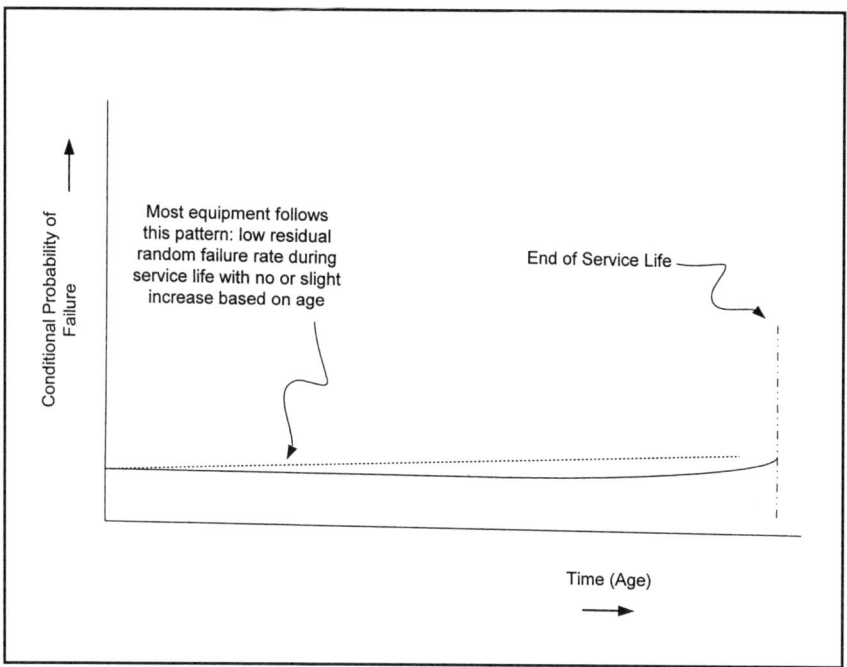

Figure 5-1: Failure Curves for 93% of Equipment Components

an engineering discipline for a long time. Dissatisfaction with the cost and availability of precipitators led to the development of fabric filter dust collection systems, affectionately known as "baghouses." Improving economics and the performance of fixed-speed drives, with the growth of power electronics, lead to development of variable-speed drives. New materials, processes, and equipment all start with the realization that equipment has inherent design limits that ultimately limit the capability, which in turn starts the search for new alternatives. Involving an entire organization at all levels with in-service performance evaluation is a profitable first step towards the next breakthrough.

New software information-management capabilities make this more feasible than ever before. More importantly, there are significance savings to be realized. Parts usage and improvement based upon service requirements and performance are one of the prime features of RCM applied in commercial aviation. In my own career, I've regularly discovered major savings opportunities by extending equipment service lifetimes based upon age exploration and redesign. The latter are usually not dramatic as engineering exercises, but in aggregate the cost savings and performance enhancement have added up to some staggering amounts.

In most plants, small parts (costing less than $10,000) and the scheduling and planning process means that parts decisions are left to the worker or planner. The planner-often in the dual role of scheduler-makes decisions when ordering or purchasing the parts based upon feedback from the shop floor. This is where part utilization improvement begins.

Stocking levels. Parts usage and processes drive stocking. Some parts need to be carried as "critical" (important) spares, most don't. Given the "spare everything" strategy of the traditional utility industry, there are ample opportunities to reduce stock—just as there are opportunities to make errors: No-parts-stocking strategies will always provide a spare on demand.

In fact, the absence of any part problems indicates over-conservative stocking practices. Lacking spares—or being unable to locate a critical spare—translates into lost production that makes for an expensive spare. Spares carried in stock for years, with virtually no movement, is no better. One expense is visible, the other is not.

To optimize stocking, operators and craft must understand and accept the parts strategy and the larger operating strategy of which it is a part. They manage failures, which requires an implicit appreciation for the plant's equipment R. Approaches differ, when using and extending the life of used parts. Some would have you toss every consumable-others, reuse everything. There's a middle ground in most cases.

When workers know that parts aren't readily available, do they exercise greater care using them? You bet! Inexpensive and consumable parts get tossed—expensive and reusable gets reused. (But, which is which?) An age exploration program can help identify them and fill the gaps. If we don't know how a part failed—find out! Many parts can be reused cost-effectively, e.g., not with a penny-pinching, refurbishment-at-any-cost philosophy. Evaluating the reuse of serviceable parts requires age exploration and training. Parts-life estimation is tied closely to in-service parts-failure evaluation.

Many companies establish stocking levels with expert software, but a conservative mentality and an awareness of parts aging and failure can suffice. It requires training and instilling a "questioning culture." There's a degree of risk, which may include not having key parts for equipment that fails in service. Like owning a car or a home, risk-management exists in several forms—financial, obsolescence, and downtime.

The traditional strategy of "carry an extra of everything" was advanced by vendors, who supplied lengthy lists of every spare imaginable, requiring warehouses with huge inventories to control and manage. Taking more risk—using parts-sharing groups and vendor-maintained parts "inventory"—reduces part costs. Truly "critical" spares can be managed with overnight delivery, parts sharing, and operational contingencies. For overall success, reduced stocking must tie to knowledge of part-failure risk, redundancy, risk profile, and equipment strategies.

Consistency. Reliable parts provide value. Seems simple enough. However, study how parts are used in any given manufacturing setting, and the strategies employed in selecting parts and vendors, and you'll come to some surprising conclusions. For instance:

- manufacturers focus on reducing parts inventory as a cost-management strategy
- general results of that strategy transfer to the generating plant environment
- parts R influences required stocking
- unreliable parts carry hidden stocking costs
- the more vendors and part sources a plant has, the greater its overall part variability
- part variability increases costs

Most engineers and planners have to deal with multiple suppliers and mixed lots. By evaluating parts service, I conclude that manufacturers of quality parts understand the cost/benefit savings that superior parts provide for users—and they price accordingly. Unfortunately, planners, maintenance managers, and buyers-those who don't have the information (or desire) to do life cycle parts-costing-make most of the buying decisions. Many utilities require firm bids when a purchase amount exceeds some nominal amount—$5,000 is common—which kicks decisions upstairs.

Special, custom-application parts increase life cycle management costs. A coal-fired plant with six mills—no two of which has the same basic hot air ports, pyrites brushes, discharge valves, rolls, and tensioning plungers—means that this individual "crafting" of equipment requirements increases parts requirement. In the absence of formal standardization policies, there are common parts-management thumb rules. SPC (borrowed from manufacturing) provides the best guidelines on how to monitor parts usage. Many "parts events" have demonstrated the folly of not following standardized "part rules." Many corporate buyers disagree with these guidelines—they buy strictly on price—but they rarely face the consequences of using and managing parts that don't fit or that break on installation (or soon after entry in service)—or that break randomly.

In general:

- know critical functions of all secondary market parts selections
- go with the OEM except for obvious substitutions
- work with the suppliers' engineering staff to understand parts in service
- "low-quality" appearing parts are usually what they are—low quality
- "high-quality" appearance doesn't guarantee performance; suppliers do
- workers have practical insights on parts performance
- parts records will surprise you

Problems. When a part problem is identified, it's best to check records. However, it's likely that records concerning parts and their use-and costs-are incomplete. Usually, the craft identifies a part problem based on service problems—or a feeling. Analysis confirms the concern and quantifies an opportunity. More capable inventory management systems and CMMSs can improve parts information and also age exploration. Systems can help, but only if there's perceived benefit to their use.

This is an area for "poke yoke"—the use of many simple tools. Simple retention-and-review "storyboards" are valuable for training and evaluation of failed-parts experiences. "Good" or "bad" examples, photographs, and failure descriptions all help improve understanding of parts.

Troubleshooting. Some basic truths:

- troubleshooting costs technical and maintenance personnel time
- the more experienced personnel are more productive trouble-shooters
- troubleshooting new equipment is harder than diagnosing a familiar machine
- a "learning curve progression" must be followed before people reach full diagnostic effectiveness in a plant environment
- new and unforeseen failures demand more resources

FMECA and a fault tree assist troubleshooting by establishing relative probabilities of what can go wrong. High-probability events can be checked first. They also provide failure symptoms that can validate actual failure causes. An FMECA indicates sources of failures, benefiting future troubleshooting.

Experienced personnel don't require these insights. Every organization has inexperienced staff and bringing new people up to speed quickly—providing all users with optional aids—is extremely beneficial. Rare failures are hard to catch and diagnostic aids such as fault trees, logic guides, and FMECAs are then very helpful.

To effectively diagnose equipment, however, the technician must understand it. The more complex the equipment is, the harder it is to diagnose. Technicians must understand the components that provide the functions that make up the equipment as well as their interactions. Focusing on functional descriptions—or even on component engineering failures—limits failure detail. (Complex equipment can also *simply* fail.)

When equipment maintenance is permitted to slide and instrumentation and redundant components reach failure, multiple failure modes begin. Failure interactions make trouble-shooting much more difficult.

A time benefit of a fully implemented PM program is that failure-identification is as simple as it can be. Fewer failures get diagnosed with complex interactions, so diagnostics are more straightforward.

One of the most difficult tasks in a complex plant is to restore abandoned equipment-just ask an engineer who has done plant restoration after a fire! It's almost as hard as startup, since everything must be checked out. Abandoned equipment also generally has multiple failed states, each of which must be separately corrected to restore function

Secondary failures drive home this point. Every secondary damage event results in many more problems. (Reconstruction of steam-damaged cable can be extremely complex and time consuming.) Multiple failures can result from flooding, fires, leaks, or a variety of other events that fundamentally change or exceed the physical environment. Some of the most severe damage (in terms of cost) comes from events leading to steam and moisture attacks on components designed for dry environments.

I prefer to deal with primary-failure prevention and simple failure modes. To do so, we should understand those primary failures that lead to common-mode, general, and expensive secondary damage.

Failure numbers

Numbers are the best way to tell the objective failure "story," yet they're missing from most traditional generation RCM analysis reports. Some RCM books go so far as to discount failure statistics and numbers altogether. In my opinion, this is a serious oversight. Implicitly or not, we use frequency and consequences to draw conclusions, and numbers tell that story. Those who don't understand this and live by the numbers wind up chasing the "trivial few." It gives RCM a black eye when analysts bog down in endless pursuit of rare or imaginary events—things that don't happen in the real world.

My approach reflects my predisposition towards measurement—I'm an engineer. In reviewing large quantities of failure data, identifying failures, summarizing statistics by failure categories, and making estimates, I work with numbers that aren't exact—but they're in the right ballpark. I view them like health physic numbers—order-of-magnitude significance. They identify sensitivity to costs that need to be understood at 10%, 100%, or 1000% payback levels over the period of interest. We need to structure our programs for value. A 10% payback on a turbine overhaul may be worth chasing, but probably not for a $20 filter replacement. A 500% savings on a $20 task clearly outweighs the same for a $2 task. Practically, this means when it comes to a tradeoff (and it will), we must give up the $2 tasks to make room for the $20 tasks.

Activities should reflect on-site statistical data and failure experience. Work environments are unique and influence what fails and when. The cultural environment influences what failures are recognized. Available levels of skill, knowledge, and other intangibles can be inferred, but are hard to measure. Just as two randomly-selected individuals will experience different success rates with the same make and model of automobile (as measured in longevity and life cycle cost), two similar plants experience distinctly unique operational outcomes. These can only be explained in process terms.

So-called "rare events"—the second aspect—pose an actuarial problem. Rare events represent the highest-value RCM learning and benefit opportunities. Most heavy production and financial losses arise from them. They are certainly worth understanding.

After many years examining major losses, I find that in most cases, a chain of events presents a history. The progression towards ultimate failure depends on systematic process weaknesses-rare events occur more frequently in the absence of process awareness and controls. They reflect random, individual, repetitive occurrences. Individually, they rarely convert to an operating event but if they happen frequently, that event will most probably occur—and, ultimately, statistics tell the story. Rare events can be managed with conscientious, complete operations. These rules, well known in theory, are well-practiced by professional operating organizations.

Safety

Direct consequences
Generating plant safety presents two challenges—maximizing safety practices and minimizing costs.

Overall, we need to apply better safety practices in many plants. Fossil generating unit accident rates are significant. High pressure steam, high voltage, coal belts, and large rotating equipment perform well under ordinary conditions, but are unforgiving. Safety awareness begins with understanding equipment and how it fails. Many safety issues develop in the course of returning failed equipment to service. This includes diagnosis, tag out, physical work, test, and return-to-service. Better understanding of equipment failure leads to better maintenance practices, more specific work plans, and planned work—and planned work is safe work.

PM emphasis is on time-based and CDM, and occasionally, on "NSM." PM work is plannable. When planned, it contributes to safe operations. Rework and repair tasks are routine and plannable. Some plants develop detailed work steps, others leave details to the skill of the craft. Whichever approach is taken, using standard PM tasks in standard blocked work formats, ensures that repetitive work is supported and contributes to safety. It's easy to make a case for safety when a higher plant-conditional readiness state is achieved—and an ARCM-based maintenance strategy in fact leads to a higher state of equipment readiness, both for critical instrumentation and redundant and lessor equipment. When work is performed in logical rank, the need to work extra hours is reduced. More efficient work practice—like equipment alignment and "on-condition" maintenance-also contributes to safety.

Operators and mechanics who learn the system review process also acquire skills helpful in understanding equipment importance and prioritization. Combined with the inherent improvement in R that occurs, it's easy to make a theoretical case that RCM favorably benefits safety. However, actual measurements, to the best of my knowledge, have not been made.

Serious injury and equipment hazard events can stem from performing on-line maintenance. As units are put under more pressure to generate, the trend to perform on-line maintenance will increase. The need for routine, high value work plans will be even greater.

Potential consequences

A second, equally significant improvement in safety can be derived from more expeditious use of a plant's safety budget. Presenting a budget case for safety modifications is a chronic problem. Conversely, many "convenience" modifications are covered in safety terms to improve their likelihood of approval.

The greatest practical safety issue in many non-nuclear plants is the state of critical instrumentation. ARCM can and has significantly improved plant awareness of instrumentation by focusing maintenance efforts on high value instruments and alarms, while discontinuing scheduled maintenance on the rest. Operations monitoring has benefited from this review. Some classes of plant equipment get taken for granted, and this practice could benefit from an ARCM review.

Wherever a station spends a substantial amount, that warrants considering ARCM analysis. Chances are tidy savings will result. All-out faith in contractors to identify plant material maintenance needs is generous, but financially risky. The chance is high that the plant will end up with a Mercedes, rather than a Ford!

In its original development phase, RCM's criterion for "critical" work was safety. Yet, safety is a routine issue at most generating stations—companies will sacrifice at least an hour a quarter to have all hands attend a "safety meeting." ARCM can neatly "operationalize" many safety issues. The airline criteria (RCM critical = direct safety consequence) meant "critical, mission terminated." *i.e.*, For air transport, a critical failure—a safety issue—is a show stopper. All critical failure modes warranted scheduled maintenance. In the event that "applicable and effective" maintenance tasks couldn't be specified, the default action was to redesign. Utilities—particularly those operating fossil plants—could greatly improve safety budget mileage by adopting simple, direct safety criteria of the airline industry. Hypothetical, possible, and theoretical safety expenses that prevail at so many stations could be a thing of the past.

Example and Case Histories

Circulating water tower

Circulating water towers (CWTs) offer a representative look at the tradeoffs in large equipment maintenance. Many possible CWT maintenance approaches are available. Towers can be capitalized, and replaced in 20 years. They can be continuously maintained, with ongoing, cell-by-cell rebuilding. The choice depends on long term operating goals and philosophy. Functions vary from cost-conserving mist-eliminators to functionally important debris screens.

Nobody loves a CWT. They're damp, dirty, infested with birds, and they smell like chlorine and biological growth. In farming areas they fill up with sediment and windblown soil. In industrial areas, they scrub everyone else's emissions. (Maintaining a CWT adjacent to an oil refinery was my personal headache.) Industrial area towers require chemistry different than those at other isolated tower sites.

Towers suffer several dominant failure effects:

- biological growth and debris accumulation
- aging of diffuser/distributor nozzles
- random environmental damage (like ice)
- cell hot water basin flow balancing adjustment
- timbers and fill aging
- structural deterioration, fastener relaxation shifting

Plant operators take towers for granted until performance is a problem. They're not complex, they're easy to understand, yet they're sophisticated in design. In a common basin-distributor design, water risers carry heated condenser cooling exhaust into cell basins through shutoff/flow control valves. Water spreads and falls through diffuser nozzles onto fill, forms droplets, and then evaporates and cools. Cool water in the basin provides a suction reservoir for the circulating water pumps (CWPs). CWPs draw cool, aerated water through large mesh screens, knocking out debris and returning it to the condenser inlet water boxes.

Cooling point depression is the measure of tower performance. Margin means there's ample cooling capacity. Ample capacity assures condenser vacuum is adequate.

Towers require seasonal maintenance. In summer they're most needed for cooling but in winter they need the most attention. Towers can be icy, slippery, dirty, bitter cold (except inside), and demand constant attention. Ice builds up, louvers tear off, fill comes down, and screens plug. Work can be miserable. Fan electrical problems, lubricating oil leaks, vibrations, and a host of other random problems make towers tough maintenance areas.

During the summer, the units are often on the raw edge of load reduction as the tower cell fans, spray patterns, distribution, fill condition, and other lesser problems make towers the key determinants of load. Every last ounce of capacity may need to be coaxed from an old and tired tower. Balancing cells on a shifting, deteriorated tower can be almost impossible.

Towers have blown over, fallen down, burned down, and rotted away. The last case is the most frequent. Deterioration of cell basin levels, or leaky distributor shutoff valves, makes cell balancing difficult or impossible. Structural sags can effect the hot water basins so that balance cannot be achieved. As towers age, their inability to balance, maintain basin temperature, and maintain condenser vacuum in summertime (during load peaks) makes replacement inevitable.

The more dramatic tower episodes in my career involved lesser subsystems that weren't appreciated until they became problems or failed outright. Before fiberglass return lines and spargers became standard, redwood staved-distribution piping was common. At one plant the staving failed, sprinkling a waterfall out away from the tower basin. Flooding resulted, and the basin went low. After the basin emptied, the circulating water pumps tripped. The unit went down on combination of low vacuum and no cooling water flow. A similar event involved a tunnel access manhole cover bolt failure on the discharge side of the circulating water pumps. The condenser tripped on low vacuum after the basin emptied. This latter case destroyed a contractor's onsite trailer. Circulating water pump head is 30 to 60 feet at rated flow, nominally— enough for an impressive waterspout!

Tower fan problems are the stuff of legends. Fans throw blades when ice damage occurs. Deicing practices aggravate this tendency. Gearbox failures, due to water lube-oil contamination, are typical as age increases. Corrective measures for throwing blades have involved creative modifications like enclosing the diffuser assemblies with heavy wire mesh. (Consider the costs for this modification for a 16 cell tower and you see the potential for RCM-based modification review! What an opportunity for root-cause analysis, too!)

Motor failures are common as towers age in service. Most new units use weather-enclosed motors. There's no work to speak of and they are essentially consumable. Sizes range from 5-75 horsepower (HP).

Secondary failures from tower fill and structural debris have been quite damaging in unique cases. We once rebuilt major tower sections (with a contractor) and wood scrap debris was left in the basin. Startup transported this debris to the water boxes and waterbox isolation valves (seats), and ultimately into the condenser tubes. Screens removed a lot of debris, but the volume and size of the debris, together with repetitive screen cleaning during adverse winter weather conditions, allowed large amounts into the condensers, where it accumulated. Silt accumulated around the packed wood debris waterboxes and flow stagnated in the partially blocked tubes. Local corrosion cells were established. The resulting tube damage prematurely required condenser retubing due to severe water conditions and the inability to control localized secondary pitting corrosion. An admiralty brass condenser that had a design life of 30 years was limping badly at 13. Production losses ran into the millions. Retubing ultimately cost around $5 million. Granted, the water was aggressive, but the condenser had been performing well until the wood debris episode.

This example reiterates the importance of such simple features as screens, and the need to do simple PM tasks-screen cleaning-very well. It must be timely, and conscientious, even in adverse weather. Our willingness to start the unit in this state of unreadiness indirectly reflected our standards. When standards are compromised at high levels, the trickle-down effect can be significant. Ultimately, workers care when they see that managers care. Standards must begin at the top.

Secondary damage of this nature is an expensive consequence of low-quality work and otherwise inconsequential failures. It is very preventable-if you're aware of the risks. Of course, this event was an infant mortality failure, but a very predictable one in light of the station's other problem areas.

RCM-based maintenance options

1. **NSM.** Replace the tower at 20 years-suffer intermediate production losses. Generally minor aging problems like lost fill, structural sagging, and dry rot add up over time until the most cost-effective choice is a complete replacement. For peaking units this may be an economic option

2. **OCM.** Use performance test and inspection to identify when to perform maintenance. Overall, balancing capability, structural integrity, and wood condition can identify selective areas for rework. Rebuild towers in part, cell by cell. (This requires skilled tower technicians.) Generally, tower fill must be rebuilt after working into a damaged section. This requires higher tower skills than a typical plant person achieves over the course of routine work. The work is difficult and requires specialists

3. **TBM.** Distributor nozzles cleaning and biological growth-control chlorination must be done repetitively on a schedule. Even with chlorine treatment, biological growth occurs in hot water cell basins where chlorine depletes. Basins need physical removal of growth with squeegees, scrapers, or other tools. Distributor nozzles require visual inspection and debris removal. Damaged nozzles require replacement.

Nozzles exemplify an area that plant staffs typically don't appreciate. Made up of simple plastic fittings, nozzles must break up and distribute water into entrained droplets to achieve cooling. Fill helps this by further dispersing the water as it splashes downward. Damaged nozzles merely dump water into the fill. With too much "waterfall"-type flow, cooling is reduced. Not everyone appreciates that the splash zone of a healthy tower effectively increases its height six to eight fold!!

We once assigned a new tender to clean tower nozzles. Although briefed, he discovered an expeditious way to perform the job—restoring nozzles plugged with flaking deck paint—by using a broomstick to "rod out" the nozzles. Unfortunately, he broke the distributor fitting on many nozzles that he cleaned! When we discovered this, months later, during the summer peak load, we faced the double whammy of forced load reductions and nozzle restoration. Unskilled people can damage even simple CWTs!

4. **NSM.** Aside from lubrication and performance, additional work can be identified, based on equipment condition. A typical multi-cell tower has installed redundancy. Other equipment strategies include:

- motors and motor starters (CM)
- lube of motor bearings (TBM). (Many are now sealed bearings.)
- balancing fans (OCM)
- distribution valve seats restoration based on shutoff (OCM)
- seasonally repair louvers (icing damage) (OCM)
- replacing damaged fill (OCM)
- correcting resonance from structural relaxation (OCM)

As perhaps you can appreciate, the distinction between maintenance strategy types starts to overlap. At an excellent operating company it almost becomes academic!

There are many tower styles. All provide the same basic cooling functions. Overall tower and individual cell performance provides good opportunities for on-condition work.

The challenge is performing inspections and tests as scheduled and then performing indicated, required, condition-directed work as demanded. Because seasonal load periods are brief, it's easy to miss the summer peak season if work can't be responsive. Cell-by-cell performance assessment, specific inspections for resonance, dry rot, and other structural deterioration are typical examples of effective OCM. Fan motor, drive, and assembly maintenance are more conventional rotating equipment OCM examples. CDM beyond the rotating equipment often fits best as a part of a cell-by-cell rework program.

Towers are greatly affected by adjunct programs, such as water chemistry. A good chemistry program helps a tower achieve its design-capable life. Problems such as erratic chlorination will result in service deterioration and premature aging.

Tower operation methods influence component life. Towers de-iced by reversing fans suffer greater ice damage and require more fill and louver rework. Fill restoration is specialist work. Replacement requires removal by working-in, and restoration working-out.

Most towers have considerable "as-new" performance margins. Many tower ratings can be adjusted as much as 25% on paper with little performance impact based on the same hardware. A typical tower will last 20 years with modest maintenance (focusing on mechanical fans and drives). Alternatives to this involve spending around $10,000/cell annually, continually servicing and rebuilding cells, or replacing the tower entirely in 20-30 years at a cost of $200,000/cell. The solution must fit with the company's long term unit operating profile. Establishing and carrying out a tower maintenance strategy in a cost-competitive environment is, in fact, a complex optimization problem. Running a tower well on a limited budget means there's little room for oversights or mistakes.

Aftermarket tower services and parts abound. Some suppliers are outstanding, others marginal. Beware of services from anyone who is not a tower specialist. The re-decking cited above was performed by a local contractor, not a reputable tower service firm. At the time, everyone saw it as a good deal, and an easy job. It turned out to be a good deal more.

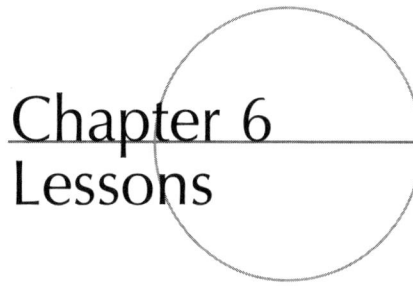

Chapter 6
Lessons

Wise men don't need advice. Fools don't take it.
Benjamin Franklin

Believe your instruments.
Pilot Adage

Calibrate your instruments.
Naval Reactors

Preventive and corrective maintenance models are built from basic assumptions—generalizations not always born out by facts. The traditional PM model assumes:

- PM is always less costly than failure
- equipment can not be run effectively with "failure-based" maintenance
- we understand and can recognize "failure"
- the conscquences of missing PM is failure

Can simple failures progress to functional failures? Can functional failures impact equipment and systems? This is what we seek to understand and, ultimately, remedy. These are the lessons.

Even the best, most optimized, experience-based designs are no more than relatively insensitive to failures. The natural progression for equipment over its production life is to increase failure resistance by design as field experience increases; ultimately, however, how people perform is what actually matters. Some organizations "gun deck" maintenance paperwork (like PMs); workers don't blindly buy into processes. Failure process ignorance, or a lack of participation in PM identification, selection, and development limits worker process commitment.

Root cause analysis can be a factor. Until actual strategy and costs are developed, it's not often clear which maintenance approaches are likely to yield the best overall task, or which combination of tasks best address potential failures at the lowest cost. Until the plan is implemented and statistically meaningful costs are collected, value cannot be established. Backing off existing, low value PM is just as tough as establishing high value programs.

Compare CDM versus TBM alternatives for a filter change-out. The accepted wisdom is that OCM is always less expensive and better than "hard time." Then consider plant pump VM-at some level, the benefit derived from monitoring pump vibration vanishes. At a $50/hr (loaded) labor rate—and with no production impact— no planned maintenance balances against monthly monitoring with a $10,000 pump capital cost and a 10 year lifetime. (i.e., if a pump lasts 10 years with no PM, costs $10,000, and carries an 8% capital cost, 10 years of monitoring to achieve a two year life extension has little or no payback.) Such numbers can be replayed with different assumptions—time value of money, and so forth—but the general range doesn't change. Somewhere in the comparison of capital cost range with projected life, PM becomes ineffective.

Until a benchmark case analysis is done, operating team assumptions about failures and costs may diverge greatly from reality—to their economic disadvantage. Note that as maintenance effectiveness increases, the break-even point drops. It's more effective to perform PM in an efficient organization—and less effective in an inefficient one.

Task Intervals

PM task intervals are based on failure rate and mean life variability. For random-failing components, special strategies may apply (within the context of the design). For a predominantly random failure mode, for instance, functional monitoring can effectively identify instrumentation failure. A check made at a fraction of the MBTF can identify instrumentation failure while minimizing overall failure risk. An instrument can be tolerated in a "failed" state for limited intervals, if effectively redundant.

Operations is charged with identifying random equipment failures in the plant. An effective round ensures that operators get through the facility often enough to identify failures—particularly random ones—without excessive monitoring. Usually, four to eight-hour round intervals are effective. If they must be made more frequently, designs should be evaluated; if they can be made less frequently, it raises questions of risk, and staffing levels. Equipment in terminal failure may require additional monitoring. Requirements to monitor terminally failing equipment can be quite substantial.

Selecting task intervals is part science, part art. Failure data alone may be inadequate for infrequent failure modes. With limited failure information, manufacturer recommendations, failure mode physics, or mode type assessment, an expert opinion may be needed to establish appropriate task intervals. In many instances, too, an exact interval is not critical. With inexact information, intervals can be over-specified (made too frequent), particularly by an unskilled analyst. An age exploration-based monitoring program can adjust intervals based on failure type and age exploration.

When craft participates in interval selection, it's a significant organizational growth step. Craft develops ownership of monitoring and parts-in-service assessment that in turn supports identification of appropriate intervals.

For expensive, age-based failures, intervals need to be conservative. Generator rotor cracking probably has an exact MTBF in excess of 30 years. Inspection on a disassemble-overhaul basis (typically, every 5-10 years) is appropriate since the equipment is at risk. Instruments monitoring for high cost equipment failures must be maintained and their failure prevented. Brief outage periods are acceptable but entail risk. Prolonged unavailability is not an option. VM on high-inertia rotating equipment must be maintained constantly, with hard-wired trips, and must have an operating limit. Anything above the limit is an automatic trip.

The problem establishing intervals for instrumentation is ambiguity. Manufacturers resist hard-wired trips to avoid spurious or undesired events. They assume that an operator can discern "inappropriate" demand trips on an instrument provided for "status," and avoid bringing down the unit. It's right in theory but wrong in practice—operations learns to ignore unreliable instruments. Regular, spurious instrument trips and alarms—as there can be at "status only" instrumented plants—means that instrument value plummets. "Status only" instruments can lead to maintenance deferral when their importance is diminished. Well-maintained, high-quality "critical" instruments are essential.

Identifying critical instruments is helpful, of course! Chasing a faulty alarm is frustrating and expensive. Sitting in the hot seat after making the wrong call in an ambiguous situation is equally trying. You don't get too many chances before the instruments are discounted and ignored. In RCM (or ARCM) a spuriously alarming instrument is considered "failed"-one of the truly great contributions to instrument maintenance programs and one that operators had been demanding for years.

It all sounds overwhelming. But in reviewing thousands of components and tens of thousands of WOs, one quickly becomes comfortable estimating task intervals. It takes experience to develop a feel for failures, but, with some quick R training, most experienced people can draw on their years of observations to make excellent judgements about parts aging—particularly when wear, abrasion, or erosion processes are at play. A failure model helps integrate a picture of the failure process with plant culture and strategy.

In practice, MTBF's are often grossly underestimated. Plant staff base their life estimates (and PM intervals) practically on a small fractional sample of failed parts. While suitable for "safe-life" interval limits, it isn't obvious that this grossly underestimates the average life. Predominantly, PMs are based on economics. What this says is that informally developed economic PM intervals are almost always grossly conservative.

Estimating organic failures is vexing. Aging is expected; however, rubbers, cloths, elastomers, and similar materials deteriorate with time and temperature. Even when visual aging evidence is missing, it's risky to assume there's been no aging. The Arhenius temperature characterizes organic aging—below it, little or no aging occurs; above it, aging increases quickly. Visual evidence can be absent in the transition range. Calculating an age—especially for components in critical applications— helps avoid gross errors. When large organic expansion joints made of reinforced cloth, rubber, and binders, reach their manufacturers' specified life, life extension is risky. During installation, the absence (or presence) of offset, vibration pulsations, and other synergistic aging phenomena complicates the picture. Only experience can determine actual in-service aging.

Remaining aware of failure processes and how they work—knowing what to look for—greatly improves setting task intervals. Fortunately, a few fundamental aging processes repeat over and over in most plant applications. Learn these, and you have the basic tools to evaluate most aging mechanisms. Developing failure mode data is a R engineering exercise. Fortunately, experience, thumb rules, and training go far.

Age Exploration

Definition

Age exploration is the systematic examination of the lifetime a component or part can support in an application in service. It's crucial to setting task intervals. The term means literally "to explore component aging, and find out what service the component can provide."

It used to be assumed that all components had finite lifetimes— equipment wore out—and needed replacement or overhaul. As first examined in air transport, it was discovered that it doesn't hold true for many components. Though powerful and intuitive, the assumption had no basis. A statistically large number of components showed virtually no deterioration during period of use, based upon early jet engine over-hauls in the late 1950s and early 60s. These included actuarial analysis of failure studies. On 90% of replaced components, life remained at the end of their specified life.

"Lifetimes" have always been based on the best available information. Aircraft turbine development—transitioning from prescribed overhauls to age study-based monitoring—summarizes the experience. When overhaul limits were eliminated and age exploration undertaken, equipment lifetimes increased significantly resulting in quick economic benefits and lower risk. Considerable actuarial analysis detailed mathematical failure analysis to quantify lifetimes and conditional probability distributions-and support this change. The concept of "conditional overhaul" gradually emerged. The results can be applied to most other industrial maintenance applications.

Conditional overhauls only address immediate failure causes and correct other necessary parts to achieve specified performance. The paradox is that conditional overhauls yield "overhauled" equipment that statistically perform the same as traditionally overhauled ones. By literally running fault-tolerant equipment with NSM until failures develop, we can use the concept of age exploration, merged with design and conditional overhauls, to give credence to the term "NSM."

Early equipment manufacturers and maintenance experts did their best to specify age-based replacements, recognizing the potential to do better through age exploration. Extending useful equipment life requires understanding how items age in service and how effective we are at discovering it—and then formulating how best to use this knowledge. Profound understanding of statistics and actuarial lessons—lessons learned from those aircraft engine overhauls, failures, and actuarial analysis—enables this conceptual leap. Evaluation of in-service performance on an ongoing basis (particularly for new equipment and components as they enter service) enables us to manage risks.

In the course of understanding jet engine aging and failures, the aircraft industry discovered that even the very best maintenance and engineering experts couldn't predict future engine performance based on overhaul data. Experts predicted the imminent failure of apparently worn-out equipment, only to have the equipment perform (statistically) identically to "acceptable" equipment. The inescapable conclusion was to extend life literally until an age-based failure limit clearly emerged and then develop lifetime limits around it. This is the lesson of the aircraft industry. And while it takes a leap of faith to adopt, it has clearly been successful for extending engine service life.

The inability of even the most qualified personnel to predict future performance with statistical accuracy-the expert-life prediction paradox-flies in the face of all common expectations. It leads us to assume that under normal circumstances, equipment has a finite life—a stochastic interpretation. The statistically correct alternative would be to assume (until proven otherwise) that all equipment has an indefinite life. This is the approach followed with age exploration.

The maintenance implications are enormous. We can finally understand why 90% of parts never fail in-service, and the powerful case to be made for CNM to establish parts life expectation. It also explains the legitimacy and validity of the case for NSM, and why the common phrase, "run to failure," is patently, statistically wrong—not only highly-biased and charged, but statistically incorrect. This is the message maintenance traditionalists need to learn.

We don't indefinitely extend life where environmental or safety losses could occur. In these cases we benchmark against our best experience to establish appropriate monitoring intervals. In practice, experience enables us to predict likely failure mechanisms and their order of magnitude. While acknowledging the paradox that says: use fault tolerant designs and then stretch, until facts point us to an age limit, we select safe-life limits for critical safety-based failure modes.

Value

For major equipment overhauls (boilers and turbines), extending an interval by even a few days can have value. Other small savings, in aggregate, also add up. Age exploration achieves its greatest potential value when a plant shutdown can be deferred.

For a nuclear BWR, replacing solenoid valves to meet an EQ often falls in the 4-6 year range. Extending the "qualified life" for control rod pilot solenoids (at 4 per control rod, or 137 control rods per unit) provides substantial savings. (This requires re-qualification testing, of

Figure 6-1: Best Value? Every nuclear plant spends time on the NRC's watch list – or so it seems. Impressive operating records lead regulators to suspect production is emphasized over safety. Maintenance expectations differ in the highly regulated industries despite the same equipment. The challenge of deregulation is to use industry-wide best practices to achieve outstanding operations at low cost.

course.) For non-essential parts or in fossil applications, this means keeping parts in continuous service until aging or in-service failure demonstrates life limitations. Obviously, this must be done in a controlled manner if failures have safety or environmental impact. Many times, however, component failures have minor impact. Age exploration may take an item out to failure to establish lifetime limitation—or to demonstrate there is none.

Solenoid valves statistically perform very well in service when control air quality is maintained. Failures are mainly "sticky operation," at which time the solenoid needs rework or replacement. However, nuclear application failures are unacceptable. It's particularly important to understand solenoid pilot actuator life in high-temperature environments (like pilot-operated relief valves). Plant shutdown can result if it's not managed. Those who work with such parts must be counted on to observe in-service performance and lifetimes that can improve parts usage. For cost-management purposes, systems used to pass this information along must be adequate to conduct effective aging studies (Fig. 6-1).

How well parts perform is information required to assess in-service aging but it's never been easy to get. New CMMS software can document failure data and retain it for future use with relative ease; raising craft awareness levels, and retaining history for future use, is a key next step. Most maintenance personnel are capable of performing and documenting very useful summary observations about parts when it's expected. Simple one- or two-line descriptions identifying unusual in-service aging performance are most useful. "Normal" aging warrants periodic assessment and descriptions. Again, when entered directly by users in their own terms, they're most beneficial for future aging studies. There's a balance between too much and too little detail that needs to be maintained. Too much detail and too many reviews of large amounts of data bog down personnel; too little, and the assessment is inadequate. Workers need guidance on what level of detail to provide.

Some technicians and mechanics systematically perform replacements using rebuild kits or available stock spares with little consideration to part in-service performance. Others assume a new part is always better than an old one. Age exploration teaches us to examine components upon replacement, using aging knowledge and observation, to assess remaining service life.

Infant mortality studies also dispel the notion that a new part is always better. This is particularly true for electronic and electrical components, and some mechanical parts. Living organisms demonstrate infant mortality-in fact, the term originated here! For service in complex equipment, proven, "performance-aged and burned-in" components have a higher probability of service than newly rebuilt ones. Everyone knows the period immediately after a startup carries the greatest risk, and that in theory, when random failures are measured, a new part doesn't always out-perform the old one.

Theory (based on experience) validates the use of age exploration to extend service intervals for new designs as well as parts. Any new component has some theoretical in-service life but until validated by performance, it remains a theoretical projection. For new, complex equipment, age-exploration is essential to establish potential life. Regulators want assurance that new parts and components will perform in service as expected. In some instances-safe-life limited equipment, for example—they may require disassembly/inspection as part of the conditional certification of a new equipment subassembly or products.

Systematic application

Age exploration in new equipment begins by removing parts from service for examination. New-failure mechanisms, premature aging, and other unanticipated failure-mode evidence requires immediate attention. Over the long term age exploration provides the basis for predicting how much service a given component can support. You can better extend life when you realize the ultimate service life limit. Done systematically, it can provide the basis for improving many plant equipment maintenance decisions.

Such age exploration principles have been known and used for years, but haven't found regular applications in electric power generation. Perhaps this simply reflects the traditional nature of power plant maintenance. The need to improve part-cost performance hasn't been a need in generating plants-until now. Legitimizing age exploration neatly resolves this cultural problem. Utilities should develop formal age exploration methods and hand the decision process back to those who actually use the parts.

Effective age exploration requires:

- awareness of part characteristics
- documentation (usually via CMMS)
- engineering assistance
- a corporate environment that encourages learning

This last element is vital. Employees must expect that plants, equipment, and systems will continually improve, and that performance will increase and costs decrease through improved utilization. Without learning, a part-aging program won't be effective.

The benefits-both obvious and subtle—afforded by age exploration include:

- steady reduction in spare part expenses, both direct and indirect
- increased awareness and understanding of the role that parts play

in overall equipment considerations—how parts age; the service they provide

- increased awareness of suppliers and manufacturers and how their parts perform in-service
- the added value provided by premium parts
- the R contribution from the "best available part"

How many mechanics understand these benefits well enough to confidently perform parts life-extension today? Many organizations buy on price. More simple case histories documenting how inferior quality parts can impact a facility are needed to overcome this bargain disposition.

Persons handling or installing parts of any sort need to assess their usual suppliers for parts condition and further service suitability. They need to do so on an ongoing basis. Once entered into corporate maintenance databases this information can be used to evaluate future supplier or vendor-supplied parts, procurements, and the suppliers themselves. Decisions to upgrade existing parts need to be based upon in-service performance, cost, and simple life-needs assessment. It's too easy to see only the "tip" of the cost iceberg—initial purchase price. The total in-service cost can remain hidden beneath many other factors. Overall costs include generation and service loss, as well as costs of replacement and parts.

Part-use problems are compounded when multiple suppliers are involved. No matter how carefully specified, manufacturers achieve different performance results. Mixed-part populations, in service, complicate age exploration. Unless a company can design and run statistical experiments—and few can—suppliers must be evaluated one at a time.

Find quality suppliers. Expect to pay more for their parts and service. Develop long term relationships. This is the lesson on parts from manufacturing.

Engineering Focus

When it's included as part of an overall corporate strategy, age exploration focuses corporate engineering on what matters. Issues of redesign, statistics, cost analysis, and new CMMS tools can help put engineering resources where they will add the greatest value.

Old and new facilities differ in their engineering design improvement needs. In the past, rapid advances in design, lowered unit costs, and load growth meant that engineering focused on construction. Plant lifetimes were short. "Disposable plants" were expected to be technically obsolete after 40 years in service. Such was the design standard. Today, plant replacement capital simply isn't available to old-line utilities. Traditional, vertically-integrated utility generating units, whether fossil or nuclear, are beset with high costs and complex processes, putting their continued existence at risk. The challenge is to redesign and re-deploy assets for competitive survival. When engineering groups lack the experience needed to effectively improve plant operations, utility engineering must turn to others-and this should not be the case.

Failure spectrums

A complete RCM review of a complex system—establishing a maintenance "spectrum"—quantifies optimum maintenance mix. At the extreme are systems that support heavy monitoring: they have a high number of random, low-consequence failures that don't (cost-effectively) support fixed-time maintenance. Personal computers—many of them controlling many complex subsystems—fit this profile. Overall, they fail randomly, but a quality machine's average age at failure—it's MTBF—is several years longer than prescribed useful life. It's highly probably the device will be technically obsolete (and taken OOS) before this point is reached. The MTBF is large—20,000 hours or more. Most "failures" are, in fact, randomly introduced software glitches or random operational losses. Hardware failure incidence is low. There are no effective tasks that will cost-effectively prevent failure so an effective strategy is one that addresses failure identification and data preservation instead.

The philosophy behind how equipment is designed also determines its approach with respect to operator intervention, monitoring, and the value placed on monitoring time. When system failure rates are low, it demonstrates integrated man-machine design success. Time (in man-hours) required to achieve failure rates may differ radically from one design to the next. If labor is valued low and capital requirements are high, an overall optimum low cost solution is labor-intensive-a CNM-intensive maintenance solution. If the cost of capital is low and manpower high, the optimum mix is little manpower and more capital. Here, the limiting case requires no man-hours for maintenance at all—the OTF case. Different cultures approach equipment maintenance differently but often apply one of these two methods.

I think of the former model as "German" and the latter as "American" because the way the design of a Porsche and a Chevrolet reflect this different thinking. European maintenance strategies lean towards more monitoring while Americans tend towards less. If equipment is capable of extended life, we should seek that approach and apply it for the optimum maintenance cost.

Operating costs can be reduced through reductions in capital expenses, provided such reductions-or an increase in unreliability-don't increase O&M. (One unplanned outage and all savings can be wiped out) The purpose of many capital expenses are performance improvement. Programs to extend life must assure against trade-offs-or, worse, bottom line losses. Invariably, production losses carry high penalties but are abstract and harder to quantify than PM. (How can we measure the cost of opportunities lost when sales are missed?) Industry faces the same opportunities—and risks—on a much larger scale. Every decision in industry is a roll of the dice-and they roll hundreds or even thousands of times a day. When we do, ineffective or incorrect strategies show up on fairly short order. (Twelve to 24 months are needed to measure the impact of a strategy change for average plant cases.)

Many technically advanced products carry specifications that assure a specific design life at a specified level of performance. Boiler tubes will last 40 years at design firing rates with specific water chemistry. (Firing rate and chemistry specifications—technical limits—assure design life.) Exceeding these specifications causes immediate "proximate" failure. Some books refer to this as engineering failure, or root cause failure. For long-lived capital equipment, understanding this relationship reflects business profound knowledge. Many companies do not (or cannot) make this relational tie. Where equipment records indicate secondary failures can be attributed to exceeding specifications, improved performance monitoring can significantly improve economics. Where expected plant life is 40 years, (with an institutionalized maintenance strategy and plan) for instance, and an ultimate pre-obsolescent life of 80 years, extending equipment life improves future returns. This reflects, for example, the future value of a sound water chemistry program or firing a boiler within limits.

Monitoring identifies those specifications that are exceeded before they convert into proximate failures. Minor adjustments and "tweaking" are required constantly spanning 40 years, but long term benefits are substantial if a unit that would otherwise be retired at 40 can sustain 80. Near-term, condensers and boiler sections re-tubed at 10-15 years can achieve 30. Circulating water towers, ready to collapse at 18 years, sustain 30.

Sophisticated maintenance practices include:

- tailored performance monitoring plans based on vendor recommendations
- identifying problems early
- addressing identified problems while minor, before the final failure phase:
  - with redundancies in place or risk managed
  - while limits aren't compromised
  - on a timely basis
- planned replacements of known age-limited parts (including lubrications)
- quality parts and competent service

Plants and equipment become uneconomic when they become unreliable. The context of R depends on the operating mission. As items age, age-based modes failure probabilities gradually increase. Eventually, failures increase. Managed carefully, the overall failure rate for equipment can be controlled.

This final phase is the most challenging for operations within any company: When do you pull out and re-capitalize or reinvest in an existing facility to restore R and performance? Effectively selecting alternatives characterizes strong engineering programs; doing so poorly (or not at all) ensures that failures will occur to make operating economics less favorable (faster), driving profits down more quickly. When costs become unpredictable, operations become less economic, and finally-uneconomic. Managing increases in aging unreliability allows a company to determine facility end-of-life by technological and mission obsolescence means.

For an obsolete facility, increasing generating R and availability with small capital investments provides attractive short-term earnings opportunities. For most aging plants, operating costs are high—as are costs of capital and risk. Such facilities remain competitive only by keeping their total costs down. This points to improved O&M.

Maintenance can be abandoned in the final operational phase when a company closes a facility. Such a decision precedes operational termination by some period, and is often irreversible, economically. Railroads strapped for cash did this in the 50s and 60s. Once suitors were found, or economics improved with "deregulation" in the 80s many properties were exhausted to a point where abandonment was the only economically viable decision. Extracting capital from a facility by means of deferred maintenance should never be done lightly, or unconsciously, as has happened in some companies strapped for cash.

By profiling typical plant systems to understand their technology and basic maintenance process, RCM-based benchmark comparisons help establish appropriate cost levels and identify effective methods to manage production and control costs.

PM Implementation Models

"Do your best"

When some companies build or acquire facilities and maintain a laissez-faire approach to facility maintenance, it's because they discovered they can run them much longer than vendor-specified intervals with no apparent loss. Ultraconservative vendor intervals partly explain "ho-hum" approaches to TBM.

Imagine, on the other hand, that missed PM intervals had (relatively) immediate and severe consequences. Time-based monitoring programs and vendors would gain credibility! I believe this would happen if manufacturers discarded the volumes of trivial, over-conservative information they provide with much of their new equipment. What's needed is a "Cliff's Notes" of PM.

The truth is, few truly grasp vendor-recommended strategies for maintenance. Vendor manuals (like company prospectus reports) are under-rated and under-read. Their technical writing styles, required skill levels, and even their basic information varies in quality and accuracy—even from the same vendor! Variation is even greater between vendors. Some fail to disclose service life information. Others don't provide service manuals. Many don't offer in-service aging and performance information. Yet, in my opinion, the more product life cycle information a vendor offers, the more competitive they become.

They may not offer product failure information because it's assumed that disclosure of this information is unwise competitively and carries legal disadvantages—particularly when the vendor recommendations aren't conservative and lawyers become involved. Vendors promote the virtues and longevity of their products in sales literature while their service manuals are conservative. Development of an O&M strategy rests heavily on the user, as a result. Vendors supply to sophisticated users but I have witnessed many disagreements between vendors and users over product usage. Several notable ones were litigated; most were resolved through user-vendor negotiations. Most were unnecessary.

Disagreements arise because maintenance requirements downplayed during the sale gain emphasis afterwards. To be economically viable, equipment must deliver a period of reasonable, reliable service with little or no maintenance. It needs to do this competitively. When expectations are met, everyone's happy. When not, accusations fly. I've personally evaluated failures as an "owner" but on some occasions, problems developed based on our owner specifications, maintenance performance, and other responsibilities. Vendors still worked with us-they must have seen weaknesses in their own equipment or services or they wouldn't have negotiated! What did everyone learn? General lessons included:

- users can successfully exclude many performance-monitoring recommendations and achieve reasonable performance
- users who fail to act on known information and problems incurred losses

- serious equipment shortcomings involved a bargain: An owner can't economically pursue an incompetent, occasional, or "in and out" supplier
- quality companies meet their obligations; most go further than owners can rightfully expect
- conscientious, quality suppliers engage in frank discussions about equipment, problems, and expectations. Often this is where large operating organizations learn about operating limitations, despite literature and training

Equipment vendors are expected to possess a high level of maintenance awareness, but this exceeds reality in many cases. Between vendor guidance and informal learning, many plant maintenance staff and managers carry on with inexact, unspecified programs. Many are merely inferred from work practices. All maintenance departments strive to be effective but only recently has "effectiveness" been defined. Maintenance models-even those structured around CMMSs and work practices-equate to defined maintenance processes. Some organizations do work in regulated environments, and still don't have process maps. ISO 9000 certification has driven maintenance process documentation and definition. In recent memory, only the NRC regulations and the "maintenance rule" has had greater impact.

I call the functional PM model the "do your best—you know what that is" model. Transferring performance responsibility from management to performers, requires performance accountability, particularly where there has been no objective performance measures. Doing "best" can be highly subjective and support radically different outcomes. Guidelines, expectations, and measurement may be vague or completely absent. Craft worker guides can be non-existent or PM tasks can lack specificity.

The old presumption, when specifying PM work, sounds like, "PM the pump—people understand what to do." This has led to three mechanics looking at three different things. RCM goes way beyond this model—too far, in fact. ARCM appropriately stops at specifying what to do.

Trust us (we know what's needed)

Indefinite PM programs are not written down yet adherents swear they exist. "Trust us" says the staff. Undocumented plans lack field performance consistency; yet bring this up and staff takes affront.

Informal audits reveal maintenance people who say they do PM tasks—they know how, and they know why-but don't have cost, performance measures, or other analysis. Since they have little experience measuring performance, they're concerned about measures. After all, measurement has never been done before! Maintenance personnel with only informal training in maintenance, PM theory, and methods gained their knowledge on the job. That knowledge can't stand formal technical adequacy or value tests in many cases.

Of course, the acid test of a maintenance program is how it performs. Continuous production increases and reasonable costs are satisfactory program indicators. The reverse of this—high failure rates and steady or increasing costs—are the warning signs of ineffective maintenance.

Traditionally, increasing or unpredictable budgets were grandfathered and the need for maintenance was universally recognized—within maintenance. General thumb rules and guidelines applied by the manager were taken as gospel. These older, authoritative shops that preceded the modern organization and CMMS produced some of the best maintained plants around. Availability was good. But was money spent effectively? We'll never know. Worse, when the central authority figure retired, it could be discovered that these leaders were the only ones who knew the maintenance program strategy. In their absence, the slate was wiped clean. Without a maintenance process, maintenance cannot be developed and grown; the risk of a major redirection when the central controlling figure changes is so high that the whole program slides back to ground zero.

Such an intensely personal program can't be benchmarked, improved, or even shared. Someone makes it work, on-site. The style doesn't lend itself to standardization, large facilities and teams. Major maintenance cost charges that were not controlled at the plant level and budgets were largely "swagged" from previous historical performance, with cost adjustments made for inflation and margin. Budgeting was based on specific activity in principle-but not always practiced.

With no competitive pressure and a cost-plus baseline budgeting systems-this maintenance strategy suited regulated utilities for many years. Many government and quasi-governmental agencies operated in this same way. Today's maintenance strategy, technology, and theory are changing expectations.

Today, the cry reflects former President Reagan's arms limitation motto: "Trust...but verify!"

When maintenance schedules are reduced to optional task lists, they accumulate backlogs. These justify routine overtime hours, budgeted across the board. Maintenance can then work all the overtime anyone wants. The only problem is that no one wants to work PMs. In part this is because planned work can be deferred—"de-prioritized"—and you can't justify overtime for it. Only emergencies warrant non-routine expenses, so planned work goes begging—and undone.

Discipline in managing the backlog is what's missing. It hasn't happened in part because of fear that regulatory authorities find backlog trimming unacceptable. Yet, it's convenient. Airlines, reactor operators, chemical refiners, boiler operators, and other risky businesses must have maintenance plans, by law. They must plan work, do TBM, and be accountable for work performance on all plant equipment.

A better prioritization method would be activity-based, as it is in accounting: All activities need to pass similar tests to be ranked for common resources. The automatic deferral of PM to crises is a fundamental maintenance paradox. Priority work needs to be value-based. ARCM provides value-based tests that can restore credibility to a maintenance system.

Maintenance organizations occasionally downplay the value of timed maintenance in a traditional hammer-and-wrench environment. "PM is not real maintenance," goes the jargon. "If you can plan it—that's cheating!" Combine this attitude with a utility's tendency to let workers self-direct (e.g., choose their work) and it's no wonder that PMs don't get done.

Typical PM implementation

In the unregulated maintenance arena, only a small number of PM tasks are performed. The tasks themselves may not specify work to be done. Review provides only vague notions about work to be done. From years of plant experience, I can infer things worth doing, but auditing work turns up many different results. The common trait is that the value of many PM tasks can't be assessed or calculated. The plant's entire maintenance strategy may be suspect. At fossil generating stations I've audited 15% of the work on the PM list is performed-on the high end. Typically it's 7-10%. Low is 3-5%.

Nuclear plants have more aggressive lists and better measurement because of regulatory requirements. Completion rates are regularly between 80-95%. I watched a BWR achieve more than 90% of scheduled PMs worked to completion month in-and-out. Those not worked were rescheduled. Failing to perform scheduled PMs had to be justified in advance. Returning a PM to a backlog list was unacceptable. The result? This unit did not suffer a plant trip in five years! Not that this was solely due to PM completion-there were many other expectations and practices that supported operations. But the culture was one of commitment, competence, and maintenance across all groups—"inspired" by the regulatory environment.

Clearly, the contextual meaning of the PM program was radically different in these two environments. To get this latter level of PM program performance requires management commitment, and PM work credibility.

Fossil plant R is a tribute to design—they run so well with so little. But if most fossil units run well without a complete maintenance plan, what's the upper performance limit? Would a more detailed plan bring down performance and raise costs? What about other aspects of PM performance, such as outages? My opinion is that more complete strategies can raise production and lower costs.

Total PM performance

Some companies develop and execute maintenance strategies based upon what I call "total PM performance" strategies. They aim to enhance production and profitability goals by centering on facility utilization. Ensuring high production, profitability, and performance facility utilization rates must be heavy and planned. The key to such projects is that maintenance must support operating and facility use plans—not drive them. Exceptional operating organizations plan their maintenance, using performance measures and production goals as feedback tools. Some of the best (based on numbers) operate in areas not traditionally associated with heavy maintenance performance. At least one prominent food service company is included.

Companies integrate these plans into their total operating strategies in an effort to see maintenance support operations. They demonstrate how maintenance can center on production-not provide a shop for people who like to take apart and reassemble big, complex equipment at their leisure!

To reinforce total PM performance, inspection, CBM, outage scheduling, and TBM priorities should derive from production schedules and goals. TBM (e.g., replace/restoration PM tasks and inspections) provides the structure for an operating facility's scheduled work. Developed week-by-week over a quarterly period, it can provide the framework for a repetitive, long-term maintenance schedule. Combined with outages and operations rounds, a comprehensive monitoring plan takes shape—one that provides the overall foundation for implementing a facility maintenance strategy. All the bases are covered.

To a traditionalist, this is backwards. "Maintenance is delivered on demand he'd say. A PM can always be deferred." But to the R engineer, applicable and effective TBM receives the top maintenance priority. PM maintenance tasks avoid future losses and expenses. PM monitoring tasks facilitate scheduling necessary maintenance. Applicability means a task is technically appropriate and effective, as intended, when performed by a qualified person. Effective means the value ratio—cost to benefit—is favorable. When PMs are tied to economics, ranking cost-effectiveness assures that high-benefit work precedes the rest.

PMs' B/C ratios rank value. PMs with economic ratios less than 1.0 (e.g., the PMs' benefit is less than its performance cost) are best performed on a NSM basis. In those cases, equipment can self-disclose maintenance requirements to operators on non-specific rounds or monitoring. This is the most effective maintenance plan there is. The challenge is to get people thinking "present value"—one day at a time.

For companies that can't "get the kids to play ball," there is the railroad approach: Take away the train set. (Railroads outsourced maintenance unflinchingly in the 80s.) In this case, it's called "corrective maintenance." Contract all repair work to specialists so that all that remains are PMs, performed in-house as the total responsibility of the remaining workforce. This could be a multidisciplinary group-electrician, technician, and mechanic skills-supplemented with operations. This approach clarifies, identifies, and prioritizes maintenance organization work. The inherent conflict-of-interest between PMs and corrective work for overtime, enjoyment, or other motivations is gone. PM is now the only game in town. Selection of PM as core work is espoused by some merchant generators.

Merchant co-generators tried this approach as an interim measure because they lacked trained, skilled crafts. Onsite plant staff performed all routine operations and light maintenance while outage and heavy work was contracted out. Plant staff-clearly focused on the plant condition-used CNM as their primary tool to identify, diagnose, prioritize, and plan outage and restorations. It was effective!

Vendor Perspective

The vendor's dilemma is twofold. He must provide a good product while generating sales. Ideally, he receives follow-up sale and service calls for training, service, parts and so forth—for each customer. When the client receives value and satisfaction from the equipment, the vendor's interest is best served when a product has a finite life. His best situation is technical, functional, or economic obsolescence before the end of useful facility life occurs. The client retires the product in-service to buy another—unless the vendor can convince him to upgrade to something better.

Vendors are also repositories for product development knowledge. In the course of their work, they must identify, understand, and remove design, production, and operating impediments that cause failures. They generally retain this information conveying it selectively to users. Unfortunately, vendors can't provide complete failure data to equipment owner/operators nor fully disclose product development and applications, as they must protect competitive positions. They need to exercise discretion in the event of legal action. In addition, plant technical staff must understand and translate the operating data they receive from the vendor. Generally, users don't need or require details about the product. They might be intimidated by too much information—weak points and total costs—or might comparison shop, or be steered towards a competitor. Lastly—and most important—vendors don't have "complete" information on actual in-service failures and aging performance. They cannot possibly understand all conceivable environmental and aging factors, applications, and uses imposed by the users and their environments.

So, our dilemma is that vendor information, while good, is incomplete. It generally gives a fair assessment of common failures and expected maintenance needs for anticipated service applications over an intended period of use but doesn't provide environmental or applications information that summarizes a product's "stretch" capacity—always the most exciting and challenging areas for users. However, vendors are always a first source to identify both expectations and reported experiences with new products.

If you can connect with a vendor's engineering staff (assuming they have one), you can resolve most questions with unpublished accounts and experience for many product use applications and most failure history. Vendor engineers are more likely to offer critical failure information over the phone than on paper.

Vendor recommendations

Vendor recommendations represent the best guide to maintenance strategies that are appropriate for the equipment they offer. The quality of vendor recommended maintenance varies greatly. Some is truly outstanding. Many don't provide any information at all. The vast majority provides useful, but incomplete or sometimes inaccurate guidance. At best, vendors provide a starting point, and so their O&M manuals, sales literature, and drawings should be reviewed while developing any maintenance strategy.

In a highly regulated environment, a vendor's guidance may carry the force of law. If the vendor specifies that a certain filter must be changed every three months, you must change it. Rarely, however, are vendors so direct-or consistent with plant time measures-in their guidance. They may recommend lubrications based on service hours while plants base it on calendar time. It's common for vendor tasks to be inserted literally into CMMS scheduled PMs at the conservative recommended interval. My review of gearbox lubes at a nuclear plant revealed that most were changed on a one year interval, even though many would not exceed their annual service hours during the entire licensed life of the plant!

My experience is that most shop people trust vendors implicitly and take their guidance without question. As a practical matter, however, reviewing thousands of vendor guidelines has led me to the conclusion that developing PMs is specialty work if you want to keep unnecessary work to a minimum. It takes several years of plant, equipment, and failure experience to learn how to interpret and apply vendor recommendations. You must understand a breadth of programs—overhauls, cals, checks—as well as traditional PM work tasks and you need to understand how maintenance shops perform work. On top of this, you must appreciate equipment functions, uses, and failures.

Failure Footprints

CMMS barriers

When performing ARCM, simple techniques can often significantly improve analytical results. Typically, RCM analysis is highly abstract, using esoteric, redundant, or even arcane statements. Many analysts focus exclusively on expert interviews to "flush out" failure modes of interest. While interviews are good, numbers tell the story as we've seen.

In my experience, the difficulty most people have—engineers included—is penetrating corporate maintenance management systems. CMMSs are difficult to learn and harder to interpret. In the role of maintenance manager-at the mercy of others to develop and interpret CMMS reports-I finally forced myself to learn how they work. Having waded through the process several additional times at several companies and plants, I highly recommend anyone involved with maintenance not yet fluent with these systems, reports, and numbers to learn at least one. If you want to evangelize, you have to learn the language.

Another CMMS barrier is report formats. Row upon row of unformatted numbers and text represent little to uninformed readers. Information is often coded and layouts must be compromised to standardize reports. Key record locations and formats must be learned. Once you are able to skim CMMS reports, recognize key information, and query interactively, you can use CMMSs effectively to understand maintenance.

Most CMMS fields are not crucial. Important fields include the component (equipment) identifier, component name, work type, and description. For a time-based PM WO, a field should identify PM scope and source. For corrective maintenance, the problem description originates and justifies the work. A basis (or analysis) that justifies doing the work initiates a PM. CM is justified by a plant problem documented by an observed symptom and problem description. Operators usually prepare CM WOs because they first notice most problems while making rounds, trying to operate equipment, reconfiguring, or just running the plant.

The "work done" field complements the "problem" or "scope description." This field is the work performer's assessment of the problem and its probable causes. Such information is only as good as the report from the original work performer author/writer. So long as descriptions are generally adequate, you can infer the scope of a problem and work completed based on two or three overview lines of "work done" text. Practically, this is the most you can expect most mechanics and technicians to write!

After participating in some CMMS redesign/rewrite efforts, I've become an advocate of commercial CMMS software products. Cost, standardization, compatibility, and response-to-changes are a few of the reasons why. Limitations imposed by in-house CMMS suppliers pose severe restrictions on end users, yet, utilities have historically taken this approach. Not surprising, few CMMSs are fully utilized in the field. Many have extensive field variations. The basic process flows can be different, making software utility less.

If basic PM/CM "work justification" information is complete, and the "problem description/work done summary" are completed faithfully by work performers, then analytical failure study results can be adequate. These two fields provide the most basic failure information necessary for these studies. For CMMS users-particularly schedulers, planners, and workers—the maintenance process flow model and CMMS must compliment one other. With the widespread application of file transfer protocols (and products like Microsoft(c) [MS] Excel), it's relatively painless to export, sort, and reformat CMMS data for analysis and presentation. (Day-to-day users are restricted by their unique systems and processes they support.)

Information displayed in a condensed report format—emphasizing work initiation and completion fields—enables a quick review of components by type. An RCM failure-sample survey (by system and by component type) provides both failures types and the work performed. Several thousand WO reports are "representative" for an average system. Typically, this encompasses one to three years of WOs for a fossil plant and at least 30 for a nuclear facility.

I read the reports, placing "tick" marks for each failure group, and then identify dominant failure modes. Based on the time period under review, I make some rough estimates of MTBF and MTTR. In this way, I can knock off a relatively large system in a day or so. My research into system problems comes next, and takes more time, but the CMMSs can be very effective when I download and sort vast quantities of information—exactly the uses CMMS promoters championed 20 years ago (Table 6-1)!

These reports tell a story. They indicate aging-based failures, random failures, and indeterminate areas that need more review. They enable me to effectively structure questions for operators, maintenance personnel, and engineering support personnel. Done well, these reports can summarize failures in visual ways for storybook analysis and problem discussions. They provide a relevant basis for types of PM tasks and their intervals (Fig. 6-2).

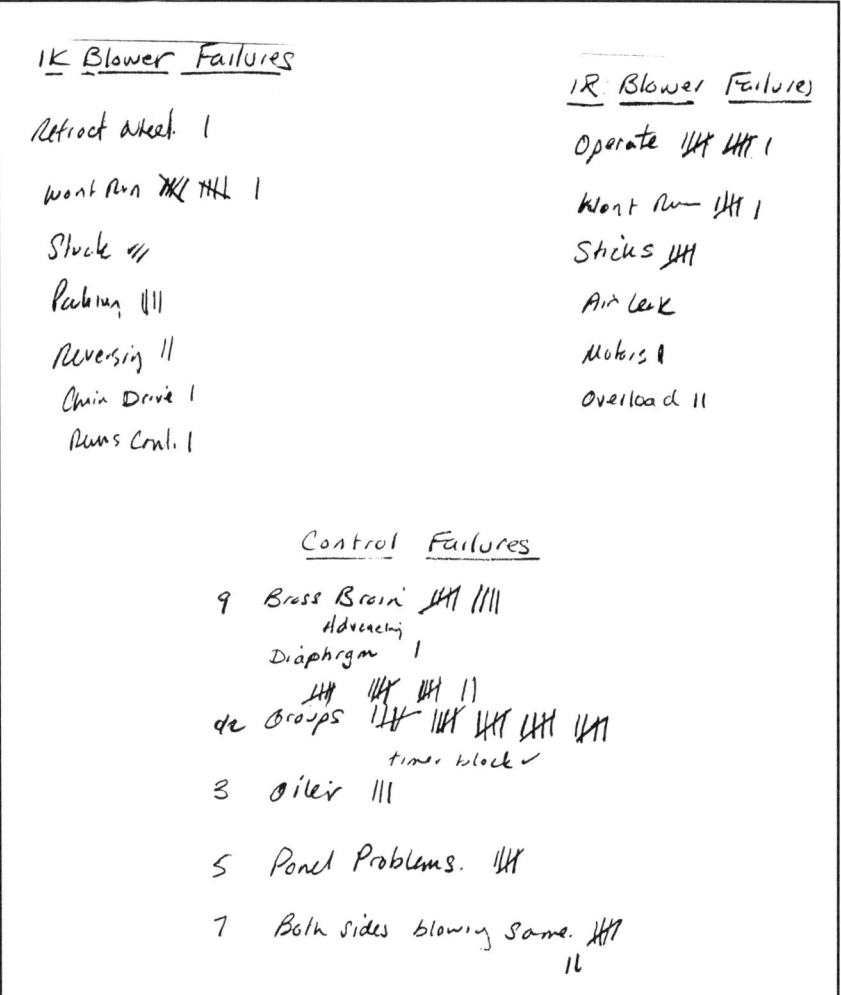

1K Blower Failures

Retract wheel. 1

Wont Run XX THL 1

Stuck ///

Packing ||||

Reversing ||

Chain Drive |

Runs Cont. |

1R Blower Failures

Operate HH HH 1

Wont Run— HH 1

Sticks HH

Air Leak

Motors 1

Overload ||

Control Failures

9 Brass Brain HH ////
      Advancing
   Diaphrgm   1

dr Groups HH HH HH HH HH HH HH HH
                timer block ✓

3 Oiler |||

5 Panel Problems. HH

7 Both sides blowing Same. HH
                               1L

Table 6-1: Tick Summary

OTF and No Planned Maintenance (NPM) in the Real World

Failure

Unplanned failures occur, although functional failures are rare. Still, signals can get missed, staff can be unaware of identified emerging failures, and instrumentation can be OOS. Failure patterns, conventional wisdom, and misinterpreted conditions can lead us to miss the obvious.

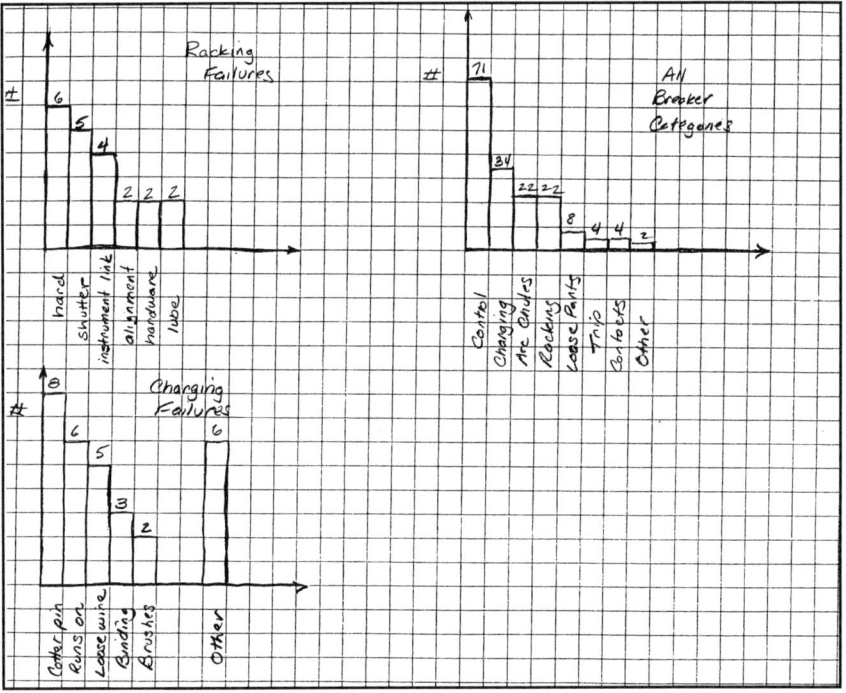

Figure 6-2: Breaker Failure Summaries

Sometimes, organization cultures are a barrier to recognition. Ignorance, fear, or inertia can combine with process barriers until a real failure develops. Failure study is both morbid—and fascinating. Any unplanned event that compromises operating goals is a failure. Unplanned events that don't compromise operating goals are the luck of the draw. Everyone gets lucky to some degree—although we usually create our own luck.

Those companies adept at avoiding failures are those that work hard at fundamentals. Those that understand failure modes and mechanisms—developing failure strategies—improve their capacity to manage risk and so favorably influence their failure rate.

Imagine two utilities with opposite workplace cleanliness standards. One staffs (and pays for) janitors—the other doesn't. One has coal units so clean you could easily mistake them for a nuclear plant. The other guarantees your hands will turn black if you hold the handrails on stairways and ladders—which you need to do for safety. One maintains current plant drawings—no engineering job is complete until the drawings are revised. The other never updates any post-construction drawings. One plant finds time and money to maintain little extras, like ventilation and air conditioning. The other can't seem to keep them up—except for the administrative offices.

Each plant's policies offer powerful cues about management expectations and equipment standards. The commitment to develop an O&M plan—in contrast to a "catch as catch can" approach- delivers greater confidence. Developing and maintaining standards takes fortitude. A line is drawn in the sand. If you skip back and forth across the line too many times, it becomes indistinct. Companies that set high standards out-perform those that do not, as reported in business books.

OTF in RCM

OTF simply means no scheduled maintenance tasks. The terms "run to failure" and "operate to failure" are similar but generate negative interpretations—and that's the least of our problems.

Regulators have come to expect that everything has a planned maintenance program and it must be understood that OTF is a planned maintenance program. The plan is no scheduled maintenance —like the mathematical null solution. The work elements and failure modes can still be virtually complete. Some nuclear plants must document their null maintenance plan, literally. The inherent robustness of design cited in the RCM classic by Nolan and Heap is, largely, beyond the grasp of the general population, regulators, and particularly the media. Those not versed in RCM—and most people aren't—simply don't understand this distinction or its basis nor do they need to.

For the NSM option, there is no "time-based" WO to kick out, but operating staff and those in the plant must identify and respond to symptoms. NSM programs depend on their personal and informal diagnostic skill and knowledge though the tasks are non-specific or not scheduled. This also illustrates why RCM-based, non-specific tasks need to be ruthlessly purged from the CMMS task list. For operators, "knowing the plan" depends on their initial condition assessment. Removing redundant CMMS tasks encourages more thoroughness in their condition-assessment learning and application.

Why does a plan of NSM work? Why, statistically, is it appropriate for 80% of the equipment in complex plants—perhaps more? Three reasons: people, design, and equipment.

People monitor. Design adds complexity and minimizes dominant failure modes. Evolutionary equipment is inherently reliable. The line between a condition-directed task performed by an operator on his rounds and a non-specific operator's CNM task gets blurry as operator skill levels increase. Highly skilled operators with many years of experience apply their knowledge in every monitoring situation, often above and beyond the call of duty. We may benefit but we haven't planned on it. And their absence has only minor impact on our program—by design.

Worker learning and knowledge buttress NSM. When the work force has a high level of equipment understanding and job commitment, informal maintenance programs like NSM work well. Operators schedule inspection tasks while on their rounds. Between rounds, they play "free-form" CNM, so to speak. What they find, through skill and attention, highlights problems that require further, specific follow-up (though equipment that qualifies for NSM has minor or mitigated plant-failure impact). Maintenance can be planned and scheduled with virtually no safety risk.

NSM should never be used to ignore equipment. Once failure identifies equipment needs, work goes into a queue. Because the work it identifies originates from non-specific rounds, NSM doesn't downplay importance—its developing failure mode establishes it! It prioritizes work with all other scheduled CM tasks, non-specific tasks, and scheduled maintenance. Although NSM-originated maintenance could be of higher priority, that's highly improbable. Rather, NSM work is generally lower priority because of the redundancy and mitigation capacity inherent in the design. NSM effectiveness, (like CNM monitoring) depends on the organization's ability to monitor to identify condition and then to include and manage CBM maintenance work with priorities in place for the planned maintenance process. Weekly reviews reshuffle the "hot and not" planned work schedules. The schedule is dynamic during the planning phase, but once a job is issued for work, reschedules should occur only very infrequently. Routine, repetitive jobs need standard "corrective" plans developed, planned, and shelved to work on demand. Sophistication improves as people gain insights into plant design depth-a depth developed with routine ARCM applications.

The term "self-identifying maintenance" also sheds light on both NSM and OTF. The bottom line of that plan is that we don't schedule formal tasks to perform maintenance - nothing more or less. Equipment maintenance needs are self-identified by the equipment. This is acceptable because of inherent R-the absence of known, effective age-based PM measures-and limited consequences of failure. No one in their right mind would do any maintenance to a car headlamp other than replace it upon failure. On failure, however, it's very important to replace promptly. It's the same concept.

Legitimate failure

Actual "failures" occur when we violate performance standards. If standards are established, it's unnecessary to discuss failure criteria. In their absence we have nothing but a discussion about what constitutes a "failure"—a discussion beneficial only to the degree that it leads to common failure definitions. The exercise is pointless-except to develop failure standards. Without standards, people cannot agree on what is important and what is "failed." For better or worse, nuclear plants have many guidance standards in place. Failures are well known. For fossil plants, the concept is abstract. Failure tends to follow a free-form definition, literally linked to a primary function failure. At plants with specifications, failures are specification-based and incremental.

An RCM-based approach to failure definition forces people to think about goals and limits, which in turn leads to earlier action. Goals can be obscure to operating staffs. Obvious limits on measured variables such as opacity and emissions, material thickness and production, can be missed. It could be argued that exceeded specifications and limits are in themselves arbitrary failures, since in most cases violation does not cause sudden, discrete events. Rather, an engineering limit has been exceeded. Real failure comes some time later and with continued loss of margin. Design specifications have margins that—if properly followed—safeguard us from real proximate failure areas. "Real" proximate failure consequences include:

- operating events that compromise company strategic goals
- unplanned, unscheduled production losses
- equipment losses that compromise design and operating margins

Risk correlates with "real" failure experience. A high risk plant experiences more proximate failures. Causes are diverse, but result from chronically pushing design margins. Plant practices—shorted time for PM and monitoring, missed CDM cues, use of design redundancy, or failing to operate within prescribed standards—lead to higher failure rates.

Examples of Group (1) failures include:

- major accidents
- excessive accident rates
- excessive environmental releases

Group (2) includes:

- unplanned unit outages
- restrictions
- unplanned plant system losses
- unacceptable loss of production equipment margins

Group (3) includes:

- failure to observe warning instruments
- overriding interlocks
- operating above tube temperature limits
- overfiring

Convergence: OTF-CBM equivalence

As much as OTF and CDM appear to be extremes in strategy, they are closely tied. The resulting "condition-directed" WO looks identical. Equipment has no memory of whether condition deterioration was caught by a formally scheduled task with a WO or by an inquisitive and alert operator. Both cases are discovered in advance of any production,

safety, or environmental impact; in both cases, the resulting maintenance can be planned and scheduled (e.g., function failure hasn't
occurred). Both tasks become common work items in the condition-
directed (formerly corrective) maintenance backlog.

The difference is in how they originated. One originates with a
time-based, OCM task. The other begins when an operator notices
something that is "not right." The former condition has a performance
limit; the latter may or may not. In maintenance performance, each type
of monitoring can provide the same outcome but the outcomes require
different levels of interpretation.

Since an OCM task that generates a CDM activity is "tuned" to a
specific failure mode, it should be easy to perform follow-up CDM. The
proximate failure should be clear and unambiguous. Functional failure—though not clearly traceable as a functional loss—should be clearly evident. On the other hand, non-specific failures are ambiguous.
Operator ability discriminates wheat from the chaff.

Indeed, non-specific monitoring depends on the operator's discretion, skill, training, and judgement. Establishing standards for monitoring, and then training the operators, can help hone their monitoring
skills. Database structures show OTF monitoring response to be virtually the same as that for CNM. The distinction is in the initiating task—
planned, scheduled, or none at all.

One could argue that identifying a "task" in the OTF bucket is arbitrary. Because the lion's share of formally scheduled monitoring tasks
are operations-oriented—failure-finding tests and specs-performance
monitoring—operations performs most ARCM-derived CNM. In
CNM, this formal leadership distinction is, theoretically, reserved for
highly skilled and experienced personnel. Maintenance-performed
time-based monitoring is much more task-oriented. Maintenance performs monitoring by WO. No WO—no work.

This can become a legal exercise—which ARCM is not—but the
key point to remember about CDM task WOs is that when a true-blue
OCM task appears:

- it's formally scheduled
- it's specific to a failure

- it's actionable (a discrete go/no-go outcome)
- it has a specific limit, margin, measurement, or attribute that unambiguously (to the trained analyst) is "go" or "no-go" failure evidence

OTF equipment self-identifies failure. We don't have to extract failure elements with sophisticated tools like vibration analysis or thermography because the benefit isn't there. Some operators possess a sixth sense for identifying failing equipment (the power plant operators' version of a green thumb?). They anticipate failures with scant physical evidence and often, their intuition is correct. They relish the diagnostic role—it's all they do. This demonstrates how the distinction between CBM- and OTF-based maintenance can be rather arbitrary.

Mathematically, OTF is simply a case of OCM without the usual time limits. The formal scheduled PM interval is taken "to infinity." Can the PM identify the failure if it happens? You bet—if it's technically effective and the failure occurs. Is it specifically worth doing? Generally not. For a facility with a 40 year lifetime, maintenance performed on even a 20 year interval is approaching plant-life limits. At such an interval, such tasks look as if they are "taken out towards infinity." They qualify strongly for non-specific monitoring if the return is small. This demonstrates how excessive amounts spent on PM can raise costs.

When all is said and done, OTF is rounds-based, non-specific operator monitoring. Everyone knows trained operators walking through the plant can accomplish a lot on experience, whether specifically directed or not. The lesson, pure and simple, is to keep it simple but keep it up! ARCM offers a streamlined method to identify the monitoring tasks expected of operations, and establish appropriate intervals to make them effective.

Complexity in failures

Failures have contexts—simple or complex. Simple failures are easier to manage so we naturally prefer them to complex incidents. Complex failures involve multiple failed items and interactions—interactions that make them hard to diagnose. Multiple hidden failures are harder to recognize, interpret, and correct. These include multiple coincident failures in complex equipment trains.

RCM analysis helps familiarize us with the spectrum of the anticipated failures. We are made aware of likely failures as well as prevention strategies and counter-measures. Task performance can be made easier by developing fault trees for complex system failures. They enable us to step through the failure analysis quickly, and because fault trees can pre-develop the probable failure modes, actual diagnostic performance take less time. When failure events and their likelihood are established up front, developing preventative measures also comes easier.

"Bootstrapping"

ARCM provides us with many ways to approach plant operations improvement. Each can add value quickly. Once specific equipment and subsystem analysis is complete, results can be transferred to similar units with little effort. Applying previous learning to similar units and systems without any formal detailed analysis is a form of "bootstrapping."

Contrary to TRCM analysis, power plant design is highly standardized. All steam turbines use a Rankine cycle, for example. Suppliers are the same in a given region of the world. Large generating facilities share considerable standardization of equipment, systems, and layouts. A few configurations have been developed and see many repetitions. Even informal standards have proven their utility over time. Thus, once a basic "repertoire" of RCM equipment and systems understanding is in place, it forms the core for many common applications.

Systems. Common system configurations abound. In fossil plants, feedwater, condensate, sootblowing air, and circulating water are very similar from unit to unit. In nuclear plants, the GE BWR and the Combustion Engineering (CE), Westinghouse, and Babcock & Wilcox (B&W) PWRs share common design elements. This supports standardized RCM analysis.

Occasionally new systems are integrated into traditional ones, and this requires system re-analysis.

Equipment. Most common equipment in power generating facilities is supplied by two or three primary suppliers. Even where there are many suppliers—as in the cases of valves and motors—there's so much functional commonality that many failure CNM tasks transfer directly. This commonality supports the development of standard maintenance strategies at the equipment level. While the relative frequency of the major failure modes shifts from plant to plant and environment to environment, the primary modes tend to be the same. Where they differ, there's often a different operational strategy or use at work.

Commonality supports standards development. Using the craft to identify and select commonly encountered failures and select strategies is an effective way to focus general information and solicit buy-in. For simple PM work formalized development and application of standards eliminates large amounts of low value work—the annual replacement of gearbox oil in a very clean plant environment, or monitoring and replacement task intervals for hoists, cranes, and other infrequently used equipment. It can lead to major labor savings.

Reduction in the frequency of parts replacements is another consequence. Small stock items are carried because manufacturers recommend it, and they're replaced on intervals to support 24 hour, round-the-clock operations. In standby, this equipment may not see the manufacturer's full year of operation over the life of the plant! To service it on the manufacturer's suggested interval is usually way too frequent.

Processes. Developing an RCM-based maintenance program invariably forces an organization to review processes. Virtually all of them must be tuned and new ones added, including cost and measurement. This stressful exercise partly explains why many RCM-based maintenance plans fail. Without a vision and commitment to change, forces naturally align to resist.

Process redefinition, however, though painful, is also the most beneficial long term aspect of an RCM-based review. Refocusing work around a PMO philosophy is fundamental. New skills, perspective, and commitment, and long term benefits—such as more directly comparable cost information between users, more useful information breakdown, and performance measurement—make the pain tolerable. Potential maintenance savings of around 40% have been cited in various electrical trade journals, studies and specific cases as the benefit opportunity derived from improved maintenance.

In contrast with another popular process redefinition method—"re-engineering"—an RCM approach offers specific methods. Organizations can use RCM approaches in discrete incremental amounts, for example. Improved PM measurement can be initiated on a CMMS with the introduction of failure criteria. New terminology can be phased within a time. Decisions about software, basis maintenance, training, and other issues can be implemented with only minor organization disruption.

The downside to incremental change is that organizations dilute lessons. More than one PMO effort has flagged and failed this way. You may be successful at introducing new maintenance methods, but when the old remains in place, an incremental approach is sure to fail.

"Critical?"

Traditional RCM definition

Traditionally, commercial air transport RCM reserves the term "critical" for those failures that have an immediate and direct safety impact. "A critical failure is any failure that could have a direct effect on safety." (Nolan & Heap) Note the word direct imposes specific qualifying criteria not imposed elsewhere. This qualification excludes failures that aren't immediately evident, or weren't single failures. For "non-evident failures," the absence of "evidence" means there is no direct impact on safety. The "immediate" requirement screens out multiple, train-redundant failures.

The first major permutation of RCM's definition of "critical" occurred in the transition from aerospace to nuclear power. NRC definitions for what are now called "essential" components—those whose failure could affect fission product environmental releases—gave "critical" a new dimension. Since many nuclear components occupy this category, critical applications grew by default. The tendency to associate the term with specific components, rather than failures, compounded confusion.

The final application of the term "critical" to non-nuclear units, produced a flow process that divided analysis into critical and non-critical. Dividing components in this way (a la nuclear) proved confusing as fossil plants struggled to abandon their historical "critical" interpretation associated with production impact. By failing to rigorously apply the original criteria, and indiscriminately applying it on the basis of production impact, "critical" scope grew. The R engineering definition from FMECA diverged, as well. "Critical" was based on a numerical value that only had meaning as a relative ranking in the context of all other equipment in the analysis.

Casual use

Many of us, occasionally use "critical" casually. Those SRCM methods that divide equipment into two broad categories-"critical" and "non-critical"-depend upon the equipment involved, and determine whether the review selected involved is a thorough failure review or a quick and dirty (cost-based) "sanity check." After evaluating thousands of components, I've found that the RCM process that identifies components for PM is not that important. Most analysts can quickly determine a component's suitability for PM, and the likelihood the PM is effective. Whether a non-technical person can follow their analysis is another matter! If your interest is solely the final product-an effective PM program-you probably don't need the extra information. If you must maintain it, you do!

One plant in which I worked developed an automated prioritization CMMS that identified all equipment as critical (or non-critical) and electronically pre-assigned priority to the equipment WO. This system ultimately produced a disproportionate number of critical, high-priority, "work-today" WOs. Because no one had the inclination to override default rank and rank any priority "low," the predictable results were that virtually everything—except PMs—were "critical." This was not only not useful, but effectively eliminated meaningful priority.

The primary work prioritization/screening tool depended on two attributes-is it emergency or deferrable? The tasks most likely to relieve the workload, long term—PMs-rarely made the cut.

The primary purpose of a priority system is to rank the importance of work—quickly. If there is no discernible priority attribute, or it's skewed, then the system-no matter how ingenuous at the software level-has little value. This system had no value. When truly "critical" equipment fails, it causes unsafe conditions. Economically "critical" equipment failures cause unit outages or other obvious design-intent and system-function failures that relate to bottom lines, not safety.

"Truly critical" failure modes. People don't think in terms of failure modes. Understanding failures does not come easily. Like any skill, it takes years of experience and missed calls to get right—most of the time. Systems understanding comes first. Many people in responsible positions haven't had the time to hone this skill but are charged with understanding plant risk and making safe, effective operate/shutdown decisions. Such managers rarely make good candidates for failure engineers—they like black and white cases, and clear-cut calls. (And, indeed, someone has to make a call.)

It's extremely important to clarify distinctions among safety, production, and economics, to best allocate scheduled maintenance resources. After 15 years of discussion and analysis, I believe that producers are inherently safer that non-producers. Plants that operate-just like cars that put on miles with no breakdowns-have to be in good hands to be able to do so. It's rare to find top performers that don't also put operational safety at the top of the list.

Just as equipment groups are similar, so unique—"different—critical" failure modes are relatively scarce. What they share in common is the fact that most equipment potentially presents life-threatening failure modes, eventually. When we buy it, get good service out of it, and experience its performance deterioration so legions of engineers have to scratch their heads arguing over whether the time for overhaul (or replacement, as the case may be) has come—that's the way we like it! Operations at this point wants a new one—pump, compressor, valves, belt, whatever—but understands the old beast well and so gets more mileage from it.

Wearout. Wearout is the desired end-state for every component in an operating organization. When equipment is worn out, evenly and well, meeting a manufacturer's promised life, it represents an ideal. It's a matter of gradual, predictable performance loss, providing lots of latitude to schedule replacement—and manage risks by scheduled maintenance. Ideally, turbines age this way between overhauls, as full turbine load gradually trends towards valves wide open (VWO) position. Centrifugal pumps show gradual loss of head, as the rotating elements deteriorate and seals wear.

Wearout is an important failure mode and the most desirable outcome. Examples include:

- loss of balance on high-speed rotating equipment
- loss of steam (or other high pressure fluid) boundary
- cracks
- gaskets
- ruptures
- packing blowout
- loss of lubrication on heavily loaded parts in relative motion
- lubrication aging
- contamination
- aging breakdown
- loading

Wearout relates to critical failure modes as one end of spectrum of concern that can be lumped under two, maybe three groups. Consider automobile tires.

Wearout—the ideal and key failure that ultimately will end the life of every tire—is balanced at the other extreme by random failures. In a car tire, it can be the unpredictable blowout or loss of air from puncture, glass, debris, or even driving over a curb. Last, there's loss of air. It's possible that a tire may not lose any air over its service life, but most will.

Tire manuals offer many examples of rare event type failures. Scalloping, uneven wear, ply separation, rubber breakdown from chemical attack, stem failure...typical drivers see one of these failures once or twice in a lifetime (assuming you maintain your car!). Secondary failures-from imbalance, improper adjustment, unintended service, suspension damage, worn ball joints, and the like-can also be induced easily.

Confusion implications. The greatest downside to design redundancy is the ambiguity it introduces into the definition of criticality. After all-using our tire analogy-is a tire critical if you have a spare (and the means to change it, and the free time to do so)? Power plants have many redundant "critical spares." Standby boiler feedpumps, redundant buses with auto-transfer breakers, spare turbine gland steam exhausters...so long as the redundant feature is available and the correct transfer sequence occurs, loss of the primary component has no impact. Hence, it can't be critical.

Or can it? What if the backup fails or the auto start sequence fails, or the backup is OOS or performance-impaired? In these cases, production or other essential functions are lost the primary is "critical."

This ambiguity is another reason why I dislike indiscriminate use of the term "critical." In some sense, every item in a plant is critical-or mis-specified. i.e., There shouldn't be any equipment in a facility that has no functional value. If this is the case, it should be abandoned with no impact. What is critical is outcome. A critical failure is when someone's hurt-a non-critical when they're not. In this context critical reflects the airlines' direct safety consequences interpretation.

Importance

Given that "critical" will probably endure as a named condition, though in confusing permutations, let's consider one more attempt to clarify and simplify its use.

Safety and cost drive all PM while safety and economics drive all plant operations. Ignoring economics fails to adequately address our sole operations purpose. For this reason, "economically critical" is an acceptable concept-provided we restrict its application to failures. The primary consequence of most safety equipment failures are operational; we must terminate operations to address a key safety function. Using our functional definition of failures, we could agree to use another term for equipment classification based upon economics, then reserve the term "critical" for safety functions and their failures.

We would then have two classes of equipment—*important* and *non-important*—and a sole criteria for classification: whether we plan to consider scheduled maintenance for the item or not. We could just as well identify these as "scheduled maintenance" and "non-scheduled maintenance." Once past this barrier, we can review equipment for applicable and effective PM tasks.

Practically, a reviewer looks at the following identifiers to discern suitability of scheduled maintenance:

- equipment size
- failure reporting frequency
- work·frequency (including PM)
- vendor recommendations
- general industry practice
- shop practice
- equipment register

Equipment size. Large equipment is expensive. Safety considerations are based upon enclosed energy and fluids. Large items are always reviewed for PM activity. In addition, since the work scope involved in opening large equipment is considerable, they're expensive to maintain. Documented PM or maintenance costs usually confirm this. Large equipment vendor manuals and other recommendations usually suggest appropriate safety and operations concerns, as well as cost-effective tasks.

Failure reports. There are two sources of failure reports: operating events and WOs. Until these are screened-ranked by criteria and evaluated-the nature of the failure is merely supposition. Practically, any equipment that carries operations impact needs careful review for potential PM tasks; if operations impact is absent, equipment with significant failure frequency warrants deeper review. Depending on the plant's equipment tag number coding detail, such WOs may be electronically hidden for some equipment. A comprehensive review will always turn them up.

Operating event records are only as good as the level and detail of the operating logs. In each case, log review conclusions should be reviewed with operators and maintenance staff to confirm trends, as well as flush out items not included in the formal records. Operating groups with a shift rotation of more than five people, providing plant coverage, should maintain a written operating log. Where logs are inadequate or lack detail, standardize them. In one event I know of, initiating new records to document the cost of operational failures was instrumental in gaining corporate support for capital investments to lower costs.

Work frequency. Systems that are costly to operate and maintain demand that much extra work-and they require PM programs, WOs, or both. Outage and contracted work not captured in plant WO systems should be reviewed for scope that reflects maintenance tasks. "Acid tests" for large work scope should be considered-is a *basis* established? Does the plant know why the work is done, and what its performance benefits are? Occasionally, capital modification work performed in off-budget areas is charged as a maintenance expense, especially when engineering groups perform maintenance support functions. This skews costs.

Vendor recommendations. Vendor literature should be reviewed to assure that vendor-identified work has been assessed. Vendors generally understand their equipment's economics better than anyone else. They also appreciate safety considerations, although they may not understand specific applications. Vendor recommendations often turn up interesting insights (and oversights) that influence large equipment costs. Small equipment that lacks obvious, integrated functions is usually NSM by definition, but reviewing this against the vendor recommendations can identify tasks the vendor thought were cost effective to perform, even for generic equipment.

Comparing recommendations from similar vendors is an effective way to establish appropriate tasks and performance intervals. Many times upgrades and enhancements influence maintenance frequency-like a superior synthetic lubricant that extends a lubrication performance interval, for instance.

Industry practice. Benchmarking equipment for standard industry practice is another effective way to establish appropriate levels of effort. Comparing one industry to another, when both run the same equipment, can provide insights. For example, how mines handle coal in different locations supports cost-effective improvements for utility coal handling operations. As with all benchmarking, understanding the methods and practices behind the numbers is essential to making appropriate choices.

Shop practice. Every shop develops techniques to manage work performance. Often some of them are unique and effective tasks that can be adapted to other areas.

Once, working at a chemical plant, a worker pointed out a simple method to determine bearing wear on a large fabric-making machine. The applicability and efficiency of this simple quick test for bearing tightness was a classic example of shop learning and creativity. The test was faster and more effective at locating bad bearings on a 17 main bearing drive than any other method I've seen since.

Equipment registers. At initial plant startup, an equipment CMMS register or hierarchy is created based upon design engineering or accounting descriptions. These lists of equipment should be considered for PM. High cost capital- and skid-mounted equipment-along with equipment that has been erected on site-goes into the register. Reviewing this list for PM candidates is an effective way to assure that everything gets considered. Design modifications that support the register also need review.

Area Checks

When operations personnel perform area checks, they're engaged in NSM. Area checks require operator rounds, to check the condition of equipment installed in the plant. This was the original intent of the hourly round.

In fossil generation, an operator's complete round can take two hours or more, even at a brisk pace. And "brisk" is not the point-you must slow down to read instruments. Often, lighting and cleanliness can make equipment monitoring additionally time consuming (especially when gauges are dusted in coal or oil mist.) When I identify hourly rounds sheets (except for the control room), I'm immediately skeptical of their effectiveness and applicability.

Yet, rounds are important. In complex plants, many failures are random, and actual system functional failures are rare. Because of this, it's essential that the operator on a round identifies failing (and failed) equipment. A log entry or CMMS trouble report are techniques to identify failures. After a complete ARCM review at a nuclear plant (reviewing approximately 100,000 components), we found that the overwhelming default PM activity, numerically, was "NSM." This plant had exceptional operating rounds. Plant operators are uniquely positioned to identify failed equipment.

This functional role was what was intended by the hourly rounds in the U.S. Navy. It's the same for commercial operators: Monitoring is the operator's principle role except during startups, shutdowns, and preparation for maintenance.

Random failure identification is an operator "value-adder." To be effective, operators must perform an adequate, failure-based round—that is, a round that meets the effectiveness criteria discussed earlier. Here's where ARCM criteria can help.

When developing monitoring requirements for systems and equipment, ARCM methodology helps develop rounds in a shorter, less-arbitrary way. The net benefit is a better, more specific, actionable, and focused rounds. Performing "round reviews" can achieve significant improvements in operating plants. An RCM system review assists operators by identifying:

- functionally important equipment (and their failure modes)
- analyzed NSM equipment checks (important failure checks to be done by operators)
- formal, defined checks (for rounds) and observed process and equipment parameters
- optimum round intervals, based on MTBF
- area checks

An area check is a general survey that integrates the senses and non-specifically identifies failures. Like functional tests at the system level, they integrate and pick up major failure indicators. They're like the "area check" that driver education courses suggest you perform before you get in your car and drive away. Airline personnel make them every time a plane takes off, confirming the absence of functional problems for various basic, yet critical components. They also enable the quick discovery of problems that could have serious consequences over time.

Area checks in plants are also cost-effective ways to identify random and general deterioration failures. Clean, well-maintained equipment provides unambiguous results. If standards drop, however, and dirty conditions develop, area check effectiveness drops. Cleanliness is essential in plant environments. Clean, organized areas enhance failure detection. A Japanese maintenance process called TQM places great emphasis on equipment cleaning as a tool to discover incipient failures.

This suggests other requirements for effective area checking. Poorly illuminated areas can't be monitored effectively, so area check value drops commensurately. Coal facilities can be dark and dirty. Aging paint reflects less light. One mechanic friend refers to his fossil plant as "the cave." It consistently underperforms and suffers from perennial operating and other performance shortcomings.

What level of cleanliness is appropriate? It's hard to say. I know of one coal-fired facility operated as a low cost producer that maintained a janitorial staff of eight exclusively for what was a (2) 550 MW unit facility. With availability well above 90% and capacity factor close to 80%, it must be doing something right!

Area check strategies also work well for complex fluid system flows where leakage can be identified by the trained eye (or the other senses), or by sense-enhancing equipment (such as ultrasonic detectors or thermograph imagers). They fit well with operator rounds. Examples of specific area check failure targets include:

- lube oil reservoir leaks
- service water leaks
- hydraulic leaks
- instrument air (or other gas) system leaks

Obviously, as operators gain experience and skill, their area checks become more effective. Innovative, as well! Operators of a large coal facility, running volatile morpheline water chemistry treatments, pointed out to me one of the most effective area checks I ever encountered: smell. It's extremely sensitive and the fastest way to identify a steam leak: morpheline has a distinctive, pungent smell. In coal handling areas, the pungent smell of burning coal is an effective tool to locate smoldering coal piles in belt galleries and bunkers. Chlorine and sulfur dioxide systems can quickly identify leakage by smell well in advance of acute safety concerns.

Instrumentation

The typical plant has a tremendous amount of instrumentation and control equipment. Much of it was installed or packaged with equipment skids, provided by the manufacturer/assembler. Many instruments have as their primary function the setup and performance of pre-operational or operational testing-lube oil skids, for example. Equipment suppliers often provide remote panels for pre-operational testing and operations. A typical plant's equipment, augmented by skid I&C, provides so much instrumentation that to tackle TBM of all of it would be a Herculean task!

Major process control loops feed a plant's DCS. This is done through drops, in modern plants. The typical DCS runs two redundant independent buses with self-checking diagnostics and the capability to swap buses, should a problem occur. Each has a fully redundant back-up with the same capability. These robust systems haven't been the focus of a detailed RCM assessment by many clients, and I&C technicians and engineers, by and large, effectively maintain DCS controls. This suggests low value added benefit here at the typical plant installation. Risk management, maintaining redundancy, and "depth" have thus far been very effective. However, there are I&C opportunities.

The first is to select and identify candidates for NSM. I&C PMs include cals, channel checks (CCs), and functional tests (FTs). Self-diagnostic equipment can reduce or eliminate the need to perform FT. Self-calibration routines can eliminate the need to calibrate. Typically, a trouble alarm sounds if a plant DCS loses a drop or channel. Periodic checking of the channel alarm is all that's required.

It's tempting to feed every instrumentation point in a plant into a new DCS during an upgrade project. However, large fossil units could have 5,000 points fed into the DCS. For important equipment (like boiler feedpumps), this enables dropping more points than original plant data logging could accommodate. More information is available on-line than previously available, or available locally at the feedpump skid (such as local hydraulic and control oil pressure and temperature) but the downside is that we must maintain instruments-including the low value instruments. Points fed into the DCS need conservative selection with an operating monitoring strategy in mind. Selection should not replicate an existing monitoring strategy. Extraneous "nice to have" instrumentation can otherwise result.

Estimating installation costs (at one man-hour per point), a new DCS installation can cost nearly 5,000 labor hours for point drops alone. If 50% of the points get no functional use, then non value added instrumentation costs $125,000 (2,500 x $50). ARCM in I&C programs not only provides guidance to control the installation, but also the maintenance costs and application selection of low- or non-value instrumentation.

Two key instrumentation applications are loop cals and functional checks. Active control loops require calibration as determined by drift. Functional checks of essential safety, production, or compliance alarms is also necessary on an interval determined by alarm failure risk. Examples include:

- safety: steam-driven BFP vibration alerts
- production: deaerator level alarms
- compliance: continuous emission monitor (CEM) flow, opacity, and species (and other alarms)

DCS systems allow taking alarms "out-of-scan"-for-nuisance status and other low-level applications. While appropriate for temporary problems, out-of-scan status can be forgotten. Routine checks for out-of-scan alarms are a necessary scheduled task.

Spurious alarms

Consider the new car whose seat belt monitor tells you to "buckle up," over and over, as you cruise down the highway. Most people can take about five minutes of that before they pull the plug-assuming they can find it. While searching for it, they're a safety hazard (unless they pull over). Nuisance alarms are more than a nuisance-they're a distraction and a potential safety problem.

In the context of RCM instrument function—providing a clear, unambiguous picture—a nuisance alarm is a failed alarm! Critical alarms should be corrected. If non-critical (e.g., you can safely tolerate their absence for long periods), remove them.

Screening I&C calibration intervals for extension based on drift experience and importance can reduce and simplify I&C work. In fossil plants, calibration frequency can be adjusted (extended) by installing new controls, but it should be considered in all plants. Many times, overly conservative intervals persist long after they've been identified. Excepting very old pneumatic control devices, there are many opportunities to calibrate less often. Newer sensors and instruments can reduce maintenance requirements by significantly reducing drift. Reducing efforts expended on non-essential control loops calibration enables more consistent focus on key control loops and essential alarms.

Critical instruments

I qualify my reservations about using the term "critical" with one exception—instrumentation. The reason for this is very simple. Key, essential safety and monitoring control instrumentation really has one single function-to provide operators with an unambiguous window on the plant world. When this is not the case, the instrumentation has failed. For this instrumentation alone-because its sole function is unambiguous safety information-its failure alone is enough to justify a plant trip.

It's like the driver with broken windshield wipers—if you can't see, it's hard to proceed safely. Or a train with no "clear" signals. To proceed with no window on the world's critical features violates the basic precepts of safety. It's like flying blind. This is why I quantify essential monitoring and control equipment as "critical." Generally, if there is an active control loop, its role is already captured as "important." No control, no operations. This interpretation is largely limited to I&C with safety status functions, and is consistent with Nolan and Heap.

Note that the vast majority of instrumentation doesn't meet these criteria—well under 1%, and maybe 0.1%. And we're not talking about "a little drift" here or there in an operating event. Although drift also has limits and boundaries, critical instrumentation that has drifted out of range is "failed."

The I&C equipment spectrum extends from non-essential to convenient controls to generation control loops to safety I&C. Equipment not directly supporting generation control provides service, convenience, time-savings, or another support function. It is non-critical.

Much of this instrumentation provides diagnostics capability, or comes installed with manufacturer packages. Once operations begin, startup instrumentation serves little or no further useful purpose. It may be useful later for diagnostics, but this instrumentation usually provides no further value.

The vast majority of such instrumentation supports NSM strategy. When an operator identifies that it's failed, it can be channel checked, calibrated, or otherwise restored. Instrumentation used exclusively for tests should be checked out just prior to test, and then revert to NSM. As tests are scheduled, cals can be incorporated.

Spare, standby equipment such as spare boiler feedpumps and I&C status equipment share a similar redundant role—they're not called into service unless a primary failure occurs in "protected" equipment. Some SRCM methodologies refer to this spared equipment as "non-critical." This explains how SRCM methods can take three virtually identical parallel components and determine that two are critical and the third is "non-critical." They've "dedicated" the third redundant item to spare service (on paper, at least). This is particularly confusing when looking for simple answers on whether or not to "PM the spare." The design symmetry is voided in operations.

Special equipment requires special consideration. That doesn't mean "no PM" or NSM as a routine maintenance plan. Spare equipment still requires periodic testing. Because it's not in normal service and can't self-identify failure, the periodic test provides a time-based OCM task. Failure can only be detected by placing it in service, so functional tests at some interval are both applicable and effective. Post-maintenance tests after restorative work assure you have a functional item available.

Instruments often provide condition alert or diagnostics. For condition alerts, reliable, high-quality instruments increase effectiveness. Ambiguous or faulty instruments can cause an inadvertent plant trip. Lost production because a protective device failed is painful; this is why plants avoid "armed" trips with such passion. If instruments that provide for status-only information show ambiguous results, that's not critical. There should be other, back-up items to resolve the ambiguity. (We expect [and pay] operators to negotiate this sea of ambiguity.) For instruments that force immediate operating decisions, there can be no ambiguity. There's no time to diagnose whether the problem is in an instrument or "real." Such instruments are critical; they have high impact and value.

Occasionally, operators aren't aware which instruments are "critical;" more commonly, "advice only" instruments are identified as critical and vice versa—and an avoidable plant trip occurs. This makes it important to understand the criteria by which I&C are identified as critical to correctly specify the type, reduce ambiguity—and avoid plant trips. The RCM paradigm provides simple, clear guidance on how to do this. Although the identification of critical instruments ties into the review of a system and its failures, it can also be performed as separate activity. A person familiar with a system's "protected" failure functions can review instrumentation and identify critical failure mode protection. They can then tag critical instrument functions, as appropriate.

Catastrophic critical equipment failure modes must not occur under any circumstance. Large fans and turbine generators are two prime examples. Fans can't be credibly "over-speeded", with fixed-speed drives. Turbines can. Unless over-speeded, turbines can't credibly shed large rotating mass parts-but fans can. Practically, avoiding these disasters is achieved through armed trip devices—overspeed and vibration trips for turbines, and imbalance trips for fans.

There are no explicit standards for armed vibration trips on large rotating equipment but industrial safety and product liability has led to informal standards that are in widespread use. With experience-particularly with large air handling fans (ID, FD, scrubber, and primary air [PA])—we've learned that high vibration amplitudes quickly lead to catastrophic failure. While most manufacturers provide armed trip device protection options, many companies choose monitoring with alarm options. This places any ambiguity's interpretation squarely on the operator's shoulders. Boiler flame scanners, ignitor permissives, main fuel trips, visual flame scanners, and other trip/alarm devices fall into this category. Such devices provide essential safety and equipment protection functions and, hence, are critical. O&M literature generally identifies the critical nature of this instrumentation by emphasizing test and maintenance requirements.

Unlike mechanical equipment, instrumentation has fewer functions. Generally, there's an active control function or a passive (hidden) monitoring function-sometimes both. Oftentimes status and monitoring are tied in the same sensing or control loop. Decisions as to what constitutes "critical" makes more intuitive sense when it's applied to I&C, because these extend the human senses and present condition information that we otherwise would not have. Operators are critically dependent on the extended sensory range these items provide! Without them, we are potentially ignorant of otherwise unprotected conditions. Personnel safety is always involved, as proven by general industry experience. Some examples are shown in Table 6-2.

Hazard	Instrumentation	Experience
Imbalance	VM	Catastrophic rotating machine missile
Combustion/explosion	Methane detector	Coal fires
Overload	Breaker overload relay	Electrical overload mechanical faults
Bearing overheating	Embedded thermocouple (TC)	Prevent bearing babbit damage and fire
Burner stability	Flame scanners	Explosions
Fires	Carbon monoxide detectors	Smoldering fires
Fire detection	Rate-of-rise detectors	Slow fire
Fire protection	Deluge system	Fast fire: transformer fire
Flame cameras	Loss of flame/rich combustion	Explosion

Table 6-2 Critical Instruments

Primary instrument functions include:

- active process control
- process measurement
- operator shutdown action (e.g., vibration)

- abnormal condition (personal/equipment safety)
- steam leak detection
- fire
- radiation
- informational/diagnostic—status

Critical instrumentation functions include:

- active main power block control process
- critical safety control processes
- electrical shedding/isolation/overload protection
- electrical emergency safety ties
- automatic important equipment protective shutdown
- protects "important" equipment
- protects non-important equipment from safety failure modes
- critical safety function
- alert to main process control limit out-of-spec
- explicit compliance role
- status action taken on any of the above
- non-critical instrumentation functions
- status information
- successful transfer
- skid diagnosis/convenience

Insurers have a vested interest in stations maintaining these devices. Insurance companies (or their representative) typically audit operating conditions independent of maintenance monitoring programs. These arrangements are generally effective. However, insurers lack the insights of plant operators and can not investigate "problem" alarms that are sources of ambiguous information. These instruments are sources of risk when operators take them "out-of-scan" and a plant suffers loss of an interlock, trip, or requires a jumpered condition. Most insurance reps won't have the authority (or the knowledge) to open breakers or control cabinets for inspection. Insurance representatives don't statistically analyze a station's MWRs—an illustrative and telling indicator of plant operations.

Vendors identify equipment failure modes that require special instruments to detect. They generally provide these instruments or identify their need. "Good" instrumentation comes with a quality product, and vendor literature is the resource to ferret out this functional information. It's essential to do an ARCM review on "important" equipment as well as critical instruments-anything large, expensive, and involving safety. Industry and insurance protection standards are a good second resource.

Critical instrumentation is not hard to identify, and it can quickly benefit from ARCM-based review to provide a quick R performance return. This is especially true at units that have never scrutinized instrument programs before. This guidance also provides insurers a focused process to improve a client's risk profile. In operating companies with functional OTF critical instrumentation programs this will mean learning new habits. They will find it difficult to change but they will also derive the largest benefit from instrumentation focus.

Redundancy

Costs and layers

Where redundancy cost can be managed, redundancy adds value. Large mechanical and electrical equipment costs are substantial. In the absence of cost containment emphasis, costs quickly escalate. With generation simply another competitive product, uneconomic costs are an added burden.

Many American families have a spare car (some two). Its value is mobility when the primary vehicle breaks down or goes into the shop. There is a cost to maintain a spare. Typically it's less than a primary vehicle because operation is limited, but there are fixed costs. Consider the pros and cons of a spare vehicle. Space, time-cost, and other less-obvious costs are incurred to have one. There are fixed and variable costs, all of which are endured for the assurance and convenience of the spare. Cost-savings are achievable if the R of one car is high enough to eliminate the spare. Herein lies the problem of redundancy: How much, before too much becomes a cost and organizational burden?

A Midwestern utility developed a system-wide blackout recovery plan-including a PM program-that caused resistance at several plants and within the substation group. After progress stalled all parties met for a problem discussion session. Applying ARCM to the blackout recovery plan revealed that it exceeded five redundancy layers. In some cases there were seven functional layers of transmission redundancy! Workers in transmission and distribution were aware of this redundancy, and had learned to use it, performing necessary work but outside the recovery plan objectives. Budgets didn't account for the sheer volume of needed work, and were constantly trimmed. Real redundancy—functional equipment in service—was slipping, as a result.

Just as redundant equipment, components, and systems increases complexity and demands resources, too much redundant equipment work can raise costs. If workers learn maintenance complacency because of redundancy, redundancy has backfired. Eventually system R suffers. In a non-competitive world, carrying redundancy is easier. Transmission and distribution R costs will be most impacted as the market opens to wheeling and competitive power sales, but the value of redundancy in generating systems will undergo review as well. If four redundant layers have been required to generate functional redundancy of two, R enhancement from maintenance improvement is in order.

"Oh my God!"

Rare failures occur with random predictability. When a plant experiences a rare event, the loss is sobering. We hope that equipment damage is the only consequence but occasionally, people are hurt. Major equipment losses are rare, averaging less than one per year per unit (based on NERC, my own experience, and other statistics) even at functionally "run-to-failure" facilities. Individual, rare failure events occur several times in the 40 year plus life of a unit. Scarcity is their problem: statistically, they're like auto speeding; many separate events are required before an event registers. But plants with events account disproportionately for overall losses. Precursor events are risk factors. Control them, and you have made a substantial impact on practical risk.

For large equipment "event" protection, instrumentation extends the human senses for failure modes (and events) that otherwise can't be detected. Fossil units are at a significant disadvantage to their nuclear brethren on this point. Not all objectively check critical instrumentation, or train personnel to follow operating guides based on what they say. Nuclear plants, on the other hand, are held accountable by license for their controls and trips, and have many of the trips hard-wired. Hard-wired trips make it absolutely necessary to maintain instruments and controls to operate reliably.

RCM recognizes that "oh, my God!" losses are often critical instrument-related. The task is to identify vendor-supplied instrumentation that provides major event loss protection, identify the essential maintenance elements, and then make sure they're maintained.

Major sources of significant event information include regulatory authorities, industry groups, insurers, technical literature, and industry conferences. Developing and maintaining a worst case "oh, my God!" failure file may be a corporate insurance or safety group function, but it requires operating and engineering input. Nuclear plants are required to maintain industry operating event reviews. Most others do so on an informal level. RCM analysis can take considerable opinion and guesswork out of the risk estimation for these losses. A complete assessment-based on industry event frequency and corporate plant risk management programs-will identify risk levels, and help focus available resources where they will do the most good.

R engineering tools that can help quantify and develop these risks are principally FMECAs. FMECAs can quantify the risk of any particular catastrophic event. Once you quantify the risk (by extracting existing designs and using industry information), O&M can manage the risks. This is where ARCM comes into play.

VM instrumentation is provided on large rotating equipment. When available, it tells a story. A plant's operating environment can use this story to prevent or mitigate a high cost failure. However, VM instrumentation has no value if not used. Typically these systems are expensive, costing about $200,000. Turbine ones cost up to $100,000. If you don't use the information, or plan to use it-why have it? Save the bucks and put them someplace else.

Instruments In Utility Cultures

Nuclear plants' operating requirements are less ambiguous than fossil. More of their world is prescribed-whether by the nuclear steam system supplier, NRC, INPO, American Nuclear Insurers (ANI), or others, almost ad infinitum. Fossil plants until very recently were at the other extreme-they could pick and choose their operating standards. To date fossil lacks consistent processes and guidelines that assure adherence to standards in practice.

Many instruments aren't reflected on fossil plants design drawings. Instruments run the gamut from irrelevant, unreliable to excellent test and monitoring equipment, from the trivial to most essential in on-line monitoring and controls. Yet, the operator is largely on his own to learn and discriminate instrument importance and consistency. This generates ambiguity.

Many utility cultures distrust instrumentation. They tolerate and support ambiguous operations. Instrumentation is considered inherently unreliable. Suppose that it is-what does this say about the instrument maintenance program, or the overall maintenance program? Instruments that cannot be maintained to performance levels required to leave them "armed" are frequently disarmed and used for status checks alone. They are jumpered, ignored, allowed to fail, not calibrated-and the operators using them cannot be held accountable for their actions because there are no firm guides or expectations on the instrument standards, functional performance levels, or maintenance programs. For engineering and I&C folks with no clear guidance on instruments' importance-they work on hundreds, if not thousands of instruments weekly-an instrument is solely an instrument. (Nuclear plants have very clear standards by contrast. "Mom," in the form of the NRC, tells them exactly what to do.)

Do you think ARCM will help this picture? You bet it will!

Critical instrumentation is so essential to reliable, safe operation that it behooves an operating company to understand the exact role of all instrumentation, and to treat it accordingly. This means that everyone with operating responsibility has to understand the processes, risks, failures, and monitoring methods. This is awareness simply not in many plants today. Outside pressure could force increased instrument aware-ness but I believe the marketplace is more efficient. Unreliable plants are uneconomic plants. Low cost producers deliver reliable operations first, and then predictable, controlled generation on demand. They will increase their market share.

Case Example

In 1998, a New Zealand transmission/distribution utility ran into the worst utility nightmare: an inability to supply loads to a major load center's customers. Most primary feeders to the downtown district in Auckland were lost. Personnel were unable to restore service in a time-ly manner. New overhead catenary had to be strung as an emergency measure, requiring approximately five weeks but ending the crisis. The event involved the predictable loss of two aging, deteriorated feeders and the additional loss of two more in rapid succession. A few facts bear scrutiny:

- The utility knew it had a load and cable problem
- The utility had attempted to obtain new underground cable access-without success
- Community support for the necessary construction and mainte-nance was lukewarm
- Management punted-failed to "go public" with the reality of the untenable situation
- The industry had recently "deregulated"
- The utility in question was a vertically integrated distribution utility in an unregulated generation environment (Institute of Electrical and Electronic Engineers (IEEE) "Spectrum," May 1998)
- The utility had made many employees redundant—what we Americans call "right-sizing"

There were other issues. When the final failures developed, the util-ity acted slowly to save their remaining good cables. The will within the company to challenge its organizational path was absent. (What occurred we call "group-think.")

In truth, it's tough to roll out these issues before a typical company's management—the "shoot the messenger" syndrome is alive and well. Those who can quantify and add substance to R issues go unrecognized in many companies.

Yet, standard pat approaches result in lukewarm results when innovation is necessary. At many plants, only the economic consequences of a plant being pulled from the rate base gets action. While residual transmission and distribution entities will be less affected and perhaps entirely untouched by deregulation, they shouldn't be from a performance viewpoint. With no nationwide network to "wheel" over, it's hard for me to see how deregulation will really work. Perhaps a franchise vote can award a system to a bidding operating entity that shows the most "R Skill". RCM principles apply to transmission and distribution also.

Criticism aside, there has been little economic pressure on regulated entities because the alternatives to the public aren't yet clear. I foresee a time in the future when the public is more aware of the nature and costs of unreliability and demand better performance from residually regulated entities and even the government! This may come solely through the market because R has tremendous value while unreliability has only cost.

Chapter 7
Fast Track

Conquest is easy. Control is not.

-James Kirk, "Mirror, Mirror," *Star Trek*

Traditional maintenance programs-those structured on a combination of PMs, overhauls, cals, and repairs-have been revolutionized by the advent of CMMSs. Maintenance planners have loaded activity identified by vendor O&M manuals and implemented them in the form of time-based activities. Where done completely-such as in nuclear power plants-the consequences were surprising.

It's always been difficult to perform all the work, work performance was often inefficient—sometimes very inefficient. Direct work request input and WO generation amplified trip and coordination time. Without a method to organize, consolidate, and perform work within the utility process environment, a complex, stagnant maintenance picture emerged. Schedulers and planners were added to cope with this workload-sometimes to little avail.

Companies with billions invested in facilities often formally budget little to maintenance plan development-even though production losses from load reductions, extended outages, and unit trips translate into millions in lost sales. This can be tolerated in the regulated environment with capital investment guaranteed returns. In competitive industries-refining, chemicals, fibers, process and manufacturing-maintenance losses cannot be tolerated.

Culturally, maintenance is an unglamorous backshop where status quo has been accepted. It hasn't achieved recognition as a strategic function supporting production. Traditional accounting treats maintenance as a variable cost of production. It is required simply to keep a facility in operating condition. But if investment in production is necessary to support revenues, then maintenance is a viable production investment. Industries squeezed by cash flow have tried to cut maintenance but have discovered their competitive position erodes as production capacity, services, and processes decline.

CMMS Strategy

To implement any maintenance plan, there must be a strategy developed on the plant's CMMS. Most CMMSs use a PM/CM work model even if they use an RCM maintenance philosophy. Work originates on a CMMS as a timed event or work request. Both are internally generated and correspond to routine (pre-scheduled, timed event) and response (requested, demand event) WOs. Pre-developed, pre-requested, timed WOs are called "PM." Everything else traditionally is "CM." This includes on-demand maintenance we prefer not to develop as routine, pre-scheduled work.

For instance, work is developed from scratch lists kept by engineers and planners and put into CMMSs several months, weeks, or even days prior to the scheduled outage as CM "demand" work requests. (Such lists may have been maintained for years in hard copy despite CMMS availability and capability.) This work looks, acts, and gets identified as CM when in fact, most outage work is preventive in nature. Equipment is operated into an outage; work is intentionally deferred into the scheduled work period. When outage lists are prepared as WOs at the last moment, potentially plannable/planned work becomes demand work, and the benefits of planning-standardization, coordination, repet-

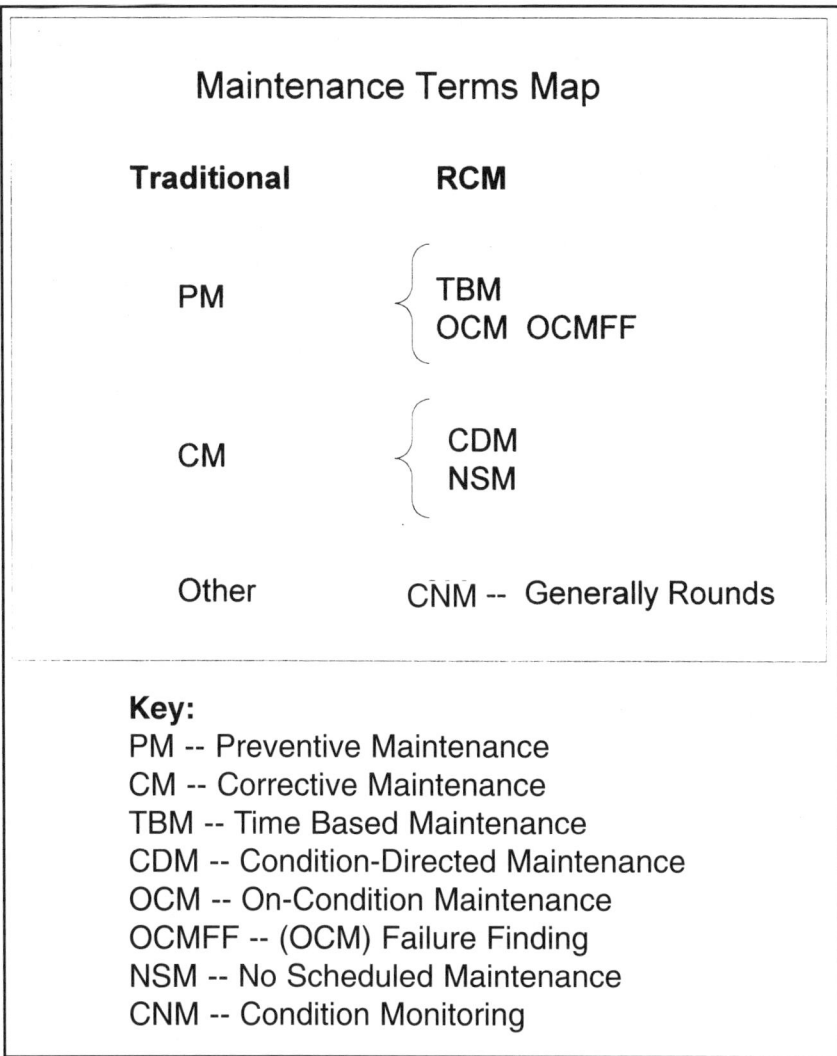

Maintenance Terms Map

Traditional   **RCM**

PM   { TBM
             OCM OCMFF

CM   { CDM
             NSM

Other   CNM -- Generally Rounds

Key:
PM -- Preventive Maintenance
CM -- Corrective Maintenance
TBM -- Time Based Maintenance
CDM -- Condition-Directed Maintenance
OCM -- On-Condition Maintenance
OCMFF -- (OCM) Failure Finding
NSM -- No Scheduled Maintenance
CNM -- Condition Monitoring

Figures 7-1: Maintenance Terms Map

itive performance, and preparation-are diminished. The CMMS strategy is to create as much known, knowable, schedulable, and organizable information as possible.

For fast-track RCM results, existing systems must be used to transition to a routine, planned work environment. To apply its concepts, we must understand and map RCM terms and interpretations onto the existing software and processes and then perform high value analysis focused on implementation. (Fig. 7-1) As this occurs, the organization develops a new maintenance paradigm focused on scheduled maintenance.

When an organization acquires another CMMS they are forced to adopt a new maintenance model. The selection of new CMMS software facilitates the transition. This time is an opportunity to introduce ARCM-based organization, planning, and scheduling methods.

Maintenance Infrastructure

Performing RCM requires a maintenance infrastructure, just as performing planned work does. Someone needs the skills, time, and commitment to do the work. The organization must have the confidence to use the results. Processes and systems grow slowly, with nurturing care. Even with focus, commitment, and expert help, learning is required. The work force grasps most RCM concepts quickly, once they perceive an organizational commitment to improve skills and manage costs. This is infrastructure and it takes a dedicated period to develop.

In some instances, infrastructure development requires new capabilities and measures. In others it requires processes-getting PM WO "change" control processes, and creating PM owner responsibility. Building infrastructure-awareness and sensitivity to a maintenance plan-requires time and nurturing.

Traditional PM Programs

Consider VM-a traditional PM program. Immediate payback comes from screening VM to limit and control scope. Only a few plant areas benefit cost-effectively from VM. Although this might at first seem like a complex task, it's by no means that hard- particularly with several benchmark VM cost/benefit studies. Developing and applying benchmark cases can quickly establish where VM will be beneficial. Using this template to quickly screen all equipment for VM can eliminate large amounts of non-productive effort for better PM paybacks elsewhere.

Other traditional PM programs can similarly be screened to get similar results. I&C programs that typically perform excessive low value equipment cals can use an ARCM filter to trim these cals to NSM. Equipment can be functionally abandoned when no value added results can be discerned to reduce daily work load. (Note that "functionally abandoned" means "no maintenance" period. This is much different than NSM.)

Substantial elimination of PM program activity can be achieved to be consistent with PMO reviews cited elsewhere. ARCM provides the value measure-every task must add value or get cut.

Managers may presume they have a PM program when in fact, they don't. Before anything else, a manager needs to evaluate the station's PM program state, acceptance, and commitment. This assessment is best done independently, and should include PM program actual performance measurement and capacity for measurement. Questions to ask include:

- Is a PM list maintained?
- What's the percentage completion rate of PMs on the list?
- Who gets PM completion rate reports?
- Who decides how to defer PMs?
- Is there engineering responsibility for PM selection?
- Is there an exception report for overdue PMs?
- How are PM priorities ranked with regard to other work?
- How are outage PMs maintained?
- What is the process to add or remove PMs from the list?
- Are the PM basic processes defined?
- Who is responsible to maintain the station's PM program?
- How does the PM program integrate (or fail to) with the CM program?

Many plants run random PM programs. That is, they have "laundry lists" of things to do as time becomes available. PM task selection, performance, and reporting are hit-or-miss. Unfortunately, it's virtually certain that plants with this PM approach will suffer R and availability losses. There is simply no credibility to this PM approach in a complex plant. Further, unless operations can provide a supporting rounds program, the culture simply doesn't give credence to PM. This program delivers a better result than nothing at all, but fails to engage the organization in the very real, exciting task of operating a plant to reduce unplanned events to virtually zero.

The adjunct to the plant's routine PM program is the plant's outage PM program. For failure prevention (and RCM) purposes they are one and the same. In plants operating with "Legacy" CMMS PM systems, it's common to find that outage work is maintained separately-even on scratch paper! Partly, this may be a CMMS design fault; partly, it's force of habit and a failure to implement an outage management PM program aspect on a software product. Outages are fleeting things-particularly in companies that experience high unplanned unit outage rates. It's the old shell game-unit X went down Monday, so defer unit Y's outage two weeks while X recovers. In the mean time Z goes down unexpectedly...and so forth. All these unscheduled scheduled outage changes mean plant schedulers can't plan for any outage window with any confidence. The lack of commitment to schedule can reach all the way to top management, where VPs juggle units outage schedules rather than bite the bullet and pay for replacement power. Nuclear units are spared because of the expense of jumping an outage around and because of NRC oversight. Getting all system generating plants onto a planned schedule is expensive-especially if unit R is low-but there is no other first step.

Corporate commitment to firm outage scheduling benefits plant outage schedulers. Outage PMs can be rescheduled, even with outage shifts, but, with a reasonable grace period—say, 25 to 33%, and, assuming application and implementation of CMMS methods—rescheduling is efficient. Having seen CMMS outage scheduling systems that are capable of this, my firm belief is that all PMs should be on the same database in the same computer. If current software doesn't allow this, find and buy some that does. It's available. Once this commitment is made, there's no excuse for missed outage PMs.

Scheduling

Once an "analysis" for scheduled maintenance is complete, the realities of making work happen takes over. Complex plants struggle to schedule necessary work by common agreement. PM programs are at greatest risk when successful scheduling methods cannot be implemented.

Consider how traditional PM programs were built. Someone-probably a maintenance manager-searched his facility's O&M manuals looking for vendor PM recommendations. Extracting these into lists, they built them into WOs-typically, one per WO-until the entire set of vendor recommendations was incorporated into the plant's CMMS. This approach featured individually applied tasks, without regard to value, service, risk, or organization. Is it the best approach?

Vendors recommendations represent a good stab at an initial program, but of necessity are greatly conservative. Elements typically missing in this approach include:

- group (operating team) assessment
- work packaging
- evaluation of the applicability and effectiveness for the recommendations
- absence of some "normalizing" routine with respect to the rest of plant equipment and their PM

This approach creates an unmanageable program-paperwork kicked out of the CMMS on a schedule that is:

- inefficiently planned
- hard to coordinate
- indeterminate or questionable value
- not ranked by some common value scale

With no planned approach to select and schedule PMs, is it any surprise that organizations schedule PMs on a lower priority than corrective maintenance, across the board? From an ARCM perspective, this is an unequal playing field. If scheduled maintenance tasks can be identified that are applicable technically, effective individually, and cost-effectively implemented—which, by the way, are key attributes of ARCM-based PMs—then PM performance should rank high on the priority scale. The cost/benefit values of these PMs are much higher than can be gained by fixing broken equipment. From a return-on-investment perspective, returns on the most effective PM tasks range from 50 to 200 times cost. Fixing broken equipment, on the other hand, has no improvement return. It merely restores status quo. From an investment perspective, then, which is more important - a 1/1, 5/1, or 50/1 benefit? Common plant priority systems are structured as appears below.

Priority	Meaning
E	Work immediately
1	Work next day
2	Work next week
3	Work when convenient

A value-based table would be inverted. Most PMs can be worked when convenient, *e.g.*, scheduled. They should also have the highest priority to complete as scheduled.

In the scheme above, failures get top billing, corrective maintenance comes next, and PMs occupy the tail end of the program. The irony is that the highest value work receives the most meager resources! In the absence of a developed system to level the playing field, priorities get skewed.

When RCM tasks are developed with the involvement and commitment of the owners, they realize value. Identified and selected this way, the priority assigned such tasks is higher than by traditional PM priority schemes. With correct ("normalizing") assessments, CDM can be ranked on the same scale with PMs and all activity value can be ranked on the same playing field as PMs. Uniform prioritization enables PM comparison and ranking with emerging CM work.

The traditional, indiscriminate incorporation of vendor manual recommendations only compounds the priority crisis. It leads to the realization that for some PMs, work had no value or was performed too frequently. The net effect is a discredited PM program, diminished commitment, and faint support. An ARCM-based PM task-selection methodology selects credible PM tasks up front to facilitate scheduling which permits less work and better commitment to selected work.

Traditional-environment PMs lack a clear tie between PMs and prevented failures. PMs focus on task performance without presenting the supporting "failure prevented value-case." That's often because questioning the basis for a task questions the competence of the entire maintenance organization. No one feels comfortable questioning the basis for the work done, so the same work continues to be done, year after year. This can lead to PMs performed on equipment abandoned in place, PMs performed that contribute to failures, and minor-value PMs performed with a work priority equal to major loss preventers. This is not good business!

ARCM instead ties the best task to one prevented failure. With this information-and the role of the failed equipment in the system-one can evaluate the consequences of the failure and assess the value of the PM activity. Each activity must stand on its own merit. This also rules out piggybacking non-value work onto an approved PM task.

Tagging "failure-prevented" to a PM helps prioritize PM and presents a set of strategy options for a given piece of equipment. For example, high-capacity rotary air compressors process sootblowing air for a coal-fired unit. A key to achieving design life is filtering particulate matter out of the incoming air stream. Compressors are installed with a filter bank of 160 individually replaceable canisters in the intake ducts. DP pressure instrumentation is installed on the blower intake head with a high DP trip for the blower motor. Filters can be replaced on time (TBM), provided filter pluggage has a time-aging characteristic. One could simply schedule a "hard-time" replacement task on either "runs hours" or for continuous loading (calendar time). Quarterly replacement is required in a moderately dusty environment. A filter set costs $500 (and time) so should be considered. This would check DP and replace elements based on the filters plugging. This latter strategy presumes:

- operator monitoring
- calibrated instrumentation
- maintenance work turnaround capacity

Item (3) requires that the organization process a OCM type WO and perform it consistently before differential pressure (DP) "pluggage" becomes a blocked filter, and functional failure. If maintenance can't deliver CDM with the necessary turnaround, and the compressor filters plug, a high DP can partially tear the fabric elements. Then, the torn filters appear the same as a new set-they pass air freely and the DP is within specs. Unfiltered air rapidly erodes the high-speed, finishing-stage blading and the compressors require premature overhaul. The obvious lesson is the KISS principle. i.e., a simple time-based task can be superior to the complex task-particularly if maintenance capacity to deliver CBM is in any doubt.

As more programs embrace the philosophy of RCM, but without the depth, the general push towards OCM will be interpreted as OCM is always better. This could add unnecessary complexity to programs. OCM/CDM presumes a sophisticated maintenance delivery system. Unless this is demonstrably available, an across-the-board push towards OCM should not be done.

Equipment priority systems are the first attempt to schedule day-to-day work. All one needs is a "relative importance code" for each piece of equipment in the plant CMMS. As WOs are generated, they can be automatically or manually sorted based on this overall priority assignment. As a first cut, this approach can be helpful, especially to a reviewer not familiar with the plant. It lacks dynamic capacity, however.

An effective program identifies degraded, not failed equipment, to provide maintenance planning with a "heads-up" about equipment that needs work. For outage work, it helps define the next "window" (Fig. 7-2). OCM identifies a work window in which we can maintain equipment prior to final failure, but doesn't guarantee work input to the CDM process on the right time schedule. To estimate a schedule window you need to understand the failure identified and the deterioration window before final failure. Many seasoned O&M people know many of these equipment windows for some equipment failures, but no one knows them all. There's also an element of chance that CM and CDM processes account for.

Some failures are best understood as a continuum that ranges from random to exact life at the extremes (Fig. 2-1). This model helps one

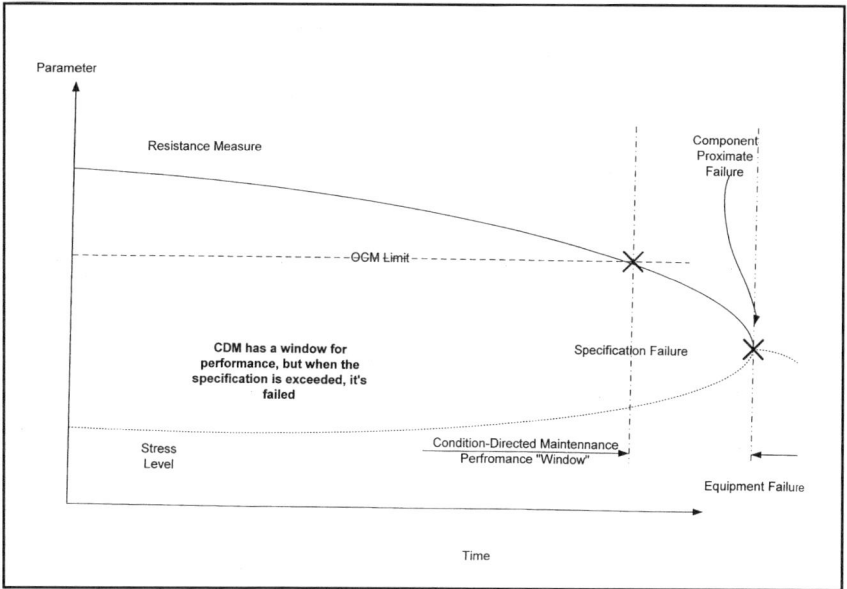

Figure 7-2: On-condition Maintenance Timing

realize that skid-mounted equipment has many sub-components, all of which have multiple monitoring tasks. To manage it well, we must understand and group components with a strategy. Experienced workers can establish a component's failure nature on this continuum. A single individual alone rarely does this well. Typically, failures have patterns and distributions. A broad, age-based failure distribution means the window for ultimate failure will be less certain. This is why we must estimate the window to prioritize each WO. Not all failures are suitable for OCM/COM stratagies. Only some exhibit a specific resistance limit, and failure window. RCM provides identification of the failure mode and identification of the failure nature-characteristic and uncertainty, associated with the mean. From these we can estimate the decision risk.

Failure modes and impacts set work priority. In a plant with two 50% boiler feedpumps, the loss of each pump causes 50% load loss. On the surface, assigning a feedpump problem a high-priority seems reasonable. However, 90% of boiler feedpump WO "failures" don't involve loss of function. Many are for instrumentation. Some identify operator confusion and misunderstandings. Some are for temporary problems at startup. Relatively few problems involve loss or direct risk to their function. Therefore, a CMMS approach that assigns WO priority solely based on equipment will not prioritize adequately. Few WOs warrant top priority but you can't even identify those without understanding the equipment failure mode, its specific risks to ultimate machine failure, and overall unit operations. This is the role operators fill.

People manage their cars intuitively-even non-maintenance people. The challenge is to transfer simple intuitive sense to complex power plants. Some prioritization systems don't support owner/operator needs. (There is a place for maintenance theory!) The sooner organizations recognize their need to "learn" maintenance, the better their maintenance returns will be.

RCM supports maintenance theory by providing on-the-job equipment failure training. One RCM-based equipment review consequence is a much greater understanding of equipment failure modes and risks based upon that facility's actual experience. This enhances intuitive prioritization-and better scheduling.

Scheduling Methods

Expedite

Traditional maintenance environments depend heavily on day-to-day expedited work. With greater coordination required by the NRC, nuclear generation has developed more routine scheduling methods. They also have more detailed PM programs, and generally, a much higher degree of PM program implementation.

An "operate to failure" perspective simplifies scheduling. Priorities are more obvious with failed equipment. Absence of daily and weekly generation "look-ahead" scheduling reflects this acceptance, and the inherent R of fossil designs. There are fewer imposed "engineering specification" failures to contend with (in contrast to nuclear). There is capacity in the traditional large generating station for operations with compromised equipment, too. As a choice, OTF offers greater opportunity to perform "real-time" maintenance on demand, as needed. So long as functional failures don't compromise production, and costs are managed, the option to use OTF is a powerful one. It also requires understanding failure context with maximum effectiveness. Indiscriminately applied, or applied by default, it leads to higher cost.

Short term (weekly)

Planned work is efficient work. Planned, efficient maintenance needs to be scheduled, simple, and standardized. CBM can virtually all be preplanned to allow workers and managers to anticipate and prepare. Weekly, look-ahead priorities should include:

- PMs (in cost/benefit ranked order)
- CBM (in priority order)
- broken equipment

Working ranked PM and CBM ahead of failed equipment means that overtime may be required for PM at reactive plants with many equipment breakdowns, or that failed equipment with no functional impacts may require operations tolerance. When ranked, the work backlog provides a rolling stack of work that can be reshuffled dynamically and scheduled to accommodate worker availability. This "stack" approach can be adjusted to reflect current priorities and needs, but once jobs are set, there should be an emphasis on avoiding schedule changes. This reflects both human and cost sensitivity awareness.

Searching for corrective and CBM in the work backlog must be understood in the value context. Operations plays a central role, but the organization needs standards to rank equipment failures. Once a system review is done-identifying key functions and operations importance-it becomes much easier to rank derived CBM and monitoring tasks. A few carefully chosen "benchmark" cases provide an anchor for comparing all work.

Ranking PM tasks can be reduced to identifying failure modes and rank.

Long term

Except for unit outages, most plants have short windows—and memories. A 12-week schedule fills the middle ground between the weekly look-ahead and the outage schedule.

The 12-week schedule was derived from surveillance tests at nuclear plants. (A colleague, Jon Anderson, developed and implemented the first 12-week schedules at Diablo Canyon with EPRI & PG&E.) It coordinates routine on-line work performance-time-based equipment change-outs, lubrications and monitoring inspections-that occur on a shorter-than-outage schedule. The 12-week schedule places routine work on a rotation (like an operations shift rotation) that comes around every 12 weeks. The important concept is the simplicity of the resulting schedule. Weeks, months, quarters, semi-annually and yearly activity support simple coordination.

The 12-week schedule provides an intermediate window for plant work unavailable in the past. This interim planning time frame suits many moderate-scope, CBM tasks. It supports work coordination and scheduling in "EGs" below the system level and so enables us to:

- manage performance risk
- improve work scheduling efficiency
- facilitate and simplify tag outs
- move "outage" work on-line
- improve overall plant R

Now CMMSs have scheduling and alignment capabilities that extend work alignment capacity substantially. Organizations with these systems should use them to maximum advantage.

Scheduling equipment groups (EGs)

My first experience with a plant PM program was deciphering a "canned" equipment maintenance program provided with a unit at start-up. Company staff, supplemented by contractors, came to the site for two years as startup proceeded. They read equipment vendor manuals, excerpted vendor PMs, and individually placed these into WOs on the vendor-recommended performance interval. The identification and performance of these start-up PMs wasn't comprehensive yet they generated some 600 PM tasks for a single 500 MW fossil unit. (These excluded lubrications, which were performed as separate "schedules") (Fig. 7-3).

As one learns quickly with the literal download/implementation of vendor PMs, many tasks are too frequent, usage assumptions are orders of magnitude off (on the "used" side) and equipment in routine service deteriorates much slower than you'd ever expect, unless you had measured it. Vendors specify common lubricants, for instance, whereas premium ones increase lifetimes. Vendor information and ARCM can jointly identify, select, and perform the right PMs-but we need both.

"Work grouping" should be intuitive to optimize performance. Minimizing overall work-the number of times an area must be entered, a tag out hung, an area cleaned-are examples of why project management is effective. Power plants often suffer the frustration of work not coordinated, equipment not cleared when crews are ready to work, or other crews working when job site space is limited. Work conflict is a very real and persistent problem. Re-entering areas for equipment "rework" is another time-wasting frustration. The fewer surprises, the fewer oversights, the fewer items falling in cracks, the less rework there is.

Organizations that measure rework are often surprised at its level as a percentage of the total. Studies I've developed and seen have indicated rework approaching 50% of all WO hours, at some plants. Unless it's measured, you'll never exactly know the loss. If we accept that rework is important to manage, then there's substantial opportunity to reduce this wasted effort.

EGs are very effective at doing this. An EG—a logical assembly of equipment identified and scheduled for work as a unit—vary in their group basis but commonly include:

- single tag out and return
- standard tag out boundaries
- one clearance (including draining or otherwise "prepping")
- one post-maintenance test and calibration for all work done
- multiple tasks performance while in the area
- coordinated scheduling of all items in the group to reduce volume of work items
- enabling of establishing standard PM work plans and schedule intervals for major trains
- coordination of work within the group
- coordination of "LCO" type license, safety, and insurance compensatory measures
- safety

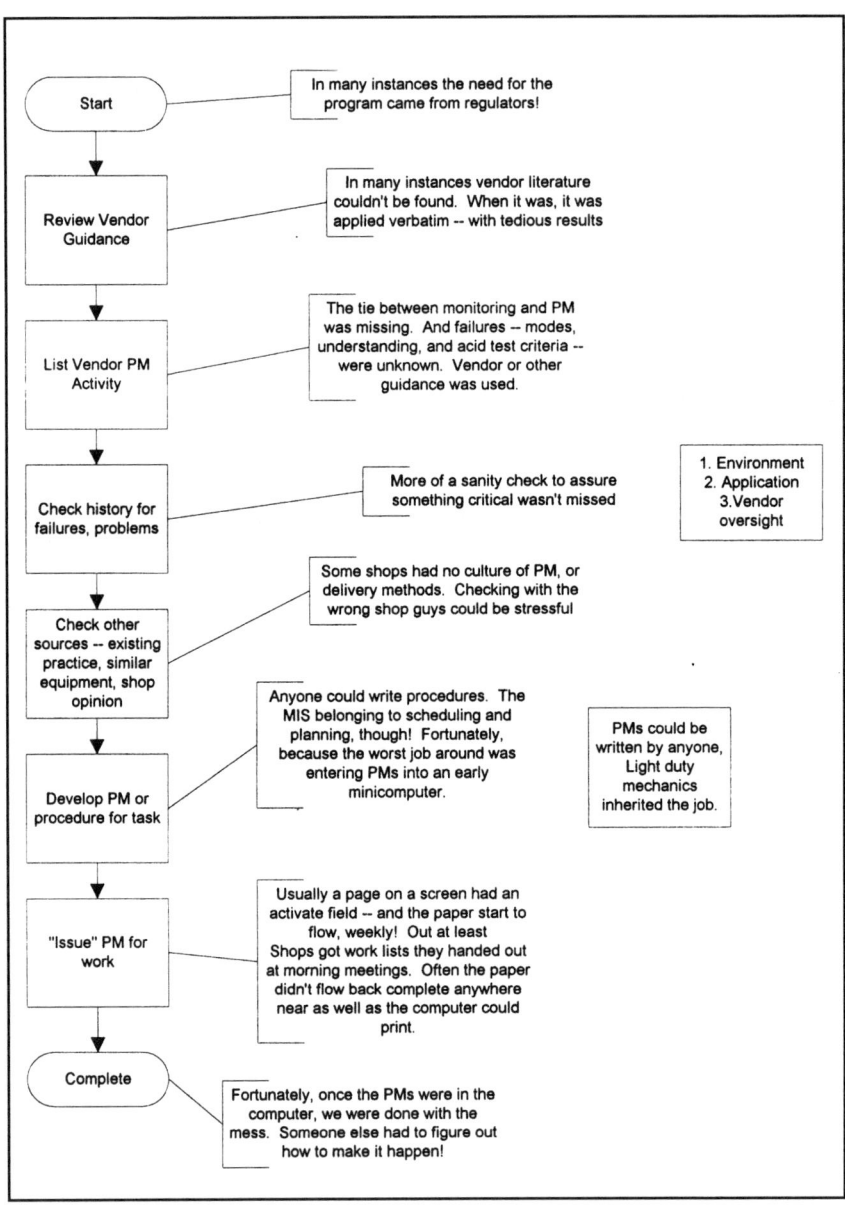

Figures 7-3: Traditional PM Development

EGs can be established in many ways, but must share common dynamics. First, they must exist for some logical purpose. While "logical" often means making all the equipment available for work under a common tag out boundary, this does not always have to be the case. Another use of an EG could be to associate many of the same equipment types into a round for common assessment under one trip. Another EG could be for common lubrications. Another could be for VM, or inspecting fire doors. There are as many group possibilities as possibilities to associate work. Newer CMMSs provide the ability to establish "parent-child" relationships, and therefore facility equipment work grouping.

EGs work best when developed as a joint operating agreement among all entities in the plant. They provide an agreed-upon standard for work performance—*i.e.*, EGs have greater value if operations can support them with a standard tag out boundary, and release the equipment for work based on the group. Any plant considering work performance at power (previously done only during scheduled outages), needs to absolutely minimize the risk of trips. This makes groups a powerful tool, especially as plants realize the value of doing more work on-line instead of in traditional outages. (OCM and CNM must be performed on-line.) Once a group is established, the risk of doing work on-line can be assessed, managed, and controlled with greater focus and certainty.

Groups fill the gap between systems and the component—*i.e.*, at the top we have hundreds, or even thousands of components per system, based upon functions and identified in the form of system drawings, descriptions, and component lists. Below the system level, there are often equipment trains, logical subsystems, and the unique organizational structures and work practices that require different groups. Groups can change dynamically over time as well. For example, an organization could move from hard time lubrications to OCM, and back to hard time for a given class of equipment, such as a coal belt (or eliminate them entirely with sealed bearings). Both types of monitoring can run concurrently. Groups simplify and standardize this practice.

Why not associate tasks within a given functional work area in a procedure and eliminate the EGs? This "fixed" grouping suffers from its frozen nature and relative difficulty in modifying procedures. Tasks are typically harder to internally reorganize. EGs retain work individually, and carry the flexibility to make new associations electronically.

They minimize the need to rework hard copy or text files, which makes them easier to use. Grouping and ARCM complement each other, since RCM identifies PM tasks to be done by functions, failures, and descriptions-all potential grouping attributes. As PM scope grows, coordination requirements are likewise greater, and this motivates grouping.

Groups also facilitate work planning. All planned work PMs and MWRs are identified and associated with a given EG by component number and coordinated for most efficient performance. Once a group window is established on the rolling 12-week schedule, time can be allotted for work based upon experience, risk, and scheduled work time. Groups can tag many CBM tasks onto one activity to improve the capability to schedule and work CBM in a controlled way.

Outage

Outage planning and scheduling benefits from the rigor and simplification introduced by ARCM methods. Outages affect unit availability and constitute the most expensive maintenance budget period. RCM reviews simplify and standardize outage work to minimize them and maximize benefits.

Getting outage workscopes, formally reviewing them for applicability, effectiveness, and cost/benefit value, greatly benefits outage scope and budget management. For units that maintain extensive outage work scopes on a routine basis, the RCM screen is virtually identical to that performed for existing PM programs. For those that develop scopes just prior to "coming down," there is opportunity to cut scope. Given that most outages slightly to moderately run over scheduled durations (based on my experience), there's great opportunity to achieve substantial returns with unit outage scope reviews.

Outages are sometimes only partially planned. Consistency and predictability of outage workloads comes with thorough, failure-based work review for value.

Project Management Techniques

A project approach helps to implement and achieve quick RCM benefits. RCM projects are "soft" programs. Support groups downplay commitments and schedules at times. Engineering must provide analytical parts life-extension and aging studies to support intervals, EGs need development and coordinated implementation. CMMS information must be downloaded-then uploaded. Without a project focus it's extremely difficult to identify problems, measure progress, and achieve success.

RCM project management is harder than many realize. It takes effort and special skills to create a core team, develop PM standards for the common equipment, refine processes and methods, and plod through the 20 to 40 systems that are typically reviewed in a major effort. Where an RCM project approach has been used, it's effective moving analysis throughout the scope of the project. Where ignored, projects become stalled, interest wanes, and results are imperiled. It's also possible to get cold feet. Groups tend to backslide on extended intervals, and there's a pre-disposition to adjust all outage PM frequencies to the lowest common denominator-the annual (or 18 month) outage.

Working to schedules

The great challenge for everyone is working to schedules. As nuclear plants discovered, adding schedulers doesn't guarantee success without somehow structurally simplifying the work. Operating groups re-schedule easily to support the operating plan, but cause severe implications for work managers. An eleventh hour operations outage housecleaning designed to maintain a schedule led to the elimination of nearly 100 PMs on one job. The limited availability of work windows, combined with the long duration between plant outages, forced a significant amount of PM work into grace periods and mandated schedule realignment. This short-perspective schedule change hurts long-term operating goals.

Once PMs are set up and the schedule is aligned, seemingly superficial changes can have long term consequences. Where PM programs are mandatory, this can force unscheduled plant outages. Accepting PM as a priority is a difficult lesson for any organization. Operations, discovering that they incurred an unplanned plant outage by "cut-and-slash" deferral of scheduled PM work, learns a sharp lesson. As workscopes shift towards a maintenance strategy, schedules become increasingly important.

Consider a plant that has 100% planned, scheduled work and "man-loads" to that schedule. There is really no latitude for change. Any rescheduled work will disrupt the schedule to some degree. For this reason, an eleventh hour PM cleanup/defer effort is likely to lead to future chaos. Clearly, the simpler the initial scheduled maintenance effort the lessor the temptation to defer scheduled PMs.

Overhaul Intervals

Basis for overhauls

Overhauls associate individual, time-based rework tasks around a common disassembly activity. Because much time can be invested in disassembling and reassembling a large machine, as much work as possible is performed with any disassembly task. The consequence of any single component wearing out before the next scheduled overhaul period is so great, it's considered cost-effective to simply replace all components, regardless of condition or cost. This is the basic overall strategy. Take it apart, replace as much as possible, and button it up.

Many traditional mechanics-as well as instrument technicians and others-follow an overhaul work philosophy. This has two disadvantages. First, costs are higher than necessary when serviceable parts are replaced. Second, by examining parts, workers learn to support a plant- or company-wide age exploration program. Without examining parts and asking the serviceability question, R engineers never get feedback on inservice performance unrelated to failures, and lose the opportunity to pursue systematic life extension based upon age exploration.

Overhaul intervals are typically based on accepted standards that represent a composite wearout picture for many components. Once established, these intervals have gone unchallenged for long periods. Only the recent drive for cost-competitiveness, and the demonstration by a few IPPs that overhaul envelopes can be stretched, has changed this perspective. Unfortunately, executive committees are too often the ones setting new turbine and boiler overhaul intervals in almost complete absence of field engineering information on equipment capability.

With today's emphasis on CNM and life-extension, plants are extending intervals, supplementing known aging problems with specific monitoring to assess the development of specific problems. Bearing wear, for example, can be monitored by a combination of oil monitoring for babbit and particulate products, visual bearing inspections (through removable caps), VM and trending, and *in situ* dimensional-wear measurement. The combination is as effective as a physical bearing examination for diagnostics. We can infer about all there is to know from these tests. Bearings can also be individually removed and inspected during light outages, of course.

Blade assessment can likewise be predicted from stage efficiency and overall performance tests. Borescope examination (on newer designs with removable ports) is also an option. But the assessment strives to extend intervals by using aging experience and intelligence to perform secondary CDM that is sensitive to real life problems and possibilities.

Optimizing strategy

Many secondary considerations go into the scheduling of a heavy outage, such as a turbine. These include availability of other units, overall load, R, scheduling of replacement power and services, and value of the deferral in present value terms. With the recognition that nominal outage intervals may have been conservative, methods to extend outage intervals (while managing risk) are considered. Methods using conditional probability have been available for years.

From an RCM perspective, a single great potential savings comes from the systematic examination of risk that comes from incrementally extending an outage period from a known benchmark. The five-year turbine standard was considered reasonably safe but companies are shedding the known safety of this interval to take overhauls out to seven, nine, and even longer nominal intervals. As they do this, they seek to manage their risk with increased use of CNM. Extending large machine outage intervals systematically is an obvious RCM capability.

Planning

Planned work. Efficient preplanning requires that work be anticipated-either because it gets performed over and over (like PM inspections) or because equipment failure modes follow statistical patterns.

For example, limit switches get loose and sticky in certain environments. Even in clean environments, the contact may oxidize. If your plant's primary experience with limit switches is that they come loose and need to be reset, that job can be anticipated and preplanned. That standard job should be planned and ready to go on a moment's notice.

A job within the "skill of the craft" requires a job plan-just not something on paper. (The shop practice and methods guide is the generic standard plan.) The planner could use a preplanned job, the standard, or none, at his discretion. But having a standard written job plan facilitates training, establishes a standard, and enables learning and revision as methods change. With electronic CMMSs, maintaining standard job plans is as simple as Windows cut and paste capability and offers the opportunity to use standard electronic plans.

Preplanned corrective maintenance. Planners and others sometimes conclude that because a failure mode occurs randomly, the work can't be preplanned. If the failure mode is predictable, and recurring, the job can be preplanned. Using the 80/20 rule, all high-frequency CM WOs can be preplanned and filed away (electronically) for immediate recall. This makes the electrician who gets called in on the backshift for an unpleasant job a little happier-he doesn't have to wait for the job plan once he gets in or plan it himself-on the fly! He has something to work from. Maintenance gets more consistent performance. When these simple aids were made available workers found them useful.

Problems: costs and rank

PM priority. PM tasks that an organization performs, properly selected, are its most important work. Based on the analysis of many failures, it's effective to establish a PM ranking system that uses three criteria.

At the top of the priority list are those PM tasks required by law and safety. Generally, these laws—like boiler codes—were put into effect in response to past disasters. Performing these are a good business practice. Very often, insurance endorsements require them. Insurance commitments maintain basic facility safety and preserve equipment from major loss. Very often this equipment work fills the same role as "critical" instrumentation. It reduces or eliminates the risk of unacceptable failures.

Environmental compliance requirements fit here. At face value, they are in the public interest. There is simply too much at risk in terms of perception and penalty to be less than completely in compliance. While some regulations can be gamed, it's not a good risk. PMs addressing failure modes that would otherwise violate these standards are the highest priorities to work. If they aren't done, it's guaranteed that an emergency results upon discovery-for someone. Examples include:

- boiler code safeties: lift-off tests
- stack lights: aeronautical safety
- code boiler inspections: boiler safety
- overspeed trip devices: turbine safety
- CEM equipment: environmental compliance

Plant outages are the second level of PM. Any PMs that directly prevent plant outage fit this criteria. Boiler tube inspections, condenser tube inspections, boiler chemistry monitoring, DCS two-channel backup configurations checks, boiler camera checks, and main steam safeties liftoff tests (where these are split, say, 3/35% for 105% total relief capacity) are examples.

These tasks assure key redundancies, backup equipment, and/or other capabilities are present. If we lose these devices or equipment, and anything else happens, we go down. Plant DCS displays are another example. By themselves, operator display consoles to the DCS control provide instrumentation. A plant can (and has) continued to operate with no active display monitors. (It's never supposed to happen, but it has—at least once!) If anything else goes wrong, the plant probably goes down because we can't respond. Redundancies for critical instruments also fit this category. "Important" equipment goes here. "Power pops" that protect code safeties fit here also.

At the third level is PM for purely economical reasons. This includes PMs that avoid reactive maintenance or large equipment replacement costs. The traditional work hours and materials B/C PM slides in here. There's no production impact at this level, but work tasks at this level are not all equal.

For favorable PMs, there are safety, production or B/C ratios greater than one. Benefit-to-cost ratios greater than one includes everything from 1.1 (marginal PMs) to 1000/1-real benefits. Total performance costs range from a few to hundreds of thousands of dollars. We want to go after high value, big ticket benefits first. We want to achieve the combination of high B/C and total payback. Unless we rank the PMs, we could be lubing a pump-well motor for a few bucks while a sootblowing air compressor filter clogs up and tears out! The former costs $20 and has a marginal benefit-the pump will still last 90% of design life without its PM. The latter costs $200-500 but has B/C of well more than 100, and close to 1000 for larger compressors. So, our PM system must identify a B/C of 1000/1 and cost of $500 (total value $500,000) over the B/C of 1.1 and cost of $20 (total value of $2). Most traditional systems can't make this distinction, and a traditional scheduler sees two hours time for either task as equal.

Standards

Every facility has highly repetitive maintenance tasks, either because of the number of identical components, or its repetitive maintenance nature. Developing standardized methods to perform work improves maintenance efficiency. Work standards should include planned NSM/OTF and CBM jobs, in addition to traditional time-based PMs. However, convincing craft people that they'll benefit from trouble shooting procedures and diagnostic guides is a challenge. Once they've developed them, they support their use. Nuclear units have procedures that provide a high degree of work conformity. Even fossil plant checklists and guides standardize work uniformity, consistency, and performance time.

Establishing maintenance standards that address common equipment classes is a preliminary step to build maintenance programs. For a fossil plant with 20 coal belts, the major components—gearboxes, take-ups, belts, drive motor, and soft start "gyrol"—have nearly the same needs. Likewise, a nuclear unit with 200 Limitorque motor operated valves (MOVs), needs an MOV standard as the first step towards overall work rationalization. A standard will need "tuning" for details, such as environmental conditions, equipment class, importance and usage, but the standard is a start. A failure review of each type ultimately identifies specific performance issues. A maintenance plan standard establishes an efficient, relevant method to examine plant equipment needs at the "big picture" level.

Standards take many conflicting issues into account. At the equipment level, in a single plant failures and wear are similar. Environment and usage factors will emphasize some failure modes while suppressing others. Yet, usage and environment will be similar within a plant. These influence failure modes.

The maintenance standard summarizes experience and establishes a plant baseline-and they can be revised with lessons learned, at any time. Standards provide a foundation for maintenance checkout lists or procedures, accommodating different requirements or classes of equipment. Taken together, maintenance standards provide the basis for a plant's planned maintenance program.

PM Reviews

PM backlog review

All plants must carry backlogged work. Very often, this is a large and mostly inactive file. WOs that have been on the list for more than a year, for example, will never be worked unless a change occurs. A quick way to establish work value is to review and screen backlogged WOs—both PM and CM—to identify the high value work. This requires equipment familiarity—knowing failure modes, the manufacturer's guidance, and industry standards. When performed by an experienced analyst, the plant can eliminate low value WOs and retain valuable ones (Fig. 7-4).

Large backlogs may mask high-risk work that fell into a crack—for example, feedwater heater tube inspections that were skipped, or missed lubrications, filters, and inspections. A CMMS review can simplify backlogs while pulling out high value work. A R engineer regularly reviewing the lists can keep backlogs short.

Reviews divide corrective maintenance into CBM and failure maintenance. The difference depends on whether an in-service failure occurs. Failures occur in all plans—even in RCM-based plans. CBM (originating mainly from operator monitoring) fails when operations isn't a full partner. Operators who monitor and manage a work backlog of CMs and PMs can recognize and extract buried important backlog activities with immediate R and cost benefits.

The review and purging of a work backlog is, in every way, like the ongoing review of problems for discrimination into CDM (FF) and "other." Most "other" is discretionary, can't have a defined scope, or needs further assessment to define the equipment state. In the absence of standards, a credible expert must perform the assessment and make the "go/no go" call. Ironically, my experience has been that the tough sells for the majority of the backlogged WOs is getting people to give up WOs that aren't ready for work! Of course, having an organization tuned to this philosophy (via ARCM) makes it easier to set discrimination levels for work and work on failed equipment.

Completed RCM-based equipment reviews, based upon standards for common, high value equipment, provide a ready reference to rank equipment value. RCM failure modes document ways in which items fail, at what relative frequency, with what equipment impact, and failure identifying method(s). This provides a tool for operator, electrician, and mechanic training and serves as the best way to get everyone on "the same page." Operators can document known failures in standard terms that O&M groups understand. When operations participates in maintenance, they improve prioritization of work resources. Maintenance is more responsive to the plant's needs.

PM list

Plants start up with OEM-based PM worklists based on specific equipment preservation. Many vendor plans assume a continuous service operating assessment. Most plant equipment is not in continuous service and sees far fewer operating cycles than estimated in vendor manuals. Adjusting vendor recommendations for these operating differences generates the first large reduction in vendor-based PMs.

Other enhancements can extend vendor intervals for continuous or difficult-to-service equipment—less frequent filter changes or lubrications, higher quality parts, lubricants, and filters, and minor modifications to improve service. These adjustments are fundamental for a

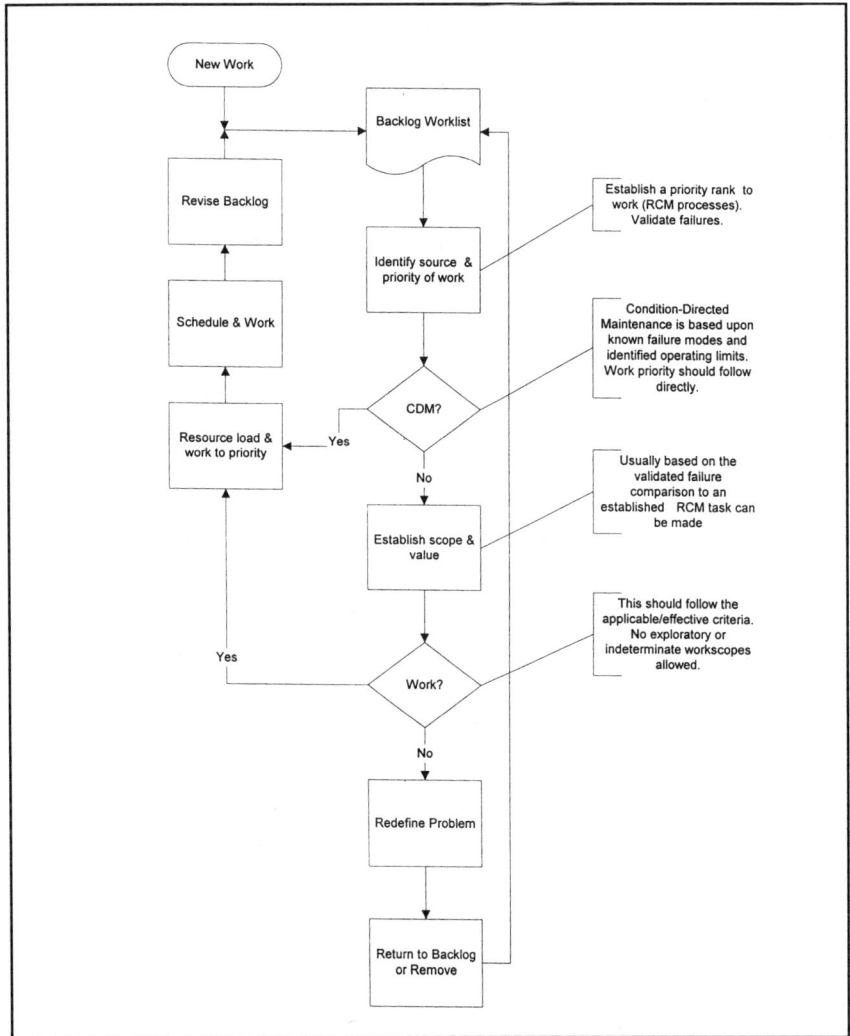

Figure 7-4: Backlog Maintenance

competent R or plant engineer. Vendors are great starting points for maintenance plans—but only starting points.

Comparing RCM- and OEM-based PM programs illustrates how RCM "screens" are effective at reducing PM hours. An immediate RCM return comes from existing PM program review using experienced maintenance engineers. Thereafter, direct vendor contact and cooperative efforts improve the use and application of vendor products. Problems should be reviewed with vendor engineers and regular contact provides insights into how vendors rate the performance effectiveness of various checks and inspections, as well as service intervals.

A surprising benefit is how control over maintenance improves when the informal maintenance plans are put on a firm basis. This establishes a baseline for age exploration and further improvements.

CMMS work backlogs

All plants suffer from WO backlogs and all WO systems operate with "in-process" work. This is a fundamental rule of differential calculus. The question is how much is right?

Absolute numbers for work backlogs (or "work in process") don't mean a lot. As we have seen, WOs can be associated, grouped, and processed in many different ways that affect the numbers. Some plants accomplish major tasks-such as entire turbine disassembly/reassembly with a single WO, and do so effectively. Others might break this job out into hundreds of tasks. There are many ways to "kluge up" a WO system. Failing to screen, manage, or group work is one! Planned and PM maintenance WOs introduce the greatest numbers problems. This is why group strategies are important.

The default standard is usually one WO per identified problem. The number of routine "demanded" WOs gives an idea of the amount of emerging maintenance work. Here, the question of WOs no longer seems academic, because one realizes a consequence of more is potentially a lot more WOs. More WOs mean more planning review, more backlog (numerically), and potentially, from a regulatory enforcement perspective, more potential of identification as having an "inadequate" maintenance program to meet the challenge.

Practically, work-in-process for degrading, but not yet inoperable equipment is necessary and ideal. With effective prioritization and lead-time to failure-such as provided by advanced monitoring technologies-there's more time to manage, control, and correct equipment degradation. The CBM backlog can rightfully be viewed as a gold mine of opportunity. The problem has been inadequate methods to screen and prioritize work. Instead, oversight group standards have drawn arbitrary absolute lines to suggest where the backlog is "too much."

With basic RCM theory and streamlined ARCM application, operating groups acquire a powerful tool to manage backlog. The RCM screen can quickly reject irrelevant and exploratory WOs and WOs where someone "thinks" there "may be" a problem. There's no basis for any WO until a problem is established. The "I think," "maybe," "please check it out..." directives represent someone dumping work onto someone else that they're unwilling or incapable of doing.

OCM gives the plant more "head's up" about work coming down the pike. More emphasis on CNM will place more work in process-logically. Mathematically, by processing more work (all else being the same) a new equilibrium backlog level will be established-higher than before. This will be one outcome of implementing RCM. This is almost certain to be viewed negatively.

Why? For maintenance, backlog means continuous maintenance "production." (This presumes that the work can be fed into the maintenance process systematically.) The traditional maintenance problem is too much "good" work to do and too little time and too few resources available to do it. It's also much easier to create WOs than to close them. Uncontrolled, unscreened, and unprioritized work requests can flood a system with requests that divert valuable resources to less-important work. A WO system with effective screening and prioritization, reduces absolute numbers of WOs going into the system, allowing more "good" work.

One spin-off benefit of the formal RCM review of failure modes and equipment impacts is the potential to simplify and standardize plant work prioritization. Most "priority" assignments are superficially based upon the equipment primary function and not on specific failure mode importance. When frustrated operators fight degrading equipment they over-emphasize their problems. Pre-assigned priorities by failures classes, based upon importance, put equipment failure MWRs on common ground. This comes through a systematic review of the system equipment using a FMEA-like assessment, which is built into the formal RCM model.

Backlog review and prioritization is an excellent way to start applying RCM concepts in the very environments that will most benefit from

RCM. Significant amounts of backlogged work may be on hold, awaiting failure, and nothing's more frustrating than to see a WO languish in a backlog while the item fails in the interim.

Equipment that fails with an outstanding MWR is a maintenance process effectiveness measure. Equipment, with a wearout period that fails with no MWR, likewise measures the health of the process and provides an effective measure of PM program performance!

Outage Work Review

Outages are major plant work. Their impact on overall plant availability and cost means that outage performance is critical.

Outages need to be planned around specific periods and staffs need to hold firm to schedules. With costs of replacement power running at two or more times the generated cost, and the cash flow penalties, there are great consequences in outages overrunning schedules. Predictable factors identify whether or not an outage schedule is realistic—a firm schedule, project management, and a realistic assessment of the work scope. Another is how the plant manages emerging work and the level of contingency (back-up) resources.

Of all these problems facing outages, the biggest is managing the scope of the work, before and during the project. Whenever you go into equipment, you discover "things"—feedwater heaters with tubesheet cracks, a deaerator with severe corrosion—things you don't expect but are predictable. Many are only interpreted as problems during inspection. Much outage work need not be done, but gets caught up in the zest for doing work. Despite the excitement, schedule driving, and momentum that accompanies an outage, however, costs are high. Plants tend to discount technical advice in outages in their zest to do work. If plants chronically disregard their technical advice, the solution is simple-let the staff engineers go. They add no value.

RCM-based outage work screens can reduce that non-value work. Reported results are favorable-up to 40% reductions in outage scope, based upon work hours. RCM outage work screens—which "pass" work based upon identifying an explicit "failure prevented"—control pre-outage work selection and work-in-progress scope. They also put

staff engineering to work on real problems to derive real value.
Finally, consider the growth of "emerging" work outage workscope.
The outage manager must manage it, using benchmarks, based on types
of plants, operating service, and other criteria. As outage performance
falls outside the benchmark, the effectiveness of emerging work control
needs to be reviewed. A substantial amount of unpredicted, unplanned
work means either the plant's program and previous outage "work done"
summaries are incomplete, or the plant's outage performance is ineffective. In either case, there's cause for concern and further investigation.

Event analysis

The best time to perform RCM failure mode identification and
analysis in an operating plant is concurrent with any major failure. Staff
is analyzing the failure; everyone's energized and focused. Now is the
time to capture the lessons for the future. I've found that RCM analysis, done concurrently in computer database format, can help focus the
failure event analysis on facts, as well as document other potential hypothetical and real failure modes discovered during investigation. It's
learning that will carry into the future.

One of the frustrations in determining root cause failure is the presumption of a single failure mode. Root cause failure analysis (RCFA)
doesn't work well for truly complex, synergistic failures-failures with
complex physical, and perhaps even organizational interactions. These
types can be decomposed to be represented as "independent failures"
on an Ishikawa drawing. Interactions, for which you lack adequate
information to determine the cause that developed into failure, are irrelevant in RCM-type process improvement analysis. The focus is learning all the mechanisms that could have lead to failure (and if you find it,
so much the better!). You need to separately address each independent
failure in your prevention strategy. In this regard, an RCM review of an
event can be more proactive and less fault-finding in nature than traditional failure analysis. This is exactly why Ishikawa diagrams help to
understand failure patterns-they seek not only the exclusive cause of a
particular event, but demonstrate the interrelationships that can lead to
failure. With this, and with process understanding, frequencies of
occurrence can be measured and action can be adjusted based on risk,

probability, and consequence.

Daily, a concurrent "significant failure" review contributes another benefit. The plant's engineer can evaluate and assess the most troublesome failures in the plant's daily meeting. In this manner, all "important" equipment in the plant will get at least a cursory SRCM review, and the plant engineer will become thoroughly versed in RCM thinking. This sort of review is enhanced by simple user-friendly software to document failures. Software, using the plant's CMMS database, can quickly create a comprehensive list of the failures that matter.

Parts and Outages

During the years, through many outages, I've observed many different parts applications in many different contexts. The ones that "stick" are the problems. In many cases, we couldn't get the desired parts on short notice and we took substitutes—sometimes "exact" aftermarket sales equivalents. Other times, we reworked parts. In still other cases, different parts were substituted. Results were not consistent! Outages and other parts-driving events are often the result in crisis procurements and so we took whatever we could get—including substandard parts.

A high cost item in any generation environment is engineering redesign—for any reason. My part substitution experience is that many engineering costs are carried off the plant's direct budget, and therefore unseen—except as a general corporate overhead. Trust me, this is expensive! It takes time to assess and reassign part functionality and specification. Even when the work is relatively straightforward, the time spent redesigning is considerable, and requires familiarity with the equipment in question to assess the true "equivalence" of the part.

Plants cannot always get equivalent parts. Even when they do, what staff remembers about their service lifetimes, they carry around in their heads. However, every part has a characteristic failure curve when used in nominal service. If we statistically group and evaluate them, we derive mean life and failure standard deviation (assuming a bell curve). What is nominal service? How legitimate is a normal distribution? (Short lifetime parts turn out to be very unpredictable, even if they fit the "error" in the curve.) In practice, with such uncertainty (and with generation riding on parts decisions) we must justify the cost of carrying excessive stock, simply because they will probably be used. Slow moving stock can easily be evaluated and parts usage improved with a joint failure and parts risk control strategy.

Focused strategy development

When a maintenance emergency develops, ARCM is an effective tool to evaluate and put the threat into a workable context. New regulations, regulatory violations, fines, or an operations upset pose challenges to continued plant operations. Likewise, near-misses due to equipment failures can be evaluated by ARCM. The potential for recurrence, need for new maintenance strategies or perhaps even redesign can be established.

RCM supports FMECA and often resolves fuzzy issues that otherwise could go beyond bounds. PM program additions can also often be evaluated in context with ARCM statistics and analysis.

Equipment Groups

EGs provide a method to make PM task performance efficient. The concept derives from "work blocking" by Nolan & Heap, and is especially critical when trip time may be a significant portion of the total job time, or a significant effort must be made to make equipment available. For a plant, it could mean the time required to tag out and return equipment to service. When equipment is available, all potential required work ideally will be performed. In nuclear applications, the NRC maintenance rule requires plants to monitor the unavailability of risk-significant systems and minimize unnecessary work. This impetus has always had value (Fig. 7-5).

The practical utility of developing and implementing EGs with a CMMS is that when a group is scheduled down, you do all planned work in the group. This requires a process of aligning all PMs in the group in such a way that they occur in the scheduled downtime slots. The other CMMS benefit is that when a group must come down, all the backlogged work for that group is immediately retrievable (by sort) for quick assessment. This greatly simplifies the job of the workscope

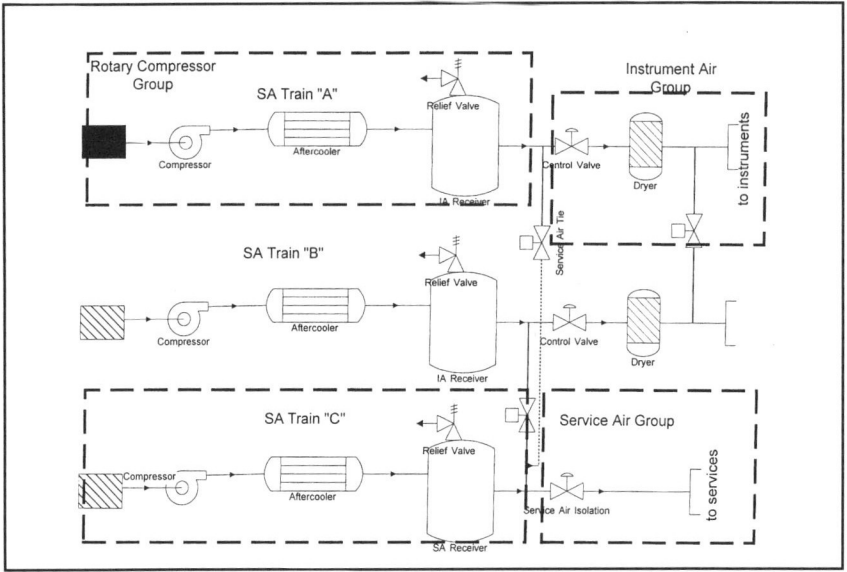

Figure 7-5: Plant Instrument Air/Service Air Equipment Group

planner and outage scheduler. EGs make it far less likely that an item will fall into a crack and be lost. The greatest benefit is work perform-ance consistency.

An operator's job includes large amounts of travel between relative-ly short periods of equipment monitoring. As much as 60% of an oper-ator's time may involve travel. Likewise, when maintenance crews must spend a large fraction of their work time getting to and returning from the job site, they need to coordinate trips for maximum time utilization. (In practice, when work isn't effectively blocked, feedback from crews is usu-ally quick and critical. The tragedy is that this doesn't always become incorporated into work as improved job planning.) PM tasks, like proj-ect activity blocks, are most effective when thought out and coordinated. The concept is like taking your car to the garage: ideally, you identify all the work and get everything taken care of with one trip. Obviously, power plants are much more complex than cars but the principle's the same (Fig. 7-6).

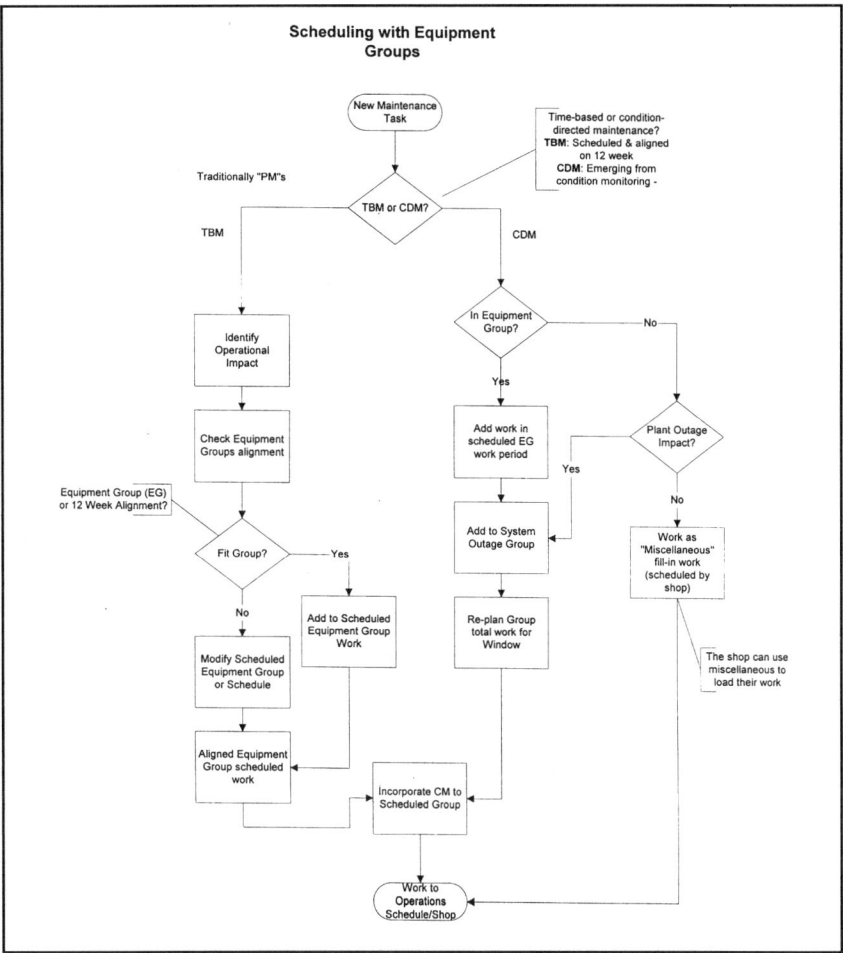

Figure 7-6 Modified Work Control Process Flow

EGs therefore, allow:

- efficient use of resources
- coordination of work
- control of work risk (especially for on-line work)
- combination of work when equipment is available

What organizations can fail to realize is that in production, most work is repetitive. *i.e.*, after any plant has been in service several years, very few work tasks are new. The bread and butter work of the organization is repetitive. Because of this, it's highly effective to spend a great deal of effort to plan and coordinate the repetitive tasks. This means EGs and blocking of some sort.

The more planned maintenance an organization works, the more important EGs become. In a purely reactive maintenance environment, things break—as they break, they come down. All that's minimally necessary is to quickly get them back into service. As a transition to planned maintenance occurs, more work originates as:

• TBM WOs and tasks
• OCM derived work tasks
• CNM/OTF operator-identified degradation work tasks

With a fully developed and effective PM program, fewer things break. With close operations-maintenance coordination (to get access to equipment to perform work), and by blocking equipment trains and components into EGs, plants greatly simplify the performance and scheduling of work. They conserve and simplify operation tag outs and standardize work.

If task blocking is so effective, why do so few organizations use it? I believe this comes back to the unanalyzed state of maintenance at too many plants. Few managers have backgrounds necessary to develop complete maintenance strategies. They lack the time to understand the detailed operating and maintenance requirements that suggest the best maintenance (and operating) strategies. Few have the maintenance support engineers to assist them. They don't always grasp the essentially repetitive nature of the work, and the need to standardize and streamline performance. Maintenance is widely regarded (with great justification) as a highly crafted skill, but that doesn't mean that every job has to be developed new, from scratch. The essential lack of maintenance processes perspective leads to this acceptance. In addition, some plants work under an outage mentality. They save all sorts of major work for outage periods when resources are more available, fewer questions are asked, and then just work like crazy for the entire outage. Where applied comprehensively, EGs can strip away up to 40% of the hours that are added onto outages and remove confusion about what outage maintenance work really is.

Residual Heat Removal
Equipment Groups - Important CIC (Identified)

	A	B	C	D
1	CIC	EG	DESCRIPTION	LOCATION
92	EE-REL-RHRP1C(52AX)	RHR-P-C	BREAKER RHRP1C AUX RELAY	
93	EE-REL-RHRP1D(30)	RHR-P-D	BREAKER RHRP1D TRIP ALARM RELAY	
94	EE-REL-RHRP1D(50-50-51)A	RHR-P-D	BREAKER RHRP1D TIME OVERCURRENT RELAY PH-A	
95	EE-REL-RHRP1D(50-50-51)C	RHR-P-D	BREAKER RHRP1D TIME OVERCURRENT RELAY PH-C	
96	EE-REL-RHRP1D(50G)	RHR-P-D	BREAKER RHRP1D GROUND RELAY	
97	EE-REL-RHRP1D(51X)	RHR-P-D	BREAKER RHRP1D OVERCURRENT/GROUND AUX RELAY	
98	EE-STR-(RHR-MO67)	RHR-RAD-B	STARTER FOR RHR-MO67	
99	EE-STR-250DIV1(MO25A)	RHR-P-A	STARTER FOR RHR-MO25A	
100	EE-STR-250DIV2(MO25B)	RHR-P-B	STARTER FOR RHR-MO25B	
101	EE-STR-250HPCI(RHR-MO17)	RHR-SDC-B	STARTER FOR RHR-MO17	
102	EE-TMR-RHRP1A	RHR-P-A	ELAPSED TIME METER FOR RHR PUMP A	
103	EE-TMR-RHRP1B	RHR-P-B	ELAPSED TIME METER FOR RHR PUMP B	
104	EE-TMR-RHRP1C	RHR-P-C	ELAPSED TIME METER FOR RHR PUMP C	
105	EE-TMR-RHRP1D	RHR-P-D	ELAPSED TIME METER FOR RHR PUMP D	
106	HV-AD-AD1405	RHR-P-C	ISOLATION DAMPER - RHR SWBP ROOM EXHAUST - DIV I	
107	HV-AD-AD1406	RHR-P-D	ISOLATION DAMPER - RHR SWBP ROOM EXHAUST - DIV II	
108	HV-AD-AD1407	RHR-P-C	ISOLATION DAMPER - RHR SWBP ROOM SUPPLY - DIV I	
109	HV-AD-AD1408	RHR-P-D	ISOLATION DAMPER - RHR SWBP ROOM SUPPLY - DIV II	
110	HV-AD-AD1409	RHR-P-C	ISOLATION DAMPER - CONTROL BLDG HVAC SUPPLY - DIV I	
111	HV-AD-AD1410	RHR-P-D	ISOLATION DAMPER - CONTROL BLDG HVAC SUPPLY - DIV II	
112	HV-COIL-(FC-R-1H)	RHR-P-B	1-FC-R-1H COOLING COIL	
113	HV-COIL-(FC-R-1H)	RHR-P-D	1-FC-R-1H COOLING COIL	
114	HV-COIL-(FC-R-1J)	RHR-P-A	1-FC-R-1J COOLING COIL	
115	HV-COIL-(FC-R-1J)	RHR-P-C	1-FC-R-1J COOLING COIL	
116	HV-FAN-(FC-C-1A)	RHR-OUTAGE	RHR SERVICE WATER BOOSTER PUMP ROOM FAN COIL UNIT	
117	HV-FAN-(FC-R-1H)	RHR-P-B	SW QUAD RECIRC FAN	
118	HV-FAN-(FC-R-1H)	RHR-P-D	SW QUAD RECIRC FAN	
119	HV-FAN-(FC-R-1J)	RHR-P-A	NW QUAD RECIRC FAN	
120	HV-FAN-(FC-R-1J)	RHR-P-C	NW QUAD RECIRC FAN	
121	HV-MOT-(FC-C-1A)	RHR-OUTAGE	RHR SERVICE WATER BOOSTER PUMP ROOM FAN COIL UNIT MOTOR	
122	HV-MOT-(FC-R-1H)	RHR-P-B	FAN COIL UNIT FC-R-1H MOTOR	
123	HV-MOT-(FC-R-1H)	RHR-P-D	FAN COIL UNIT FC-R-1H MOTOR	
124	HV-MOT-(FC-R-1J)	RHR-P-A	FAN COIL UNIT FC-R-1J MOTOR	
125	HV-MOT-(FC-R-1J)	RHR-P-C	FAN COIL UNIT FC-R-1J MOTOR	
126	HV-RI-862	RHR-P-B	AREA TEMP OF SW QUAD	
127	HV-RI-862	RHR-P-D	AREA TEMP OF SW QUAD	
128	HV-RI-865	RHR-P-A	AREA TEMP OF NW QUAD	
129	HV-RI-865	RHR-P-C	AREA TEMP OF NW QUAD	
130	HV-TE-865	RHR-P-A	AREA TEMP FOR NW QUAD RHR PUMP ROOM	
131	HV-TE-865	RHR-P-C	AREA TEMP FOR NW QUAD RHR PUMP ROOM	
132	HV-TI-862	RHR-P-B	AREA TEMP OF SW QUAD	
133	HV-TI-862	RHR-P-D	AREA TEMP OF SW QUAD	
134	HV-TI-865	RHR-P-A	AREA TEMP OF NW QUAD	
135	HV-TI-865	RHR-P-C	AREA TEMP OF NW QUAD	
136	HV-TS-862	RHR-P-B	AREA TEMP FOR SW QUAD RHR PUMP RM	
137	HV-TS-862	RHR-P-D	AREA TEMP FOR SW QUAD RHR PUMP RM	
138	HV-TS-865	RHR-P-A	AREA TEMP FOR NW QUAD RHR PUMP RM	

Figure 7-7: Residual Heat Removal-Equipment Group Registar

New CMMSs promise to help classify and coordinate maintenance performance. With respect to work grouping and scheduling, new CMMS hierarchies have the functional tools to attain this promise. As simple as the concept of blocking is, the actual development and implementation of processes that block and work based upon EGs is rather

Fossil	Nuclear-BWR
Baghouses	Minor work Reactor core isolation cooling High pressure coolant injection
Coal mills	Residual heat removal
Coal belting	Radwaste
Dust Suppression	Off gas and augmented off gas
Feedwater heaters/boiler feed and condensate pumps	Service water
Circulating water tower cells	Instrument air
Condenser waterboxes	Turbine light maintenance [blocked by As Low As Reasonably Achievable (ALARA)]
Fire control	

Table 7-1: Areas Not Worked Online

sophisticated. Groups must be developed. PM tasks for equipment must be identified and blocked into convenient work activities. These activities must then be tagged with and scheduled onto EGs. Without a computerized CMMS and the availability of an implemented, functioning PM program, there is no structure to support "working" EGs.

On the other hand, where these elements are present—as in commercial aviation and large power plants—there's substantial opportunity to use computer-scheduling tools to simplify work. This opportunity has been overlooked in some older plants. Several large integrated coal-fired sites that I've worked with have made easy work gains by EG development. (Fig. 7-7)

Available plant design redundancy is used ineffectively because most plants have never crossed the threshold of performing all the controllable work on-line that's available. Some examples of things not worked on-line (with the potential to be) are higlighted in Table 7-1.

One common reason things aren't worked on-line is that isolation valves won't allow it by leaking through. Another is that plants usually won't isolate tie buses to facilitate work. This means that one com-

Figure 7-8: Can It Be Worked On-Line? Can 40% coal-burning efficiency hold the line? What level of reliability must back this up? These plants require staffs of 100. Forty maintenance workers stretch to get all the work done. Yet plants in Australia use two-shift operations with idle shutdown periods on automatic startup sequencing – unheard of in the US. Although unit outages will always be required, systems and equipment can support more online work performance -- provided maintenance is carefully coordinated with operations. In fact, as more maintenance work is initiated by on-condition maintenance/condition monitoring, online work fraction increases. This increases revenue.

mitment to performing on-line work involves raising the perceived value of maintenance so that valves, dampers, operators, and switchgear are available and clean tag out isolations are possible.

To test for on-line work availability, ask, "if it fails, do you always take the unit down?" If the answer is, "take the unit down," on-line work may truly be too risky. But in those cases in which plants work failures routinely on-line, something may be missing. Generally, my experience is that planned work is much less risky, and more thought out than the emergencies. They really entail little risk when planned and supported by sound tag out and maintenance processes. Obviously, every plant environment and feasibility differs, but with more CNM, more on-line inspection is a necessity (Fig. 7-8).

EG development steps

EGs can be developed in two fundamental ways. One is based on design—P&IDs, trains, and equipment layout. The second is based on existing operation tag out boundaries. Each has its advantages. The designer's intended EGs are often described in A-E system operating procedures, plant operating modes, and system descriptions. These materials can occasionally provide surprises—like the fact that designers anticipated isolating equipment for maintenance on-line that plant managers never envisioned possible.

In essence, every designer had EGs in mind as they laid out their plant. This is often the reason, for example, for the check and isolation valves in standby lines. It's obvious that any redundant standby train must incorporate the means to both bring it on-line as well as isolate it for work. Designers have learned standard layouts and methods for incorporating redundancy into designs and have used them for years.

The problems occur once the design leaves the designer's control. Although many design intentions get faithfully reproduced in the real plant, as-built systems occasionally don't function as expected, either due to design oversight or quality problems. If contractors use substandard materials or undersized components to manage costs, equipment doesn't perform in service as expected. That's not the designer's fault.

The greater problem occurs, however, when plant operators and maintenance staff do not fully "own'" a design at the time they become responsible for it. Designers usually provide training, but there's no assurance operators will comply with the designers' intentions. The design is too often compromised as it's incorporated into operating rounds, PMs, monitoring strategies, and allocations for maintenance. Existing operating cultures have powerful influence on planned operation regardless of equipment or system capabilities. Given these facts, and the haphazard methods by which we bring new systems on-line, it's inevitable that some design compromises occur.

Because the primary reason for forming EGs is to perform maintenance, EGs don't need to be based upon a physical boundary. An EG should provide a unique identifier in the plant CMMS for sorting.

Other group possibilities include:

• operator rounds
• VM round
• calibration round
• loop calibration
• thermography round
• lubrication round
• fire door inspections
• fire alarm inspections

In fact, any logical grouping of equipment based on inspection or monitoring criteria, location—or both—can be effective criteria for an EG. While the first criteria for developing rounds—performing on-line work—is more powerful from a work-leveling perspective, the concept of EGs has great potential to make many more work associations possible. In fact, an operator round is essentially an on-line EG structured around a fixed amount of time and a route. The on-line work group is hardly different except that the work "belongs" to maintenance and has an accompanying WO.

An obvious concern among many managers today is workload leveling. To control the tremendous augmentation and overtime worked for regular outages, more plants today are performing on-line work. Yet the predisposition to lump all heavy, major work into outages continues. Coal handling is an example. How many coal plants take coal systems down to avoid performing heavy coal handling work in outages? Few, I imagine.

Certainly, doing on-line work requires coordination. If you fool around the units could go down. But controlling risk by equipment grouping was an intended use in the first place.

An EG strategy is essential in nuclear plants, where more armed trips and license-LCOs come into play. Nuclear units have larger amounts of on-line surveillance testing, failure-finding instrumentation, alarm checks, and standby equipment tests. Nuclear plants—with their large staffs for work coordination—offer the opportunity to "align" PM work into groups once and then keep it there indefinitely.

EG types

One method to develop EGs is using plant A-E P&IDs to identify multiple trains, logical sub-groupings, and other sub-units that so naturally form work organization skeletons. These EGs are a physical type, based upon the tangible layout for installed equipment. Such groupings are most often based upon designer intent—*i.e.*, the check valves in a feedwater pump loop were installed to facilitate on-line work. A second common grouping is the maintenance round. Thermography, VM, even fire door inspections, logically fit into this kind of special EG—a "do it all at once" group. The key to this group is the equipment availability on-line.

The first EG type includes a physical, energy, or pressure boundary—areas where steam, water, voltage, air, hydraulic, or special gases (CO_2, H_2, He) are common. Usually, these can be isolated on-line by train, using block and/or root valves, breakers, and other isolation devices. The boundary points for EGs on a P&ID are also typically tag out points. The second group type is a convenience group. These facilitate the use of a checklist to perform a PM round on a routine basis.

A combination of the boundary and convenience groups results, when a system has separate trains that can be conveniently grouped for simple work, even though trains can be crossed. Reactor water cleanup at a BWR involves two pump trains on the front end and two filter demineralizer trains at the back end, separated by common regenerative and non-regenerative heat exchangers. These can be grouped for simplicity and convenience, even though the A/B pumps and A/B demineralizers could be cross-connected. We can thus construct an arbitrary group, but a simplifying and convenient one, within the pressure boundary. Similarly, the BWR augmented off gas system has two similar yet independent front- and rear-end trains, which grouping simplifies.

Implicit grouping results when routine work within a system is "aligned" on a 12 week schedule. Once equipment PMs are placed on the schedule, and worked as scheduled, the work groupings naturally stay together as time goes on because they're aligned to the same work window. This grouping, developed as a natural outcome of alignment, minimizes work periods, system outage time, and the other negative work impacts. On a system with non-intrusive PMs (such as lubrica-

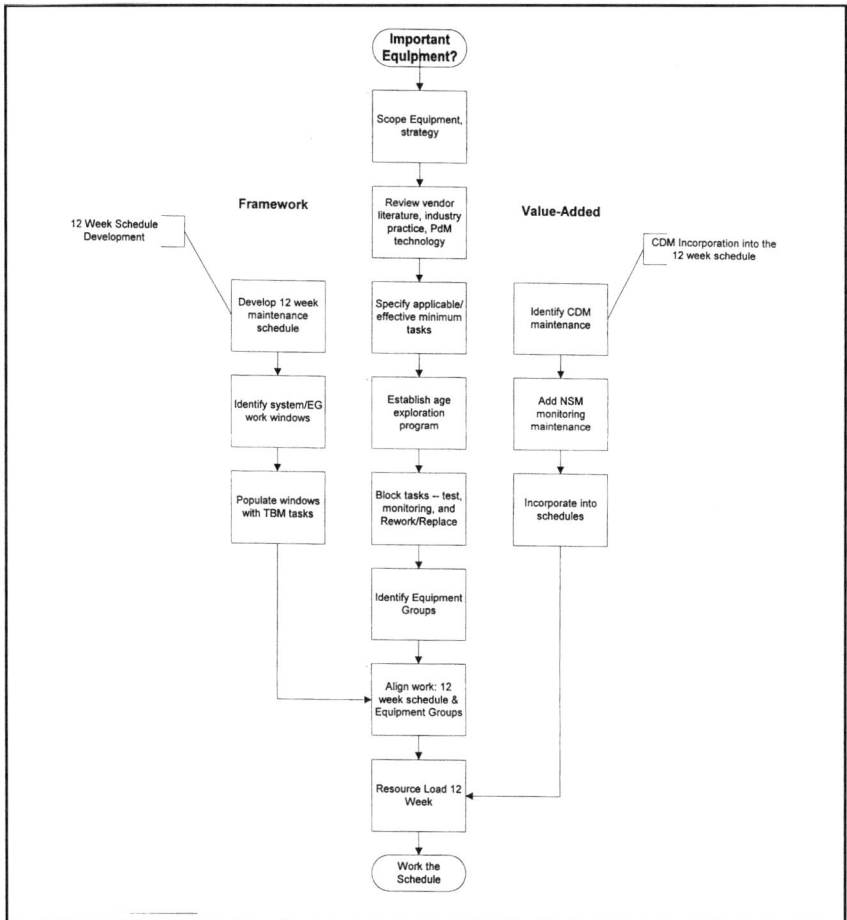

Figure 7-9: Modified Maintenance Process

tion, calibration, and electrical switchgear checks), alignment helps associate, and keep associated, all work naturally occurring in a PM group. The alignment group and boundary group may be the same, but this is not required. External valve operator PMs, for example, line up with other non-intrusive work that may be scheduled separately from the boundary group.

Grouping also occurs at the skid and equipment level. It's natural to group electrical, mechanical, and operator stroke tests of motor operated valves together at nuclear power plants. Here groups are based on skills required. Valves and their operators—motor operated, air operated with solenoid pilots, and hydraulically operated—form equipment level groups. Groups can also be developed around similar technology, such as VM, thermography, or air-operated valve testing. Groups can then be associated with fixed work-order scopes for performance in convenient work hour blocks.

The primary goal in developing groups is improved work performance. Consequently, as WO lists for equipment are displayed, they may suggest other unique groupings that offer convenience, simplicity, and time savings (Fig. 7-9).

Operations

Operations copes day in and out with residual, random equipment problems. They run complex facilities, and a general trend is for complex equipment to fail randomly. Outstanding operations reduce randomness. Mediocre operations introduce it. What factors separate outstanding from mediocre to control randomness of operations?

Factors that have been identified by risk analysis and good operating practices for years include "poke yoke" methods and devices. Some are:

- simplicity
- procedures
- standards
- marking
- lighting
- cleanliness
- training

These are obvious. But consider some not-so-obvious methods. Using the knowledge of the body's natural clock to schedule shift rotations makes for more alert operations. This improves response to events. (We used to have a joke-"All bad things happen on graveyards". In fact, they frequently did as often as not because the response of an un-alert operator to an event is just about random.)

RCM applications can be extended to the operator interface. An alarm may correctly notify an equipment problem or exceeded limit, but if the operator doesn't recognize the alarm, it might as well have failed. Any factors that increase the eligibility and predictability of operator response to mitigate random events have great value.

Modification reviews

Ranking unit modification capital allocation requests is an annual budget exercise that is becoming more complex. Safety and environmental improvements often are handled as separate line items, leaving a limited amount of money to be divvied up among many needed improvements. Ideally, improvements have high paybacks—certainly high enough to pay their way. Instrumentation upgrades—such as fossil's transition to distributed controls—may be needed to continue economic operations. Every modification should have a projected benefit.

RCM thinking has identified modifications and upgrades that made no sense, and, once identified and cast in a R perspective, could be deferred or eliminated entirely. Several of these discoveries and ARCM has paid its way.

A single unit PRB coal unit was originally sited for two units, with coal handling service sized and built to allow a two-unit operation. All major belts except the transfer and tripper belt, along with the crusher, and dust collection system had completely redundant spares. When one of the long, inclined-yard belts became an aging concern, the coal handling people requested to replace the belt as a part of the annual capital budget request. The belt replacement (with a cost more than $100,000) made the cut, even though there was no generation improvement to be derived from the upgrade.

This improvement could easily be deferred by simply "spacing out" the aged belt-using it as a backup for the otherwise redundant paired belt. This was possible at virtually no production risk. Of course, long term, the operating strategy in this case requires that everyone understand the modified approach and accept the marginal increase in risk.

In another case, a $1 million dollar capital rail loop construction project was replaced by $50,000 of capital improvements, and an overall greater operating monitoring program. This capacity to focus capital improvements is an obvious ARCM benefit.

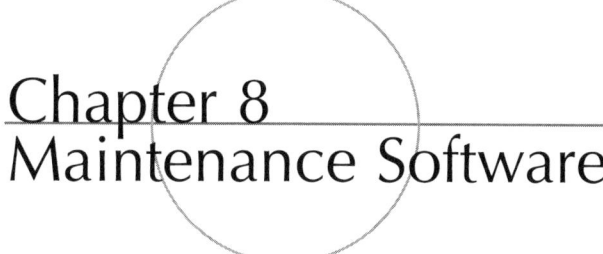

Chapter 8
Maintenance Software

There's a right way, a wrong way, and a Navy way. We do it the Navy way.

-Master Chief, USN

Goals

Why do we need software to help us perform maintenance? People and organizations have performed maintenance for years without it. On the other hand, the primary purpose of software is to improve maintenance productivity and work performance. With so many prescribed statutory rules, it's hard to imagine organizations doing work without software and many have now used it for 20 years. Some do without, however, often very effectively.

It's instructive to remember the promise of maintenance information system (MIS) maintenance software as we review the changes necessary to successfully implement RCM. The software's original objective was to vastly improve the use of maintenance information. Did this promised benefit occur? In many situations, it didn't. Access to computers was functionally limited to the "front offices." Software never

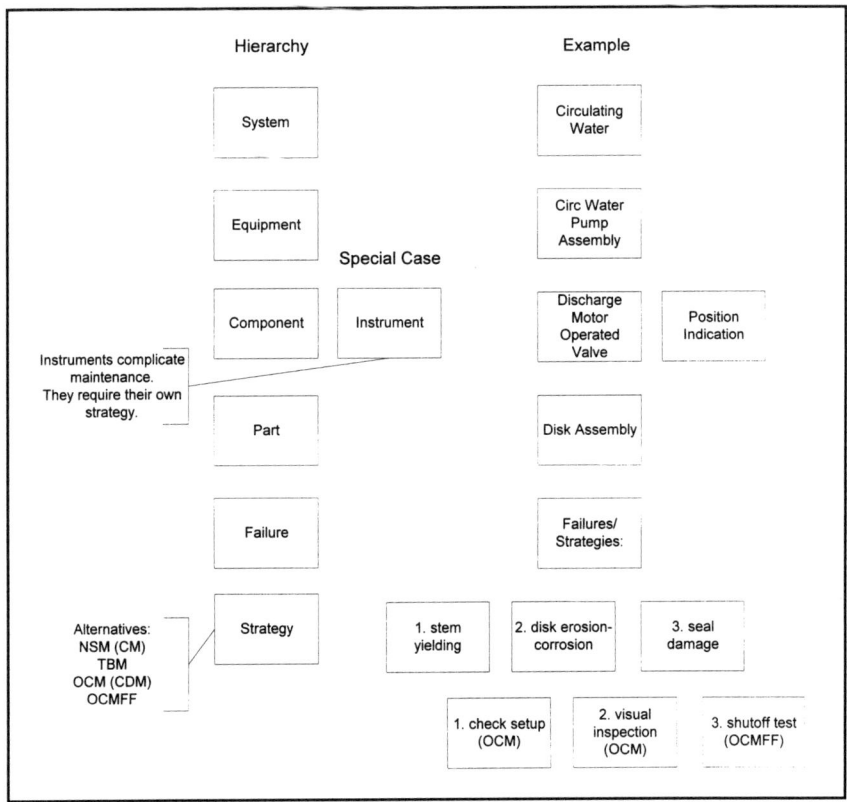

Figure 8-1: CMMS Hierarchy of Equipment

accurately modeled the work-control process and in some cases, the process did not lend itself to modeling, since it was indeterminate.

What was involved with goal setting, and where did it fail to miss the mark?

Hierarchy

Software was developed to facilitate the whole of maintenance planning—including the entire equipment hierarchy.

Systems are the highest unit level—parts that can be removed from service and replaced singly are at the lowest. In between we have equipment trains, logical equipment groupings, major equipment assemblies, and components, all of which can be tagged, isolated, and worked at one time. (I think of equipment as component assemblies providing major functions within systems.) Components—items that are built into equipment and replaceable as units "out of a box"—are the discrete building blocks. Logical groupings are available between the component and system levels to better coordinate, schedule, and facilitate work (Fig. 8-1).

EGs simplify tag out work. Groups once established advance planned work performance by tag-out processes to a consistent, predictably scheduled effort. EGs with standardized tag-out routines have less potential for error and facilitate the coordinated scheduling of large work blocks in single packages. Once a group's work is aligned, blocks of work stay together until they are realigned.

At the lowest work coordination level equipment can remain in service while it's monitored and maintained. VM routines can be placed in "miscellaneous" groups and be scheduled to work anytime. They have no impact on operations. Shops may schedule this work for their own convenience.

At the other extreme is equipment that requires unit outage to perform work. They go into outage groups. Boundary groups have these two general default categories-on-line or outage work. The balance of the EGs is made up of trains and associations that facilitate tag out, on-line work, and standardization. Physical boundary groups depend on design, but they form equipment associations that can be worked together. A plant coal-mill or baghouse fly ash train could constitute a group.

Convenience groups facilitate working activities such as fire door inspections. Rather than bog down the CMMS with 50 different WOs, a checklist can support one. The PM now involves one MWR and the checklist.

Grouping success depends on organizing for simplicity and impact. Maintenance success depends on grouping.

Coding levels

Equipment can be coded to any detail level. The key requirement is to be able to uniquely identify equipment with no ambiguity for maintenance. Inconsistently coded equipment systems pose problems. The primary reason to code and identify equipment in a CMMS is to facilitate maintenance and so the coding must support this.

Informal coding doesn't work well at large plants with many maintenance workers. Coding consistency leads to work coordination, fewer mix-ups, and a usable maintenance history. Accurate logical coding supports not only monitoring and maintenance, but clear tag out boundaries.

Standardize

Equipment coding structure reflects how you do maintenance. Coding must fit the CMMS and tie into any equipment grouping scheme the plant intends to use. Equipment grouping is a powerful tool that enables the free association of equipment for the primary purpose of accomplishing maintenance. A CMMS should facilitate the development and use of arbitrary EGs.

Applications

The value of EGs increases where equipment coding systems are detailed. (At fossil plants, a boiler feedpump could be the smallest coded unit in a group. A nuclear plant might have 100 coded identifiers for the same equipment.) This concept supports natural grouping. The trick is to uniquely locate the skid-identified components. For a plant coded to the component level, group associations coordinate work. Newer CMMS systems that support hierarchies automatically provide grouping logic.

CMMS Computer Software

The maintenance process

The maintenance process used at a plant determines how a CMMS needs to be structured. There are many different flow paths for maintenance performance that lead to similar outcomes. But if measurement and trending are the goals, the path needs to be consistent with data and paper flows.

For example, a unit with a CMMS but an operator who can't make computer entries means he'll have to delegate MWO entry to a clerk. Quality and accuracy of problem descriptions will drop. It's important to use plant staff for credibility of work as well as to validate MWO requests. Any CMMS with inaccurate MWO entries that aren't quickly purged will suffer credibility problems.

Unique CMMS

Companies used to maintain large information services groups. These groups are vanishing, victims of outsourcing and cost/benefit assessment. Their legacy is custom software products—the most important aspect of the CMMS. These large legacy mainframe codes must be mastered to extract data, generate reports, and otherwise interact with and manage the day to day work of the organization. Few engineers and managers have learned these systems. CMMS systems are largely the software tool of planners, schedulers, and maintenance staff. However, all the functions of maintenance at most stations—for better or worse— are computerized, with most permanent records maintained this way. CMMS systems offer operators great power to access and interrogate information. It's kept maintenance managers either in control (or in the dark, as the case may be) the past 20 years. Those who could access and generate their own reports had an inherent advantage.

The advent of second- or even third-generation CMMS products offers greater flexibility-though at the expense of tailored applications. These products, with their Graphical User Interface (GUI), Windows-based environments are truly exciting. From a standardization perspective, you will likely see the kinds of evolutionary paths that accompanied the adoption of Word and WordPerfect as document software standards in the PC world—*i.e.*, much greater exchange of documents and other information developed in the same Windows-based applications.

As this transition occurs, maintenance organizations must learn to adapt systems they grew up with. Just as industry-wide application of word processors has developed some incredibly powerful and common routines, CMMS capabilities depend on having and learning the software. Since a CMMS installation for a modest-sized utility runs well into millions of dollars, some won't be able to afford the transition until market pressures force it. These organizations will continue to struggle with their specific software application. In addition, newer systems are capable of more efficient import and export routines, which will facilitate strategic steps such as standardized work plans and procedures. Improved sorting and information-management capabilities, as well as "user friendliness," should put CMMS applications much closer to the worker and prepare for a transition to a paperless work environment. This is an attractive advance. Real-time maintenance information is a reality for some plants.

From an ARCM perspective, one of the greatest benefits of new CMMS technology and software will be the access and availability of failure and performance data at the system, group, or component level. This enables system level management of common equipment. In the nuclear industry, "system engineers" perform this role. When system level data (particularly cost) can be reviewed, and there's ownership for system management, I anticipate real performance advances. For the first time, operating groups will be able to see how their maintenance approaches roll up to the system performance and cost levels. With this information, in integrated formats, it will be easier for professional maintenance personnel to manage cost and production.

System level measurement

In ARCM, the functional level is most natural for performance evaluation. Integrated system performance includes functional failures, costs, and unit availability/R impact. Other measures identify the maintenance strategy, system health, and results. For example, what's the relative breakdown between time-based and CBM? Or the proportion of hours and work orders attributable to "emergency" work? Or percentage of all work that is planned? What is the relationship between estimated and actual worked hours across the entire system? These are some of the questions and measures that we've used in the past to uncover major areas of opportunity. The key to achieving these results was the ability to access performance numbers in the CMMS at the system level.

In the absence of this capability, some or all of this information could require manual manipulation and presentation. With the wide availability of systems such as MS Access, a database can often be attached and data manipulated to extract information, even though the primary database is on an older mainframe computer.

Age exploration

To perform age exploration, there must be a convenient way to capture the performance date of in-service parts. For instrumentation and controls, age exploration is as simple as looking at calibration data after some period of service, and estimating the allowable drift before the instrument goes out of range. Obviously, many assumptions and a lot of skill are needed-not the least of which is familiarity with the equipment. The principle is the same with mechanical parts, but there may be multi-dimensional requirements to consider. Opinion and judgment are still a key factor.

The tendency with age exploration is to underestimate mean life. An interpreter sees the first few instruments drift or datapoints out-of-range and they estimate the mean life at this age. In this manner, extremely short mean life estimates result.

RCM Software Development

Process standardization

Software speeds and standardizes RCM analysis. If you need a basis for every plant component (the case at nuclear facilities), then software is indispensable. Software packages are available that cover the full spectrum of capabilities, including at least one that faithfully reproduces TRCM, including the classic LTA approach. Most perform some sort of SRCM/PMO simplification necessary for simplicity and speed. Each supports different raw data sources, ranging from user input to electronic download data at the two extremes.

Having developed bases by hand, in word processors and on spreadsheets, my opinion is that software documentation packages add value (Figs. 8-2a and b).

Identifying common PM tasks for standard components, with easily reproducible tasks and adjustable inputs, is a desirable software feature. If your organization is also considering an RCM-type basis for your PM program, you'll need to task group features. Before you consider software, ask whether you need a documented basis at all. Some companies don't document their PM work tasks, much less their task basis. Some are very effective in this way, but it's because their people

Fixed Time Maintenance

No	Task	Freq	Performer	Hours
1	fill lubricator	sh	O	0.3
2	blowdown intercooler	sh	O	0.1
3	blowdown AC oil/water separator	sh	O	0.1
4	blowdown jackets	wk	O	0.3
5	replace oil filter (5 only)	6m	L	0.2
6	replace crankcase breather	6m	L	0.2
7	clean glasses	3y	M	4
8	calibrate key instruments	3y	R	4

No	Task	Freq	Performer	Hours	Annualize	Annualized Hrs
7	turn commutator	5y	E	8		1.7
8	adj lubricator	3m	M	1		4
9	change oil	6m	L	4		8
10	acid clean jackets	2y	M	16		8
11	check IC/AC liftoff	12y	M	2	0.4	0.3
12	replace air filter	6m	L	2	0.4	0.3
13	replace packing	5y	M	5	1.3	1
14	replace brushes	3m	E	1	1.3	4

Condition-Directed Maintenance

No	Task	Freq	Performer	Hours	Response Action(s)	Est. Freq	Performer Hours	Subs	Total
1	measure temps discharge	sh	O		isolate/inspect valves	1y	8		
2	measure SW inlet-outlet	sh	O		unplug/restore flow correct overheat	0y	M	0	0
3	measure IC pressure	sh	O		diagnose/correct leakage	4y	MO	16	4
4	inspect jackets	1y	O		(clean compressor jackets)				
5	test performance	12w	O		see 10 above				
6	measure piston	1y	M	2	rework valves/seats replace rings	6m	MO	16	32
7	check crosshead	1y	M	2	rework cylinders	3y			0.7
8	check commutator excentricity	1y	E	4	rework piston/rings	3y	M	8	2.7
					adjust crosshead	3y		8	
					turn commutator	5y		0	0
					correct unloader solenoid correct				

Condition Monitoring

No	Task	Freq	Performer	Hours	Response	Est. Freq	Performer Hours	Subs	Total
9	test unloading	1y	O	2	unloader valve(s)	4y	R	2	0.5
10	motor diagnostic	5y	M	1	rework motor	10y	E	40	4
11	overload trip check	5y	E	1	adjust/replace overload elements	10y	E	2	0.2
12	high temperature trip check	5y	R	1	(SBAC 5 only)	20y		1	0.25
13	check oiler drops	6m	M	0.5	adjust drops	1y	M	0.5	0

Condition Monitoring	Freq		Key Instruments
check commutator brushes	sh		Pressure
check air, water, oil leaks	sh		Intercooler
check service water bubbles, flow	sh		Discharge
check instruments T, P	sh		Receiver
check touch pad temps T	sh	Temperature	
check noises (loading, knocks, leaks)	sh		Inlet
check oil levels	sh		Outlet
check unloading	wk		Differential
check belts (5 only)	wk		
check bearing oil glass	wk		

Subs Total: 3.4 / 3.1 — Total: 34.4

Total: 52.85
Grand Total: 52.85 87.25

Key: Performer
O - Operator
M - Mechanic
R - Results Tech
E - Electrician
L - Lubrication

Intervals:
sh - shift
wk - week
m - month
y - year

Estimated Annual Cost Sootblowing Air Compressors X 4 = $30,000

Figure 8-2a: Sootblowing Air Compressors Task Summary

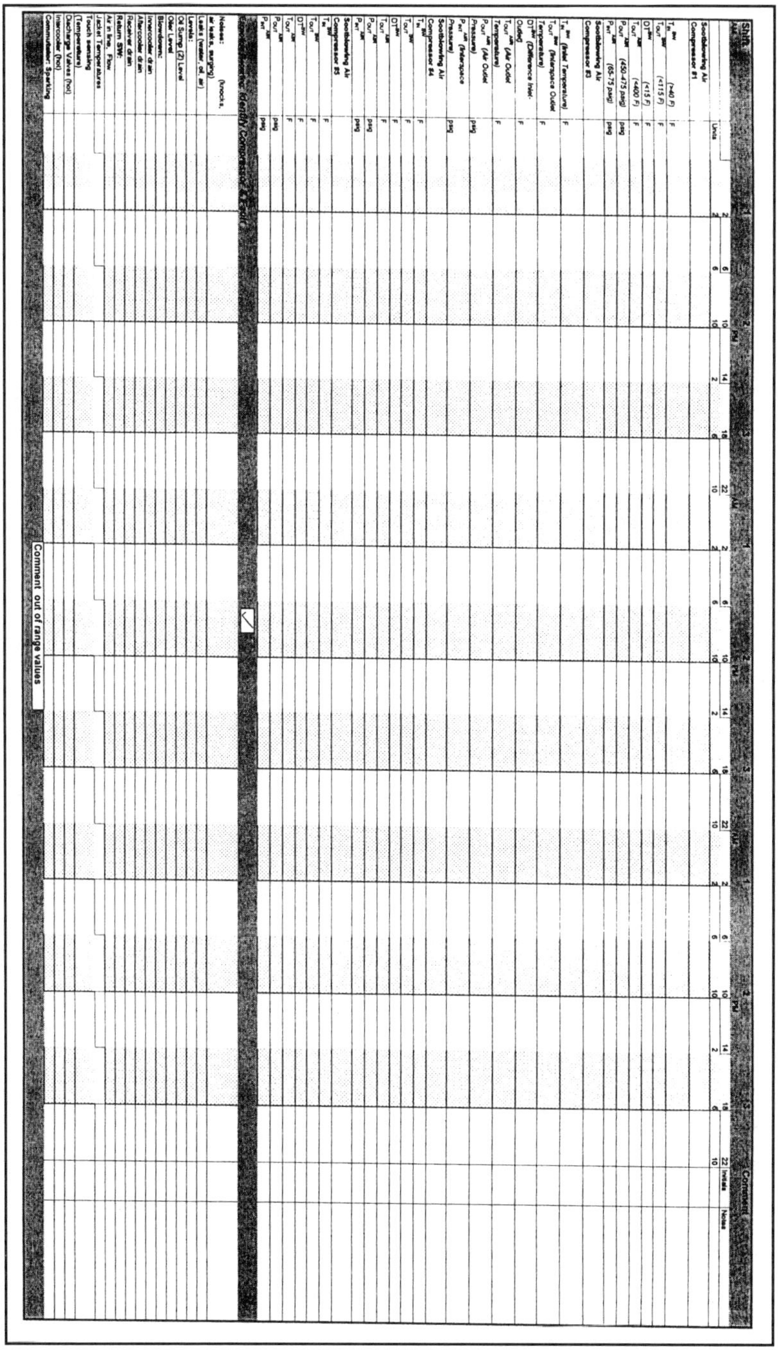

Figure 8-2b: Compressors Loaded Round

are highly skilled and motivated. Even when there are lapses, the simplicity is unmistakable.

A basis-to-PM relationship reflects the analogy of an architect to provide the basis for the builder's construction. Most builders would not consider building without a plan, but this wasn't always the case. Is it possible to build a house without a plan? It sure is—my granddad built at least two that way. Is it effective? Perhaps marginally. Is it competitive? No. Existing buildings constructed on an "as you go basis" cost more.

Maintenance programs built as you go can likewise be expensive—anywhere from 20% to 40% more, and they don't develop as much inherent equipment R based on post-implementation PMO/RCM reviews. Existing facilities must have some plan. Making adjustments to the plan as you go is often the most cost-effective approach. On the other hand for a new facility, LCM costs can be a major opportunity to achieve long term production benefits and lower cost. Is there value to increase production from the same capital asset? Ask any banker.

Most current RCM software are user friendly and save efforts. Value doesn't come from documentation of every plant component's PM case, however, although such software is available for nuclear plants. Rather, value comes from building effective maintenance strategies. Building and maintaining standards that reproduce failure mechanisms and identify effective preventive tasks is most cost effective in the long run. The organization that can retain strategies over time to support a living maintenance program will maintain lower production costs.

Ultimately, there's a tie between basis software and a plant's CMMS. Basis software provides the justification for the tasks performed. Since PM tasks (as a part of a work performance package) address individual failures, PMs get scatter-gunned, as well. Software that facilitates grouping is helpful. Groups statistically increase the odds of PM on-time performance. More advanced RCM packages provide task grouping and sort capabilities. In a seamless way, the things you do and how you do them can be imported into CMMSs to provide the PM tasks and frequencies. No single software has this functionality at this time.

Ideally RCM software should be simple and intuitive to use. Some general goals include:

- facilitate ground-up basis construction
- facilitate existing PM program reviews and basis creation
- facilitate PM task upload to the mainframe or CMMS upon completion
- provide real-time inquiry support to users
- support growth of new applications and technology
- support gradual trimming of legacy PM systems with RCM principles over time

"Basis"

The basis for any PM task is the reason why the PM task makes sense to do (or not do, for the NSM case). In nuclear applications, there is great emphasis on developing and maintaining a basis. In non-nuclear applications, the justification for a basis is underlying task value. Implicit basis requirements can be lost during time. Plants have abandoned very good PM tasks because they forgot why they did them. The task's very success may have been its downfall—tasks effective at preventing failures weren't recognized as adding value, and were dropped and a painful learning process begins anew.

This occurred once at a two-unit coal unit in a dramatic way. The plant manager eliminated virtually the entire sootblower PM program, based on low failure rate. For the better part of a year his costs dropped with no apparent consequences. Then sidewall boiler tube blower tube cutting developed in a major way. Within three months they'd experienced two cut boiler tube leak outages. In a year they'd had six. This was eventually traced back to corrective maintenance by untrained mechanics. After two years it was clear that the net gain had actually become a significant loss—more than five times the promised operating savings, at the plant's wholesale production rate. One outage saved could have paid for a whole year of sootblower PMs.

When you don't know why you're performing an activity, there's temptation to change. For PMs, you must know why you're doing them and what the underlying value is. From the perspective of a living program, that's when a task basis is most valuable. Changes can be evaluated with more focus, clarity, and with less regulatory or safety risk. Developing a basis for an activity is like keeping a log. Every time you change or note something, you log it, and file the result where it's retrievable for later use.

For large plants and companies, this means reliance on computers and their networks. The value of a basis is underscored in this age of cost-trimming, when many young Masters of Business Administration (MBAs) with virtually no practical experience are tempted to make career-serving cost reductions. These only reveal their true impact later. A documented basis gives staff more armor to defeat such moves.

Analysis

Software enables us to perform detailed failure and cost analysis quickly. Several software products have this capability. Where generation could be lost, and expensive corrective action is a necessity or option, cost-calculating software can establish B/C ranges and total costs to enable us to focus on high value activity. Exact costs aren't necessary, but we need ranges to know if we're talking B/C ratios of 1/1, 100/1 or 1000/1. Ideally, we would like bound our approximate upper and lower PM and failure cost estimates.

In non-regulated environments, cost is the primary focus of all PM activity and naturally forms the primary basis for any activity. Practically, even with regulation cost is the basis for most PM hours. You don't have to cost out every PM case in detail. Rather, you need enough benchmark cases-on the order of 5 to 10, so that you can quickly assess any new case by comparison-for cost analysis and priority rank purposes. Standard development has an obvious cost tie.

The development of benchmark cases aids in regulated PM activities, as well. Regulated PMs are usually easy to identify and because a law or license specifically requires PM tasks, these tasks often record their legal source in the basis. Rules for reports concerning continuous emissions-monitoring performance are a straightforward example. Under the older discharge permits or licenses, PM requirements were non-prescriptive except at the brush-stroke level. They once went no further than general mandates for "appropriate" PM programs. (What's an appropriate program? An outcome-based answer is obvious—one that's applicable and effective!) Nonetheless, there's considerable latitude for interpretation. However, a rule for annual lift-off tests of main steam safety valves has no such latitude. Both rules require compliance. The latter is more explicit! The challenge is (always) to operationalize maintenance strategies. Laws just make this cursory. (CEM requirements are now explicit and detailed.)

The RCM process was originally developed to resolve problems in a regulatory environment, so it's a naturally effective fit. Regulators often have trouble with maintenance engineering's finer points that are not so obvious. (One is OTF.) Many regulatory bodies work with a TBM mentality because it's concrete and simple. It's obvious when the job is done. (The irony is that the FAA nurtured the development of RCM in the '60s and '70s) Until regulators see more direct applications of RCM-based inspections and age exploration by competent organizations, they won't understand it. Another challenge is streamlining existing programs. There are many regulatory proponents of TRCM maintenance program development methods, but few SRCM advocates. Yet SRCM is just as valid in regulated environments as any other. Seeking a closed-form solution is simply forgone in full keeping with the spirit of original RCM.

If an empirical approach is effective then use it. ARCM takes an empirical approach. TRCM advocates will say that TRCM provides a complete solution, but no one can guarantee that there won't be improvements. No one can guarantee that all failure modes are enumerated, though reams of paper analysis give some confidence. New technologies are always developing, and the better you understand any failure, the more options and ideas you gain to manage it. Working with plants over time into implementation, we always learned and refined our RCM results as we gained more experience and learned more about costs, techniques, and methods. We often compromised exact RCM results to achieve a middle ground that supported plant personnel. Software must accommodate such learning; it must be a part of a living system.

PMO streamlines PM programs. While PMO (the fossil equivalent of SRCM) isn't as thorough as ARCM from a basis perspective, it offers attractive maintenance returns by eliminating ineffective PM activity and extending otherwise conservative task intervals. There's a bias in any maintenance program to increase work in the form of ineffective

PMs to fix a host of partially examined problems.

Addressing this problem—whether by PMO, RCM, or ARCM techniques—requires discipline. It's stressful. Those charged with making judgments, such as extending intervals, often have little formal training or background to do so. (They do so incrementally.) Management rarely can spare the time to train. This is specialist work that often calls for a contractor. It's difficult to achieve appropriate intervals based on actuarial data alone. Until a person gains confidence with equipment and the process of routine PM interval adjustment, he's reluctant to make any substantial changes in tasks or intervals—even those that would greatly benefit the existing program. Maintaining schedules and management's expectation is likewise difficult.

One solution is a "tiger team" task force on PM task selection and intervals, headed by a central R maintenance engineer with the authority to make calls. Nuclear plants have a R engineer role for the interpretation of maintenance rule compliance and measures. But theoretical compliance and practical support of PM activities are at extremes. My experience with maintenance rule engineers is that they don't offer useful hands-on experience and support developing appropriate maintenance tasks and intervals. They are immersed in the arcane world of nuclear licensing.

Even with something as innovative as a "tiger team," however, traditional CMMS processes start at a problem report from operations. Other work tasks aren't the primary focus of the CMMS. WOs originated from OCM need a follow-on work tag. PM follow-on condition-directed tasks head off equipment failures. RCM associates secondary condition-directed work to primary time-based monitoring. The resulting process is particularly simple to model. A RCM-designed CMMS would be different from current failure-based models that presume maintenance starts with operator-identified problems.

Tracking and measuring time-based work could be handled more easily in a CMMS that was designed in an RCM format. Any new PM task (as well as any proposed changes) would need justification. Basis maintenance is a maintenance engineering function either way. When PM "originators, assignees, or deferrers" sign for a change decision, there are fewer problems with deferred PMs (such as safety/relief valve tests) that somehow just disappear! This discipline diffuses through the PM program so that it gains credibility.

Air transport, hazardous material handling, and nuclear generation are areas of public concern that will need to continue to document and justify maintenance programs from a regulatory perspective. This, too, is an opportunity for software "heavyweights" who develop an interest in maintenance performance models. A CMMS ultimately manages work—any work—that must be performed. A CMMS that provides a seamless tie to an RCM-based work development system will always have an inherent advantage. This transition to RCM software will follow the implementation of standard CMMS programs.

Backing-off an existing program (one with too high frequency or where regulation is relaxed) requires a basis document. PMO or RCM software that provides basis maintenance as a feature has a "one up" on other methods.

A non-regulated market environment really doesn't care why a change occurs, in contrast. It is more than sufficient to justify where you are at a point in time, based on cost. In this case, a well-prepared basis document must present a case for a PM activity for economic reasons. RCM development software and it's related CMMS database must have work efficiencies as a goal whether the environment is regulatory, economic—or whatever.

Many PM programs developed informally and were subsequently "grandfathered" by regulators. A basis is implicit. We presume that at one time there was a good reason to do all the PM work specified. Justifying a specific change is referred to as a partial basis. It documents the purpose of one change. A complete PM program justification, on the other hand, is comprehensive and includes all relevant program documentation. These are known as "full bases." Many programs survive on partial basis PM changes.

Documentation

Basis development in a non-regulated environment only needs to suit the company and economic conditions. At a very basic level, all PM is based on cost. Regulation-mandated PMs means that the potential "cost" is the risk of being shut down for not doing legally required PM tasks. Or worse—the costs associated with injury to the public or employees from an event.

Typically these involve rules, agreements and understandings with state health departments, boiler inspectors, federal agencies—FAA (for stacks), EPA (emissions), OSHA (industrial safety), Department of Transportation (DOT) (gas transport) and occasionally others. It's common to have a regulatory body endorse a professional group's standard, like the NRC and state endorsements of the ASME's Boiler and Pressure Vessel (BPV) Code. Occasionally, rules or standards conflict, like the "mixed-waste" jurisdiction issue only recently settled between the NRC and EPA. Overall, these standards are the first levels of compliance that need to be assured in a PM program. Their source documents are voluminous.

Many companies also separately endorse building codes, the National Electric Code (NEC), and many of the supporting ANSI, American Society for Quality Control (ASQC), ASME, IEEE, American Society of Civil Engineers (ASCE), SAE, ASCLE, and other technical body codes. These codes are impressive in size. Codes represent the best effort of a group of knowledgeable and interested people to provide guidance on how to do something. They're often general, vague, confusing, and subject to interpretation and change, but they're also the best source of information on any subject for which you aren't already an expert. Occasionally they're dated, or organized based on changes.

Insurers develop and maintain inspection standards in addition to codes. Boiler and fire protection requirements are two examples, but there are many others. Fossil boiler insurers and their representatives often want to know specific ways that an owner implements a code requirement. Occasionally an authority designates an implementing agency for a code requirement. (I worked in a plant that had this arrangement with the state boiler inspector. The state recognized that industrial insurance agent and his engineer's recommendations had the authority of the state boiler inspector behind them.)

Industrial insurers may identify risk areas they would like addressed. Sometimes this has to do with a facility. More often it's for new monitoring or other risk management equipment. It could be fire detectors, wet pipe deluge, or an upgrade in physical plant. These agreements carry the force of contract behind them, and should be tracked like regulatory commitments, when made. Being able to track and assure that insurance commitments are in effect has a favorable impression on insurers. Obviously, they don't carry the force of law and can be more easily negotiated. But if management agrees that something is really good to do, we can only assume we should do it until directed otherwise. We should provide the means to performance within our processes.

At that point, procedures and checklists should be based on ARCM-derived work activity.

Products

The products of RCM basis documentation are the tasks done to avoid failures. Organizations that have traditionally performed them, as well as the "products" themselves, can summarize what these tasks are. Most nuclear organizations would recognize the "on condition-failure finding" RCM task as the rough equivalent of their SP. (The SP is based on the literal technical specifications of the nuclear plant, so the correspondence isn't exact.)

This illustrates why the RCM paradigm is so useful. It's helpful in cutting through the organizational muck, and getting to the basics. Sound engineering, sound maintenance, appropriate to the situation.

Round checklists

Checklists for rounds are the staple of the roving operator. They provide guidance on what to check, how often, limits, and other incidental information. For the control operator, they are summarized in software as screen pop-ups that require entry from a DCS.

Rounds are being modernized with hand-held wands and monitors. Portable-monitoring devices can provide a seamless tie from the rover, reading non—DCS data, into the CMMS or even DCS through a download. This in turn supports trending.

Obviously out-of-limit equipment is a candidate for immediate remediation—classic RCM-based "CDM." The largest fraction of rounds is non-specific checks, where the operator's judgment determines whether an item is serviceable or not.

PM tasks

Scheduled work activity that includes tasks implemented by WOs using personnel assigned to maintenance is the traditional scheduled PM program. Two primary task groupings comprise the TBM plan-the traditional PM and on-condition-maintenance. While OCM has no exact equivalent in a traditional program, it does reflect the intentions of the traditional program PdM. The distintion is the level of diagnostic task scheduling (OCM) and follow-up task performance (CDM).

Organization

Organization of PM activity into useful sub-categories is a prime benefit of RCM. The distinction between a non-specific CNM task and an on-condition one is the simplification of routine scheduling and work priority this provides. In so many words, the mature program schedules failing-condition equipment for maintenance before indeterminate-condition equipment.

CMMS integration

The best RCM systems integrate cleanly with the CMMS. On the front end, they use similar coding and system definitions to organize strategies. They tie in instrument plans. They easily upload completed grouped plans into the CMMS.

They should not require repetitive entry of CMMS WO plans for the maintenance work plans. They should allow the later addition of work plan detail from reference documents. They should support standard work plans for repetitive work-the case for virtually all PM.

RCM/CMMS idealization

What would the ideal RCM-based CMMS/RCM maintenance strategy development and implementation system look like?

It would probably have a very different emphasis than traditional CMMS systems that are based on the concept of broken equipment.

(Someone sees broken equipment and makes a trouble report/work request.)

Practically, this latter model is not only highly reactive, it's not a good model of how world-class maintenance works. The very best maintenance organizations maintain a continual process of monitoring equipment and anticipating its failure. They don't wait to notice problems. In this regard, the traditional model lags workers and is more hindrance than help.

An ideal system would provide ongoing monitoring guidance, as well as authorization to correct degraded equipment on an ongoing basis. This requires very highly skilled, proactive maintenance, and operating staff.

Configurations Simulation

What maintenance work is best in a given situation? With a simulation model, we can take the maintenance plan, enter all the factors, perform a Monte Carlo simulation, and see what R answers result. Simulation is available today on PC to help facilities test strategies for redundant trains and instruments to see which, in fact, is best. This can help avoid learning lessons the hard way. Very often minor design changes can have significant R paybacks. Maintenance routines are also likely to benefit.

Software can simulate the availability impact of proposed modifications, before they're made. Abandoned modifications resulted in some cases only after the plant saw the impact of the modification in a plant trip. Plant trips at even a modest-sized plant cost well upwards of tens of thousands of dollars. For large base-loaded facilities, they may approach six figures quickly. The age of trial-and-error design changes in the utility industry is rapidly coming to a close. It's simply too expensive!

Simplicity

One risk of simplified, computerized, RCM analysis capability is analysis-paralysis—documenting every potential failure possible.

Having been down this path myself, it is necessary for me to exercise great discipline to avoid it on any given project. As an expert, you can easily go into hypothetical mind dump. Clients may encourage it. In reality there's rarely more than several important failures, and these occur only on larger subsystems or components. We need to remind ourselves of the potential benefits of effective maintenance as well as what maintenance cannot do. Most components have one to two dominant failures that occur commonly. Some might prefer exhaustive analysis done as a matter of course. But it's not necessary to paint a clear R picture. For those who disagree with this position, consider how you maintain your car or house. Chances are you plan around the one or two things that do happen.

Policy

Some corporations may find it useful to establish company-wide RCM standards. After an analysis has been completed, it represents a significant amount of learning, and it's logical to apply this at similar facilities.

In the final analysis, the system that makes the work elegantly simple is the best one. Where learning is transferable with existing processes, it should be transferred.

Chapter 9
Measures

We don't see things as they are, we see things as we are.

-Anais Nin

Measurement

Global

Accepting the challenge of an RCM/ARCM program is an example of a process shift. When a process shift occurs, what precisely happens? At the plant or company level, it doesn't mean that we instantaneously have a new process, with new results. Any major organizational change requires time, effort, and resource dedication to implement. But what if we could change a system or its inputs instantaneously? We check a control system response by feeding in a new signal. (Fig. 9-1) Treating a system in the same way, we should see a new response (with some dynamic delay). If we could model in this way, what would the response level be?

Theoretically, output would start to respond once the change

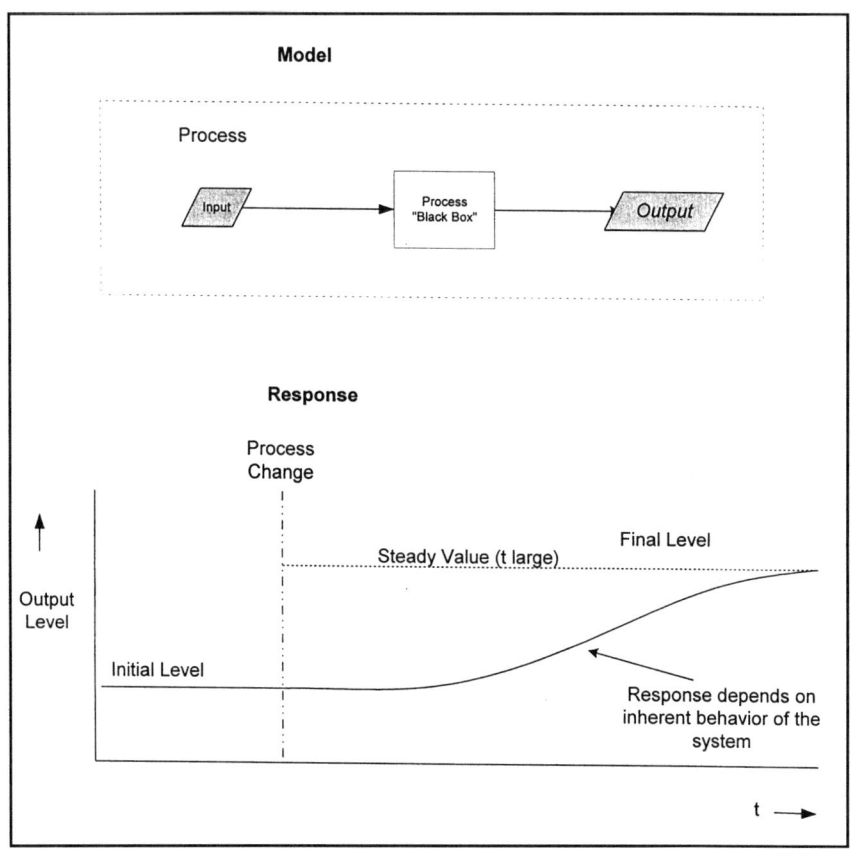

Figure 9-1: Process Change

occurred. When control system input takes a step change, the process output instantaneously has a new equilibrium value. It just takes time for the process to get to the new value. Taking the system to be "the maintenance process," the input as "maintenance selection," what are some suitable output measures? Based on theory and our projections, what do we expect to change? Our outputs are maintenance costs, unit production cost, and R. To see change, we must measure their response. Ideally, we achieve an appropriate level (Fig. 9-2).

We change a maintenance or operating plan to either increase production, reduce costs, or both. R is a byproduct—it's difficult to measure until we drop to the more sensitive system level. At the system level, responses are easier to see—if we have system level measures. We seek

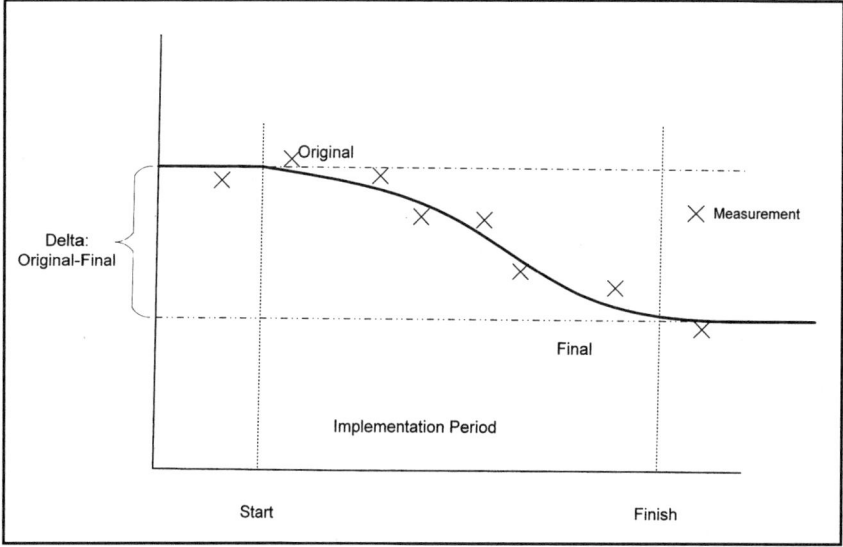

Figure 9-2: Maintenance Cost

measures that tell us we whether our change influenced performance as expected. The questions are, what do we measure? How can we measure projected benefits?

The process and inputs are:

• maintenance planning (input)
• maintenance performance (the process)
• measures (output)

No change can occur until the maintenance plan changes. Thus, the initial effort after an ARCM effort must be implementation of the plan. This takes time, because ARCM is implemented at the system level. To speed the results and identification process, those systems with the largest potential for improvement need to be selected. "Problem" system selection usually isn't difficult, but depending on the level of plant measurement awareness and sophistication, it can require some time to identify the potential value-adders. Sometimes secondary failures aren't accurately reflected by their root causes.

The kinds of system problems reported in NERC statistics are generally the same for a given class of units. For example, coal-fired boilers tend to have a high rate of boiler tube failures induced by fly ash-erosion. Boiler and turbine losses typically represent the top two loss contributors. To understand the spread of losses at a particular unit requires understanding "root cause" losses in depth. This starts with loss reports, but doesn't end until the loss "drivers" are understood. This identifies the low hanging fruit. Reviews require an up-front assessment-an intermediate step to assure the effort focuses on the best improvement targets.

When a maintenance management system provides accurate numbers up front, they're most helpful. Many CMMSs can identify these statistics, but only to the degree data allows. Sometimes a surrogate statistic must be sought when a key field or other indicator isn't available, or available but unreliable. For example, CMMSs that record hours independent from time cards are suspect for time accounting data accuracy. CMMS reports of emergency WOs may be suspect based on the uniformity and control of the category "emergency."

Systems with performance problems often have multiple problems. Systems with low availability are also typically high cost systems. More hours are worked on these systems, much of it overtime, on short notice.

By focusing on the half-dozen highest cost systems, the effort has much greater probability of success. There are often masked secondary failures in the measurement, so analytical review of system costs is required. Three factors that trend together are availability, R (evidenced by forced outage rate), and costs/work hours. By simple thumb rules of estimating, non-labor cost and work hours tend to roughly approximate each other in total cost terms. i.e., a staff with an annual payroll of approximately $5 million spends $5 million on parts and services.

Focused

Every system has inherent cost profiles, based upon designs showing their inherent R with regard to cost, availability, and man-hours needed to support a given level of production. Benchmark comparison figures for similar plants are very helpful to understand where nominal levels should be. After selecting improvement areas for focused effort, several change iterations may be needed to achieve the desired result.

A detailed performance measurement system is necessary to measure the effectiveness of the effort. Competing measurement factors can pose difficult alternatives. The three major measures are:

- production
- R
- cost

Workers today are more aware of plant level production costs. Value added varies by day, season, and conditions, and is highly variable. Plants can't plan to market projections. They must focus on well-known production factors. Market projections are at best an estimation of expectations.

Benchmarking processes have gained credibility as a viable method to compare costs. Widespread distribution of operating and maintenance generation methods makes process benchmarking a useful tool to understand competitive position. However, access to useful benchmark information is also difficult to attain! Potential competitive threats make benchmarking partners less likely to provide useful inside information that may point towards fundamental weaknesses. Third-party vendors, contractors, and consultants provide another way to obtain competitive information. Benchmarks outside the industry are also useful. Comparison with manufacturing and service providers often points to innovative opportunities.

There are, for given classes of generation, intrinsic costs. Nuclear, coal and gas all have intrinsic factors that fundamentally shape their cost profiles. These profiles change slowly during time, but can sometimes be influenced fundamentally. Three Mile Island upped the nuclear ante, just as the energy crises of the 1970s impacted fossil generation. Greenhouse gas remediation may eventually increase the cost and utility of high-carbon fuel combustion. While the burden of regulation has been born most heavily by nuclear, indicators point towards steady increases in the fossil generation areas.

Regulatory costs vary dramatically site to site and may be the most significant hidden cost factor. Certainly, few nuclear planners dreamed that the impact of licenses and construction delays would so strongly influence their competitive position. Some plants racked up licensing delays of nearly a decade in high-interest finance costs added to their fundamental capitalization. Surely, in hindsight, those plants would have never been built. This explains a hurdle to new base load generating units today-too many risks in one site.

My favorite ARCM effectiveness measures are based upon financial performance supported by subjective interview assessment. Financial costs include, by rank:

- unit cost of production
- system production cost
- major equipment production cost
- major department costs
- work type costs
- services and parts costs

Effective CMMS data sorts are indispensable to account for costs. Where these CMMSs are not available, cost collection becomes dependent upon FERC reports and in-plant techniques.

Performance measures include:

- unit forced outage rate
- unit availability
- market-weighted availability
- system forced outage rate
- system availability
- major equipment forced outage rate
- emergency work
- overtime work

Employee interviews help establish perceptions as well as problems. Where there are discrepancies between perceptions and facts, perceptions establish the organizational perspective, beliefs, and focus.

Perception at one plant was that coal handling had no impact on production despite a severely degraded coal handling system. Aside from high cost, coal handling functions had been severely compromised. One obvious deficiency was that the plant had no means to remove iron or debris from the coal feed stream. All such material was fed into the crushers, bunkers, feeders, and mills, where it was either pulverized or caused random trips that required isolated, unplanned entry to remove it from the equipment. All the while generation was lost-up to 1/2% availability due to these events alone. At the wholesale generation value added rate, this added up to a cool $486,000 annually for that base loaded unit. Year after year the capital budget request for tramp iron removal equipment and metal detection upgrades (budgeted between $100-200,000) was edged out by more glamorous projects.

Focus can be redirected once interrelationships are understood. But it shows-from a RCM perspective-how design basis system functions have gradually eroded and even vanished over a unit's in-service life-to the unit's cost-competitive detriment. Nuclear units do not suffer "design memory loss," but they pay dearly to maintain that memory. Training and documentation expenses are correspondingly higher.

All measures start with operating goals. Awareness of the competitive profile, as well as industry standards and capabilities, are helpful in establishing meaningful goals. The pursuit of meaningful goals is exhaustively covered in literature on TQM process methods.

The key is to find a parameter that provides improvement focus. Even in those rare cases where systems or equipment don't show obvious stratification, there are ample opportunities to focus on performance.

Changes and Measures

As deregulation progresses, generation R, and cost per net MW hour, take on greater importance. These integrating measures give you health overview, but they don't focus on improvement opportunity at the plant level. For individual units, opportunities must be evaluated in terms of specific mission goals. Measurement capacity depends on the company (and station's) CMMS and other management system capabilities. My experience has been that within the same company, different units, and different departments within units, will use CMMS software differently so that comparisons are both hard to make and take lots of manual data manipulation. Ideally this would not be the case.

Measures

Consider traditional system level unit cost measures. Units can also be measured at the system's level with an FERC-like cost reporting system. FERC categories are:

• boiler steam supply
• turbine generator
• auxiliary

FERC measures don't roll up as a system-oriented hierarchy and FERC categories don't match A-E system boundaries and descriptions. Information is ideally accessible at both the systems and equipment level so that a cost hierarchy can be rolled up for coded equipment to any appropriate level. Many legacy CMMSs today can't meet this need because measurement hasn't been essential.

Measures need to focus on risk and economic importance. Safety (accidents) or lost generation measures are most difficult. Maintenance costs are known at the generating unit levels, but their allocation downward to equipment is typically unavailable. To quantify risks, consider such things as:

• unit forced outage data
• equipment emergency work orders
• overtime
• special part usage
• insurance claims and audits
• regulatory information or audit findings
• industry experience

Extracting data requires reading computer and summary reports. These are accessible from the right sources—the insurance department, for example. There are also simple default measures and subjective staff interviews. Operators and their direct maintenance support, have excellent risk perception. Interviews can confirm other operating data. Industry experience is also an excellent tool to quantify risks. Experience around plants for many years, knowing how they are run, and what fails, helps interpret risk patterns. Operating risks can be managed as long as they're known.

At an operating level, economics factors (corporate wage rates, historical costs, and trends) are known. Total systems/equipment hours worked, maintenance strategy mixes, equipment CM/PM (by hours), emergencies, and total costs are also known. To interpret these numbers, you must also develop system/equipment risk profiles. There's nothing inherently good or bad about working 60% corrective/40% preventive work on a system, unless you compare it to a known benchmark. Knowing that a system's ratio of reactive maintenance and its competitive operating costs are high suggests finding out how competitors operate similar systems or performing general benchmark studies.

One of the most exciting aspects of ARCM implementation is the opportunity to view performance data from an entirely new perspective. Historically, plants followed costs and work hours. Some further broke these down into CM and PM. But if maintenance is better off performed "on demand," PM/CM categories have no meaning. New RCM maintenance categories enable consideration of measurement-and what can be measured. Most of these categories can't be measured directly with existing CMMSs. But, new CMMSs can.

System measures

System level cost measurement is the minimum level to ensure that unit performance expectations are measurable. Usually a system must meet minimum safety standards and support pre-set production levels. There are two broad systems categories. The first directly supports production, the latter provides production support service. A loss of service system functionality has a delay before production halts. Only a few production systems directly support production.

Examples of production systems include:

- fuel system
- primary coolant system
- main steam
- reheat steam
- feedwater

- circulating water
- flue gas (boiler)
- electric conversion

Examples of service systems include:

- waste-water
- ash handling
- service air
- coal handling
- domestic water

There's always a level of risk with production. A coal-fired unit operating without flame-proofing boiler protection is taking a grave risk. Companies (and their insurers) apply grace periods to restore a "failed" system of this sort to service. Instrument air (IA) systems could fall into either the service or production category depending on equipment. IA usually supports feedwater-regulating control valves, and turbine extraction valves-important equipment at startup.

Minimum availability performance for production systems is nearly 100% at the unit system level when the plant is scheduled to be available. For support systems, a predefined service level is based upon historical performance. A coal unit with a coal handling system availability of 85% (defined as the capability to run coal to the bunkers) can run indefinitely at full load. Another unit, with fewer redundant belts, found that 95% availability is necessary to assure continuous production with the same maintenance approach.

When scheduled maintenance periods are considered, system availability will shift. But pre-planned, on-line maintenance periods can usually be planned and managed so there's less risk of unit outage.

Systems that impact production—whether it's sootblower-induced boiler tube erosion or feedwater upset that trips the unit on high deaerator level—demonstrate explicit system functional failures. These should be displayed in Pareto fashion by availability loss contribution. Doing so accurately requires careful record reviews and RCA. If sootblower-induced boiler tube failure is never identified as a secondary failure, then the boiler is charged for the tube failure. This bias misdirects our effort on the wrong systems. Inconsistent failure reporting makes this doubly difficult. There's just no way to avoid becoming familiar with the actual performance numbers.

One refrain heard many times over from station managers has been, "That failure will never happen here again." It won't—at their unit—for five to ten years. After that the lessons learned are forgotten and the potential for failure reoccurs. Monitoring one company's "fleet" of 20 large generating units, I found that fleet problems such as winding failures *do* recur. Unless they change some aspect of basic operation, overall risk levels remain the same.

Statistical failures can be compared to traffic tickets. Speeding tickets, accidents, and insurance losses correlate. Eliminating risks from an insurer's portfolio begins with elimination of speeders—charging that risk category a higher premium to cover the higher risk. At the plant level, few stations keep "speeding tickets." Major loss precursors often go unnoticed, but at the system level, it's much easier to identify and track precursor "near misses" and use them to predict future risk and system level performance.

In the absence of a "near-miss" program, how can you identify system level risk performance? One method is to track two measures that correlate system risk—system equipment emergency WOs and overtime. These indicate the degree to which unplanned events influence system performance. These indicators can serve as red flags. Of course, the absence of a system management plan, a system owner, and operational awareness are the big warning signs. With one or more of these factors present, loss factors decrease.

Cost measures including total man-hours worked, total costs, and how these are allocated between and among various work categories need to be followed at the system level. Remember that a PM hour is an effective work hour-the work is planned, predictable, and the value added has been pre-identified-while an emergency hour is most ineffective. Given alternatives the PM hour is preferred.

After system outage work is controlled, emergency work should be addressed. A system profile of planned CNM, TBM, and OCM offers the lowest cost. Superficially similar systems can have very different cost characteristics. A system maintenance plan compares and contrasts cost factors for future needs and reference.

Failures

Another exciting aspect of RCM is the new perspective it provides on failures. This is true both in regulated and unregulated environments. In place of the easy and conservative (and 100% bulletproof) position of, "When in doubt, or questioned, call it inoperable!" we have proactive, pre-planned assessment engineering and failure descriptions. A confident pre-assessment can be provided to those who must make shutdown decisions.

Failures influence economics, and are therefore worth measuring. Some failures have license or emissions impacts, as well as production loss value. There needs to be a consistent basis for measuring failures at the system level. The NRC imposed this by regulation with the Maintenance Rule at nuclear facilities. Nuclear plants must track system availability and MPFFs at the system level for "risk-significant" systems. Risk-significant functional failures can be identified from operating logs, and "E" WOs.

Two difficult failure types involve redundant equipment. The first is redundant trains, like a standby feedwater pump train. The second is protective devices. Their function is redundant in the sense that they serve to alert or prevent another primary function failure. Protective devices whose function isn't required (nor typically desired) until an event occurs spend the majority of their lifetime in standby waiting for an event. Like redundant equipment, no backup need occurs until we lose the primary. An unintended transfer constitutes a control transfer functional failure. In the case of the redundant train, inadvertent transfer doesn't constitute functional failure-it merely swaps trains. But for a protective device, the component functional failure as an inadvertent activation very often creates an unplanned event and system level functional failure results. So, an unplanned and "incorrect" feedwater level trip is an unintended event-and failure. A spurious high-vibration alert trip on a steam-driven boiler feedpump is a failure.

Sometimes a functional test, a near miss, or a "demand-event" uncovers a protective device failure. Because safety devices have multiple backups, demands rarely utilize all devices and trains. Events typically are near misses, and most of them reveal loss of protective device redundancy and greater risk of functional failure. Nuclear power plants are effective at testing and identifying device functionality under nuclear technical specifications, safety programs, and general public safety considerations. Fossil unit rules aren't as structured.

System functional failures provide an objective measure to track failure performance. Though subject to interpretation—more so where the functional requirements haven't been formally identified—they are still very useful. Total WO numbers are arbitrary as a system performance measure, but system work hours and costs are not. For nuclear plants, MPFFs remain a convenient measure. For fossil, large unplanned expenses may provide the "failure" measure. Because failures themselves are hard to track, I developed two measures that correlate with failures and are simpler to use: system emergency work orders as a total number and percentage system overtime. These are easy to extract from CMMSs at most plants. Both are indicators of functional failures.

Costs

System operating costs are the obvious summary performance measure. Some systems are more cost intensive than others. A soot-blowing air system is an inherently high cost system in a PRB-fired boiler. Cost performance per standard cubic foot per minute of air produced is one measure of this system's performance. Unit efficiency and boiler pluggage events are others.

Because CMMS systems don't always allow cost monitoring below the plant or unit level, system costs—combined labor and material—provide meaningful cost data. As important as total system costs are, other numbers can tell more. The cost of overtime hours worked per system, or costs of irregular part expense, are examples.

At one plant, we arbitrarily selected unplanned failures costing above $25,000 to measure for overall performance. This selective analysis required manually tracking CMMS entries (subject to interpretation) but the results told a subtle story. Quantified in this way, eight major failures a year dropped to five because of our efforts. Stratifying measures must be performed with great care, and the advice of a statistician.

PM Hours

Real hours

Although in an ideal world all planned PMs get worked, in the real world this is never the case. Auditing PM jobs—like other CMMS-reported jobs—provides insights into how completely a program is implemented.

Many "Legacy" CMMS systems ran redundant time accounting systems separate from the time cards submitted for pay. WO completers could report any amount of time on WOs—the CMMS couldn't check or enforce simple time accounting rules. One aspect of this was that workers could work (according to CMMS time reports) any amount of time. In my experience, workers are biased by work planning time estimates and management expectations. Without an independent time accounting system, they aren't grounded by real world limitations. Measurement helps to provide this!

For accuracy, a system should charge time concurrently from time cards to WOs (or jobs, as appropriate). Fractional time charges down to the decimal hour are needed for PM time measurement. Many PMs are brief jobs—short enough that several may be worked in a morning or an afternoon. Accurate cost accounting is necessary to understand where time costs are allocated. In a typical plant, only 40-45% of work time gets charged against work jobs. The challenge for maintenance is to put in wrench time. Other things (like training) are important but all compete for limited available time. When wrench time drops below 40% the plant needs to worry about time usage competitors. These include safety meetings, extracurricular duties, and other supernumerary tasks. Mechanics who don't turn wrenches have little value.

WO time accounting must be controlled like checks. Paid time must be managed. Major time charges should be organized by Pareto chart, be tracked, and audited. Routine time consumers—rework, tool and parts shagging delays, tag-out delays, engineering support delays, and job planning (by the workman)—should be given charge numbers, so detailed time charges can be allocated. Bottlenecks, delays, and other losses can be identified for organizational review. The goal must be to continually increase work time charged to jobs in spite of the steady stream of worthy programs and other distractions calling for worker time and support. Without a wrench time productivity increase, corporate programs won't justify their costs.

Cost of a maintenance hour

A maintenance hour costs much more than a maintenance line item budget indicates. The range is several times more. (Fig. 9-3) This is why maintenance is more expensive than most people assume. Estimators peg job costs at two times the labor hour rate. When the time charged to hours and actual time on the job are factored in, one quickly concludes it's hard to get hours worked in plants. The cards are stacked against the poor maintenance worker. He has to do many things—especially if he plans and coordinates his own work. This makes expensive maintenance.

Why perform PM? A PM hour is a leveraged hour. It provides a payback (when properly developed) that is many times the invested time. Corrective maintenance isn't so leveraged. Unfortunately, many personnel fail to grasp the value of these different kinds of work hours.

Emergency. Emergency work hours are underestimated, based on detailed correlation with actual time worked. An emergency work hour is between five and ten times underestimated, by estimating bias. This estimate is based on numerical studies of work categories. (Many organizations fail to evaluate and learn their estimating biases.) If the estimate goes unmeasured and uncorrected, and it changes the capital hurdle rate costs over a factor of five-to make a modification financially successful, you must get at least five times more than the estimated return!

PM/CM Mix: Maintenance Process Measures

Effectiveness

No discussion of PM is complete without considering the measures of the entire maintenance process. While there are many measures required—FERC, NERC, INPO, Edison Electric Institute (EEI), EPRI, and others—utilities vary in their measurement performance. Regulatory-based need measures fall short when used for other purpos-

The Real Cost of Maintenance

Item	Cost	Description	Percentage	Additional	Running Time-cost Total	Range
1	Base Rate	The straight time base that workers are paid, hourly.		$17.00	$23.00	$20-40
2	Overheads	Normal corporate overhead charges.				$15-20
3	Material Cost	One-for-one assumption. A dollar spent on direct labor is matched by a dollar materials. Based upon experience.		$ 23.00		
4	On-the-Job Time	Time workers actually charge to work orders	35%	$ 42.71	$65.71	35-45%
5	Wrench Time	Time charged to job less transit, getting parts, waiting for clearances, getting tools, doing paper, consulting for help	50%		$131.43	45-55%
6	Rework	Percentage of work redone due to unsatisfactory completion	15%		$154.62	10-20%
7	Other Costs	Operator support time, based upon base rate.	20%	$ 4.60	$154.62	15-25%
Subtotal	Add-ons Sub & Total	Materials & direct support, items 2 & 6		$44.60	$154.62	
	Direct Time Cost	*Cost of one "good" hour of maintenance*			$199.22	

			Total Hours Straight Time	Annual Budget	Average Rate, Hour Charged	
	Overall Calculation	For a 540 Mwe Coal Plant with a maintenance budget of $8 million, and 40 maintenance workers:	91520	$8,000,000	$87.41	

			Hourly Incom	Annual Income		
Production Cost	Lost	At a production valuation of $18 – the 1997 corporate marginal rate.	$9,720.00	$ 85,147,200.00	Note: Production Value by far outweighs maintenance cost – as shown here.	(assuming no lost generation)
	Generation	540Mwe net @ $18				

Conclusion: Maintenance is expensive. Every hour of maintenance optimized pays back roughly 10 times the apparent budget line item savings.

Figure 9-3: Maintenance Work Cost

es, such as cost accounting. Measures based on first principles have more direct use.

Responsiveness

As maintenance programs evolve towards CNM, measurement balance needs to be sought. Appropriate levels of CNM depend on system type and strategy. Many strategies can achieve the same operating objectives though at different cost and complexity levels. A strategy must fit an organization.

Failure measurement is not possible without accepted failure standards. A traditional program lacks explicit definitions. Even nuclear maintenance programs lack function-based failure criteria, as relatively minor events—the charging motor run-on in a 4160 breaker—get described as failures. Major failures can go entirely unrecognized. The secondary failure that results is often the focus of investigation. A breaker fails to trip on overload, causing a fire, or an alarm fails to annunciate an unsafe condition, like methane or carbon monoxide gas. Events that should have been excluded by operation in fact become the focus of investigation.

Total hours/system

Total work hours per system—broken down by PM (TBM + OCM), CM (CBM + OTF) and functional failures—are a meaningful RCM-based measure. The key ratios are the percentage of each ARCM category. These profile the system. The continuous improvement goal is to reduce required hours and costs. Where systems lack total cost measurement capacity, tracking total work hours is a second useful measure.

Trends

System downtime and functional failures are important performance measures. Identifying functional failures is a challenge when functional-failure definition and perspective is absent. Having these measures requires that a company has engaged in goal setting for the unit. This establishes relevant failures. Many haven't.

Aging studies

WO populations should show continuous progression to completion. If we examine progressively "aged" WO group snapshots over time, we should see incomplete WO numbers decline. Mathematically, WOs are worked proportional to number and age. When this doesn't happen, the system's in trouble. Regular aging reports are useful for telling management whether their maintenance system fundamentally works.

Costs

Change should generate improved performance, lower unit costs, more flexible operations, or all of the above. Historically, plants have never had income benefits allocated at the unit level and so higher income generating units have requests buried in with peer units. Utility cost and income aggregation are to blame.

Merchant-independents stand alone from a financial perspective. Even then, unit costs allocate downward to systems and equipment by tedious manual methods. The before/after snapshot of any significant process change can be obscured. Utilities, as vertically integrated structures, suffer incomplete cost accounting at the unit level.

New CMMS systems greatly improve cost tracking. They're dependent on data entry, but use hierarchies that are interact with and extract data easier. They promise better information capture and presentation.

It's difficult to tackle more than one plant system improvement at a time. System cost trends—total, routine, outage, emergency, and modification costs—are major interest categories. Some cost expenditures are most important. PM time and expense are among them. These need tracking categories. For the cost-driving systems, these cost trends will be important.

Ratios

Maintenance ratios tell a story. The emergency to routine maintenance ratio-by hours-reveals how a system's work is managed. One CMMS coded work priority on a continuum range from E to "3"s (E-1-2-3). Es were unscheduled, unplanned WOs; "3" were planned and scheduled. High E/3s reflected reactive maintenance. (High and low are relative.) Compared to other systems, units, plants, and companies-these tell a story.

A high E/3 ratio may be suitable and effective for some systems. Generally, for high intensity mechanical equipment, it's not.

Companies pay maintenance workers overtime for emergency (E) and unplanned work. It requires additional support, lacks parts, and when all is said and done, is the least effective work an organization does. A plant should avoid E work. Routine, planned, repetitive work is at the other extreme. It's efficient and low cost. More routine planning should be sought. The ratio emphasizes this relationship.

Rework

Rework is a significant cost-contributor because maintenance is expensive. If you can track rework causes, you can reduce them with significant benefits. Manufacturers follow rework and "scrap cost" in depth. Manufacturers neither want to make scrap nor send it out, incurring warranty or other cost charge-backs. Maintenance is a process—but tracking rework is like tracking scrap. WOs need to identify reworked equipment and jobs for trending and root-cause assessment.

Workers, generally readily identify rework on jobs. With their participation, rework maintenance can be measured. Things you can measure can be improved. Sources of rework should be identified for process improvement.

Screening for effectiveness

Effective maintenance screens new WO requests as they're input. Different systems generate different work values. Many utility systems focus on capturing all WOs—legitimate, undefined, and speculative. Nuclear plants in particular allow the documentation of incompletely specified work. This is a great burden to a WO system, particularly when it could be screened.

Maintenance "screens" quickly identify and reject inappropriate work. WO screening is improved with a developed RCM process because operator non-specific work requests are identified as "CNM" type work orders. These can't be worked directly until someone—usually a work planner or engineer—has defined their scope.

Contrast this with a developed TBM or OCM (or OCMFF) type WO. These are exactly defined and have exact follow-up work plans. They have well-defined workscopes and failure criteria. The adoption of an RCM model allows schedulers to quickly separate the known, defined work for immediate work, from the vague, conjectural, ill-defined and speculative work that constitutes OTF. These latter work requests need specification, or return to originators until a failure can be defined. A failure you can't specify, you can't fix!

Chapter 10
Conclusions

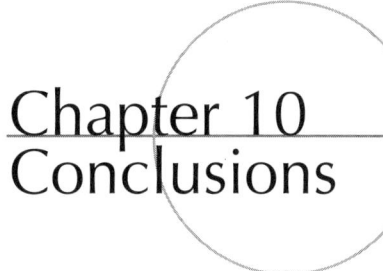

Learn from the mistakes of others. You won't get to make them all yourself.
-Eleanor Roosevelt
Just because something's old doesn't mean you throw it away.
- Scotty, "Relics," Star Trek
God integrates empirically.
-Albert Einstein

Organizational Entropy

Entropy explains why our natural order trends with time towards disorder. Entropy explains why thermodynamic cycles have limits, heat flows in a single direction, and why temperature and time have meaning. Entropy is a powerful concept and one of the three laws of thermodynamics, (paraphrased):

- You can't win (conservation of energy)
- You can't break even (entropy)
- You can't quit the game (thermodynamics provides the rule book)

Figure 10-1: The Fort Saint Vrain power plant in Platteville, CO., was operated as a conventional nuclear plant until its sporadic operations and high cost caused its shutdown and decomissioning. Today, it's operated as a combined-cycle gas generator.

Recent organizational models apply thermodynamic principles to organizations and processes. Entropy helps to explain the apparent confusion and disorder among some large organizations as they do so. Entropy can help us understand operating environments. We might view a "situation normal all fouled up" (SNAFU) as an individual fault—a person messing up the job—but an entropy model suggests it is the nature of the system. Things will malfunction without continuous addition of energy and intelligence to the process (Fig. 10-1).

This explains accepted aspects of operations that have never been theoretically considered before. Operations demand intelligence and energy. Every conscientious worker in an operating environment knows this. Outstanding operations demand more! The assumption that "order" is the normal state of affairs is simply founded on idealism. Complex operations need information and control to produce value. This only happens if "intelligence" offsets the system's natural tendency to unwind.

Management provides the framework that provides order.

Traditional management techniques—strong-arming, intimidation, shooting from the hip, power plays, scapegoating—are inherently weaker than techniques founded on facts, processes, and scientific methods. TQM provides a general management model for process improvement. Two distinct characteristics are organizational systems and processes with "feedback"—precepts similar to ARCM.

What are some "intelligence" inputs?

- operating goals
- operating plans
- training
- staff selection
- work processes
- standards

How can these be used to achieve consistency and predictability of operations?

Organizational process intelligence is, in part, the shared knowledge of the organization's members. Randomness results when workers don't understand that individual effort matters. Successful organizations increasingly tap workers to help create value, control entropy, and to retain and increase market share. Making a product—any product— is a difficult task, even in a simple shop. Doing this profitably demands creativity in a competitive environment. Electricity is a product and generating it demands these same traits.

Maintenance is a complex organizational process with different levels and perspectives. Through the years, organizations have developed methods of performing maintenance based on experience, and applied these with success. Thumb rules reflect fundamental rules and laws. We don't need to understand theory to apply them successfully. On the other hand, understanding the theoretical basis may provide insights to enable us to apply the rules more broadly, and ultimately provide a competitive edge.

Natural laws also govern business. Top performing organizations are those that embrace fundamental processes and principles, those that actively search out new theory and technology to further define their processes, and so maintain a cutting edge.

What is ARCM?

The general focus of RCM is identifying and preventing functional failures. ARCM goes one step further. It throws out the dogmatic styles in favor of pragmatics. In effect, "If it works, use it." ARCM retains the unique, fundamental principles introduced by Nolan and Heap, Matteson, and all the other RCM pioneers—the factual basis, applied statistics, applied engineering methods, benchmarks, and applications based upon basic logic principles. Basics that validate "NSM" in a complex equipment strategy. Basics that can effectively control high-impact, large equipment outages with on-condition/CDM, and use CNM as a general operations strategy. Basics that faithfully apply and operationalize "on-condition" limits that so uniquely delineate Nolan and Heap's published work from others. Methods to schedule time-based and OCM equitably with the balance of non-specific work, with assurances that CDM is "worked." Methods that accept and apply risk management to substantially improve performance, and lower risks and cost (Fig. 10-2).

The bottom line is improved R (reducing functional failures), reduced costs, improved quality, and well-supported corporate missions.

ARCM shares basic similarities with TQM. Both are founded on statistics. TQM summarizes generalized lessons from early SPC applications that became tools for present day managers. Some conclusions still apply. Others provide insights, but must be taken in context.

Embracing ARCM—developing strategies to reduce costs—is what facility operations are all about. A facility or company can practice RCM and still not know about LTA or other detailed TRCM methods. Does an understanding of RCM methodology help? Absolutely!

Years ago, corporate cost-cutters—accountants—trimmed plant budgets and cleaned out the shops. Experienced people left, planned maintenance and training programs were cut. The opportunity to improve processes was lost. Workers obviously disliked the top-down cutting, but no one understood the value of the losses, cuts, or future costs. Several years later, R was down, production down, unit costs up, and maintenance was more reactive than ever.

Maintenance is an inherent cost of production, a fundamental con-

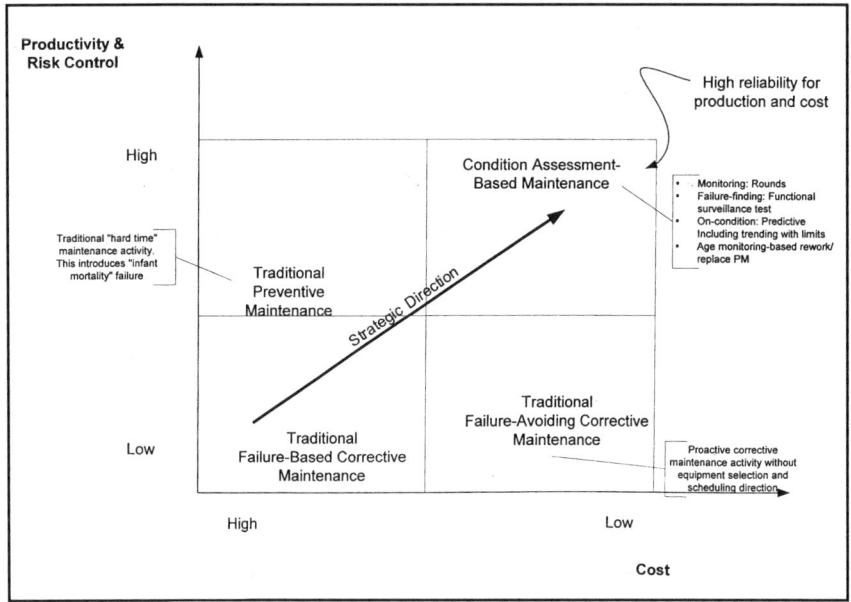

Figure 10-2: Maintenance Strategy Map

sequence of the second law of thermodynamics. You can trim maintenance costs, but you can't eliminate the randomness or time elements of failure. Improving maintenance requires process improvement. You can play accounting tricks to flavor costs, but ultimately equipment and system failures tell the bigger story. Creative numbers can't subdue entropy or reality.

For those people who enjoy maintenance (just as others love design and still others like to operate), who understand that the three roles interconnect, and who pursue maintenance theory and technology in large facilities hell-bent on the being the best, ARCM can help bring order to a crowded, complex field.

Japanese authors, Masaji Tajiri and Fumio Gotoh describe a maintenance strategy called TPM. TPM shares some aspects of ARCM in the final result. The end product—tasks scheduled and worked—are similar. Different equipment and cost optimization schemes can lead to similar ends. However, based on the published descriptions, TPM and RCM have paths that are radically different. TPM presumes a large amount of time available for group learning and fails to explain how the infrastructure that supports maintenance performance gets built. (Yet, infrastructure is required.)

TQM has supporting techniques such as SPC that share a great deal in common with RCM, as originally defined. Statistical analysis, factual basis, analytical methods, processes, and ways of deducing causes and focusing on the "critical few" are common threads. TPM (the Japanese approach) and TQM depend on an intense pursuit of work that, in my experience, isn't a part of today's American industrial culture. Not that people in American industry aren't very proud of their work and work efforts—they are. They just generally don't work 60-or-more-hour weeks without extra compensation as other cultures do!

In addition, TPM and TQM are iterative approaches. In the absence of another specific strategy, either (or both) appear to be good ways to iterate general process improvements. ARCM, on the other hand, has the capacity to generate the "end-run"—the substantial leap that companies strive to find in benchmarking outside their own industries.

As a maintenance process, ARCM provides an improvement path. ARCM is an objective, rational, statistically-based engineering strategy that can be applied theoretically to new facilities to help initiate programs. Different means can achieve the same end but one approach may support the end better based on culture, cost, and commitment. There can be multiple paths but ARCM best integrates two functions that have for years been culturally disjointed in plant maintenance in North America-engineering and maintenance.

Consider baseball. You can commit an error here and there and still win the game. You can't commit more than a few, though. Given the tools available, developing a near error-free maintenance plan has never been easier than it is today. Planned maintenance performance is very controllable. To suffer losses because the maintenance plan isn't developed, implemented, or followed represents "inadequate fielding." In baseball, players need to field consistently. A similar standard is being raised for facility managers today. Those who can manage maintenance will step forward, others may need to step aside. Competitive markets will force the changes.

Statistical maintenance

Maintenance itself is inexact, with many strategies, no single one of which is "right." Many work. Statistically, we must:

- understand basic equipment failure types and frequencies
- put controlled processes into place to manage failures
- use statistical tools to improve actions taken to address an identified equipment deficiency
- learn and improve

A simple feedback loop goes a long way toward improving performance. Consider the metaphor of the basic operational amplifier. Feedback allows this relatively insensitive but high gain electronic device to be transformed into a powerfully accurate, simple signal amplifier. Sure, you sacrifice a lot of the theoretical gain, but you now have a smart device locked on the input signal with plenty of practical gain left. Feedback into the maintenance process can likewise greatly help to reduce output variability and improve results.

In hindsight, most high impact events in my experience had obvious monitoring precursors. Some were so pathetically obvious we had to tell ourselves (at the time) that nothing could have been done to overcome glaring equipment deficiencies. These organizational blinders were maintained even as outsiders repeatedly pointed them out. A common thread was our failure as operators to consciously appreciate our decisions and their risks.

There is a cultural weakness in the engineering transfer of intelligent designs to operating groups. In many cases, the designers anticipate eventual failures and provide the means to monitor the high-impact ones. Operators, on the other hand, systematically ignore essential instrumentation in the vast sea of available equipment and fail to act based upon instrumentation and senses when major failures are imminent.

Our ability to monitor for failure is too often compromised. ARCM provides a substantial opportunity for designers and engineers to develop better operator guidance for anticipated events over the facility's economic life. I envision a day when, along with the vendor manual, operators receive a detailed ARCM-based optimized vendor maintenance program. Operators, through their A-Es, will have the foundation for a cost-effective maintenance strategy from startup. This in turn will support better staffing, life cycle decision-making, and, ultimately, lower overall facility maintenance costs. Performance levels will be known for operators to benchmark. In short, a step-advance in facility O&M performance is dawning.

Maintenance is to some degree an art. There are no panaceas that suddenly make all maintenance decisions simple and clear. Even with the very best RCM plans in hand, hard diagnostics, interpretations, and choices are and will be required. But, armed with a maintenance plan, operators will have better tools to interpret equipment, plan for maintenance, and perform work in a reduced cost fashion. This has been the lesson of the commercial aviation industry. The challenge is to introduce the appropriate degree of rigor into an ambiguous environment to improve managing risk and cost.

A general, repetitive RCM lesson has been identifying a basic essential instrumentation strategy to help manage operating risks. The opportunity suggested here is obvious: if the reader grasps an appreciation for the need to quantify, understand, maintain, and manage essential instrumentation from reading this book, they will gain great value. The flipside of the coin—learning to manage non-essential instruments—is a corollary. In this world of expanding hardware capacity, it is particularly important to control the "vital few" versus the "trivial many." Operators must learn to discriminate and act on essential instruments. This is the "low hanging" fruit that many maintenance managers should grasp. Unknown or inadequate instrument maintenance plans cost dearly in production, cost, and (rarely) in employee and public safety.

The general lesson of RCM is CNM. Organizations with a CNM philosophy are reliable. It's possible to go overboard but generally the other case occurs. Little or no CNM, and absence of follow-through on the insights provided by monitoring, are the trademarks of unreliable and unsafe operators. Like the person who feels ill but is afraid to visit the doctor for fear of having a worst fear confirmed, failure to act on CNM adds risk. Understanding problems and alternatives enables us to select options. Rarely are we saddled by "Hobson's choice:" take what's offered or take nothing at all. Which organization is more successful in the long run—one that avoids reality and flees from information? Or the one that embraces the future and its risks, and makes the best possible decisions with the available information? From a plant perspective, a "heads up" of a few days (or even hours) can often significantly reduce operating impacts of imminent failures. The very best plant operators demonstrate that unplanned production outages and costs can be safety put to rest with a plan.

Applied to existing facilities, the ARCM approach can zero in on areas of maximum value. Facilities are often inhibited from changing existing practices. Change is expensive and hard to sell. There is a cultural foundation at every plant that's hard to move, even in light of compelling cases for change. New operators are more receptive to new processes, and an owner's desires—but it's still difficult.

Are there alternatives?

Twenty-five years of maintenance have led me to look for new methods that could fundamentally improve maintenance performance. I've reviewed TQM and TPM materials, both of which offer valuable insights. Yet, I don't believe they are "technology" as much as philosophy. Tajiri and Gotoh (TPM) outline methods that, in their final form offer some of the same general insights as ARCM. If a solution is unique, then differing methods should point towards the same solution. This is a comforting validation.

What some maintenance optimization approaches ignore are specific methods of technical approach and process. This book has avoided detailed RCM process theory tedium or general strategy but readers can readily find suitable materials in the references that will guide them through RCM subjects. We've focused instead upon the lessons and practical elements of maintenance performance as they relate to operations.

Maintenance is a sophisticated process mostly taken for granted—especially by corporate financial staffs, managers, and executives, who presume they understand the finer details when in fact that takes years in the trenches. The good news for them is that there are many opportunities to improve maintenance while holding the line or even reducing costs. Many organizations can and have positioned themselves to take advantage of this. Many more will join them with existing staffs and processes largely in place. External forces will redirect some others. For traditional utilities, competition is a powerful outside force, and most will need to make a conscious decision to find new maintenance approaches.

For innovators, these are exciting times. A tremendous amount of theory, technology, and process understanding has led to practical problem approaches. OEMs are more aware of user needs. Software suppliers are finally grasping software friendliness and accessibility needs. New monitoring technologies have been perfected and become accepted. Corporate utility staffs can now visualize profit and loss differences in a competitive environment in plant availability and R terms. Some players are getting out of generation altogether, creating more opportunity for others staying in. Competitive forces are forcing serious players to wake up. For operating staffs, there's a better chance that management will fully engage them to develop methods that deliver performance. They will be supported with tools and training. For the traditional "mom and pop" utility, the writing is on the wall. Their family approaches can't be competitive with best performers. In the absence of a regulated economic protection, they too will pass away.

To understand maintenance, you must understand operations, engineering, and statistics. Outage management, computer scheduling, and data management are also needed. But finally, in this world of change, a profound knowledge of equipment design, it's functionality and weaknesses are necessary to support safe, cost-effective decisions. The excitement is that a relatively new maintenance theory is still available to reshape and simplify fundamental plant operations thinking. And it can still transform your way of viewing and performing maintenance!

Appendices

1. Glossary of Terms
2. Further Readings
3. RCM Software Applications
4. References

Glossary

80/20 Rule
A "rule" attributed to the Italian statistician Pareto based upon his study of 18th century economic wealth distribution in Italy. For our purposes, it attributes 80% of the problems to 20% of the equipment. Generally the Pareto rule can be found in many skewed distributions.

Abandoned-in-place
Equipment removed from service and left in its location in a plant because the cost of removal exceeds the scrap value. Equipment that adds marginal value to a process compared to cost, may be left "unmaintained" in place with no cost or production impact.

Acceptance criteria
Specific limits for acceptance. A term with general meaning, but nuclear origin. Also used for operationalized failure criteria, generally describing a test. Time-based PM will sometimes have acceptance criteria. Large turbine bearings will be reused as is, *i.e.*, unless wear exceeds so many thousandths.

Actuarial
Mathematical statistical failure analysis using practices accepted by the insurance industry and Society of Actuaries (SOA) or other professional groups (as appropriate) to measure aging, risk, and mortality (based on study for human populations).

A-E
Architect-engineer. The facility designer usually hired as a contractor to provide a facility design. Occasionally also the constructor ("design-construct")

After-market
The parts and services supply market for equipment other than through the OEM. After-market parts can be superior to OEM parts, but there's much greater risk in the after market for parts reliability and quality. This risk is generally borne by the buyer.

Age

Time correlation factor. The accumulated time since the equipment was placed in service. In general, time equivalent for the aging process in question. This could be tonnage for wearing parts, for example.

Age exploration

A systematic process of using conditional overhauls and opportunity samples, with formal cost analysis, to evaluate and improve designs.

Age parameter

The measure of aging, which correlates with resistance to failure for parts that age (dependent on the part, application, and use). Used as the general time basis for maintenance.

Aging

A process that can reduce a part or component's resistance to failure over time. To grow old or show signs of growing old. Synonyms: deteriorate, fatigue, waste, tire, exhaust, flag, droop, ply-out, drained, spent, depleted, obsolete, erode, consume, fret, rub, fray, erode, weather, corrode, oxidized, rust, disintegrate, spoil, decay, decompose, break.

ALARA

"As low as reasonably achievable:" a program of minimizing radiation exposure required for all nuclear license holders. The basis is to avoid unnecessary radiation exposure and its long-term damaging somatic and genetic effects.

Align

Block: put multiple PM tasks together for performance at one time when an equipment train, unit, or even plant is available. Aligning intrusive PMs into a unit outage window is an obvious example. For scheduled work, once work is aligned, it facilitates the performance of scheduled maintenance.

ANI

American Nuclear Insurers: a consortium of insurers providing nuclear insurance backed by law.

ANSI

American National Standards Institute: a standards organization that certifies and maintains standards. These include many industrial and power generation standards. Common ones for nuclear plants are ANSI N45.2 and N18.7, for procedure use and maintenance management.

Applicability
In traditional RCM use, the requirement that assures a PM activity is technically and statistically effective—*i.e.*, it actually prevents failures.

Applicable
Prevents failure.

ARCM
Applied RCM: one abbreviated version of RCM that retains the fundamental salient elements of RCM (as described in the document by Nolan and Heap). Includes maintenance strategies TBM, OCM, CDM, OCMFF, and NSM, but simplifies analysis.

Area inspection
A general walk-around inspection that checks for random or other failures. For an airplane or car, pre-service visual check for "obvious" problems.

ASCE
American Society of Civil Engineers.

ASME
American Society of Mechanical Engineers.

ASQC
American Society for Quality Control.

ASTM
American Society for Testing and Materials

Availability
Defined exactly by NERC. The period of time a unit is available to be dispatched for generation, whether it is or not. Expressed as a fraction of calendar time.

Base load
Loading a unit to full rating between scheduled down periods. Typical for nuclear and low-cost generators. The opposite load term is "peak load."

Basic interval

"Prime interval"—the most fundamental interval when aligning a PM task that fits the frequency, and provides reasonable multiples for the overall maintenance program. A car requiring PMs at 12, 24, 30, 48 and 60 months has a basic interval of 12. Often the interval that also carries a fundamental condition-directed maintenance program inspection task. Missing a basic interval PM in a completed program carries a serious consequence.

Basis

The justification—the reason "why." Usually at least partially implicit.

B/C

Benefit/cost: benefit-to-cost ratio, commonly represented backwards as cost/benefit ratio. For instance: replacing the oil has a cost/benefit ratio of 10/1. It is really the other way but most people understand and state it this way.

Benchmark

Comparing costs for similar processes, facilities, or equipment. Often performed within an industry to validate competitive position, and outside to identify world-class performers.

Bit map

An image in a computerized format literally imported as a map of image "bits."

Block

Aggregate, group, or align for performance.

Blower

A low head fan that supplies gas (usually air).

Blue blush

Deep blue-hued carbon deposits on high-pressure, high-temperature steam valves. Through time, these build up to where the valve may bind during stroke. These require periodic removal.

Bootstrap

In computer jargon, a preliminary short software routine that allows loading the main operating system. A system that gets the machine up to minimum "smarts" to run.

Breakdown
Fail suddenly, with little or no warning. A failure that impacts operations and production schedules.

Breaker
Circuit breaker

Burn-up
Slang: in nuclear work, reaching the radiation exposure administrative limit. It pulls a person off the available worker list; they are referred to as "burned up" based on reaching their maximum weekly or monthly radiation exposure limit.

Bus
Electrical bus

Bus bar
The output bus that connects the generator (stepped-up) output to the transmission grid. Often used in the context of "bus bar cost"—the cost to generate at the grid connection.

B&W
Babcock and Wilcox: a large industry supplier of boiler and nuclear steam supply systems.

BWR
Boiling water reactor

Calibrate
To adjust an instrument for zero and span due to drift. A basic PM activity that is often time-based.

Call out
Calling out a person for work after normal work hours end, usually in response to a plant need.

CBM
Condition-based maintenance: the same as condition-directed maintenance. Sometimes used to maintain a distinction between condition-monitoring program derived maintenance tasks and formal on-condition-derived condition-directed maintenance.

CDM

Condition-directed maintenance: maintenance directed by condition. Obligatory maintenance based upon defined failure limits exceeded. CDM is the fundamental differentiator of a firmly RCM-based PM program. To use effectively, it must be reserved for those on-condition tasks with explicitly defined failure limits.

CDM (FF)

Failure finding condition-directed maintenance. A special type of condition directed maintenance in which the acceptance criteria constitutes satisfactory test performance. For instance, a diesel generator in standby mode could be required to demonstrate that it can meet its design specifications by starting and loading to 650 KWe within 10 seconds as the test criteria.

CE

Combustion Engineering: a supplier of power generating plants and equipment. Now ABB CE.

CFR

Code of federal regulations.

Chargeable loss

A loss that can be charged to a specific cause. Used in particular for forced outage measurements. A restriction due to water chemistry during startup is a chargeable loss to water chemistry.

CIC

Component identification code: a unique equipment identification code.

Clearance

Tag out: a method to control equipment for work. A tag out or clearance is required to isolate equipment from energy sources for personnel safety.

CNM

Condition monitoring: operations-implemented general equipment monitoring for failure. Tasks for which no exact "on-condition" limit can be determined. When there's agreement that a benefit exists these tasks are implemented as non-specific condition monitoring, *i.e.*, CNM tasks for a coal belt walk-around check includes lighting, leaks, oil and water, and coal accumulation. All have distinct benefits, but it's hard to create an "on-condition" task for each.

Caution: Some writers use the term synonymously with "on-condition" used in this book.

CMMS
Computerized maintenance management system (from 1990s onward). Called maintenance information system (MIS) in the 1980s. A computerized WO initiation, tracking, planning, scheduling, approval, and archive system. A sophisticated computer software system that's at the core of maintenance management in complex operating environments.

CO
Conditional overhaul: an overhaul that corrects the proximate cause of failure, secondary failures, and restores equipment to performance specification. CO does not completely disassemble nor replace all replaceable parts, although it does replace any aging components prior to the next scheduled overhaul period. CO "zero-times" the equipment. (See also Control Operator).

Cogen
Cogeneration: a type of generation authorized by law to allow non-utility participation in the generating market. Being phased out in favor of independent power producers and separation of generating and transmission and distribution assets.

Common mode failure
A failure mode that compromises redundant train, equipment, or component independence. Because of this, it changes assumptions of independent failures in FTA. Common mode failures significantly change the overall probability of accident events and are of great concern to regulators. Practically, maintenance practice has the potential to introduce common mode failures systematically, so performers need to be sensitive to this. For instance, one person using inappropriate grease lubrication could inadvertently set up a common mode failure on all the equipment greased incorrectly with that lubricant. This is what happened with motor operated valve operators and the use of incompatible grease in the 1980s at nuclear plants.

Complex Item
An item that fails to exhibit dominant failure modes and thus fails randomly in service with no known age. Practically, an item that fails to show aging.

Conditional probability of failure
The probability of failure in the next operating time interval, given survival into the current one. Also, one of several mathematical failure probability curves.

Conditional Overhaul
An overhaul that conditionally addresses on the observed discrepencies, returning the cycle-dependent time measure to zero. An overhaul that addresses all proximate failure causes, as well as any aging to restore the unit's post-overhaul aging parameter to zero. A maintenance activity that corrects and "zero-times" a piece of equipment.

Condition-based
Condition-directed, including the non-specific results of condition monitoring, which are separate from condition-directed in that no commonly agreed failure resistance has necessarily been exceeded.

Conservatism
The tendency to prefer an existing situation to change; safe. In engineering, the provision of design margins to accomodate uncertainty.

Control Operator
The operator who manages the control room boards or DCS CRT. Practically, the operator who is running the plant.

Constructor
The facility builder. Occasionally the same as the A-E.

Corrective maintenance
In former days, work "on demand" to correct failed equipment. A term gradually falling out of favor due to its limitations and bias. Most corrective maintenance is a combination of condition-directed, condition-based no-scheduled maintenance but has never been differentiated. Old MIS systems categorized work two ways—preventive, and corrective.

Cost

Maintenance cost: combination of hourly cost, material cost, services cost, and overhead. Typically five to six times the hourly cost. Excludes opportunity cost of lost generation.

Cost effective

Worthwhile based upon cost/benefit perspective in a general sense. Since PM implicitly considers the time difference from performance cost incurred to benefit received, PM cost effectiveness implies using the time value of money.

Critical

Immediate and direct safety consequences, usually unacceptable. Critical is used in common language to identify any important failure. Readers of the literature and particularly in RCM should skim to discern the author's context for the use of "critical."

Critical failure

A failure with an immediate and direct safety consequence. A failure whose risk is unacceptable based upon accepted safety standards. Usually an immediate and direct threat to personnel, public, or (in rare cases) the environment.

Critical few

The statistical few that predominantly drives the totals presented in Poreto format by category. This comes from statistical analysis of data in TQM and Process Improvement Technology. As an example: one finds that the combination of bearing failures and insulation resistance failure account for most motor failures—statistically.

Critical instruments

Instruments that identify critical failures, such as excessive vibration for a large rotating machine.

Criticality

Criticality analysis is analysis done in some streamlined RCM approaches to identify how important a failure mode really is. It is subjective, based on interview and conjecture, and therefore of limited use for assessment and management of experiential risk.

CT
Combustion turbine

DCS
Distributed control system: a plant-wide control system common in many non-nuclear applications. The state of the art in control system technology at this writing.

Design
Involving the specification, selection, and layout of equipment and materials, in contrast with maintenance. Maintenance and design often overlap in indistinct ways. Maintenance in fossil environments often performs "light" design roles—sometimes without being aware of the design role.

Direct cost
Cost at the point of application—in contrast with indirect cost or overhead. Direct labor, material, and services costs show up as plant expenses in traditional utilities. Indirect costs (in my experience) don't—although they ultimately effect the bus bar generation cost to the consumer.

Discard
A type of PM task where a part is replaced based on time.

Dissimilar metal welds
Welds in boiler tubing used where the tube makes a transition from steel to stainless or Inconel, and vice versa. Typically, the design objective is to use expensive alloy only where necessary—in the high-temperature area. This introduces a DSM transition—which is a weak point.

DOE
Department of Energy

Dominant failure mode
A predominant failure mode. Often used in reference to complex equipment and equipment with multiple failure modes. Dominant failure modes are based on statistical experience. Their utility is that they help focus failure management effort.

DOT
Department of Transportation

Economic dispatch
Dispatch of the next unit of available generation based upon marginal cost as the system load increases. Economic dispatch calculates the cost of the next available MWe of generation and dispatches that unit that provides it. Most PUCs require public utilities operating distribution systems to follow such an economic dispatch model. The automatic generation control (AGC) will identify which units are to be loaded next (or removed first) as the load varies over the course of the shift and day. In reverse, as the load drops, the most expensive generation is shut down first.

EEI
Edison Electric Institute: a generating utility trade group.

Effective
Used casually in two contexts: technically effective and cost-effective. RCM reserves the term to address cost-effectiveness and uses "applicable" for technical effectiveness.

EFOR
Equivalent forced outage rate: forced outage rate adjusted to reflect the equivalent effect of forced load restrictions. Defined exactly by NERC.

Empirical
Based on trial and error—experiential.

Engineering cause
The local cause responsible for failure. A bearing experiencing a wipe with no lubrication has fretting, spalling, or wipe as the engineering cause. A plugged filter, mixed-up round, or broken supply line is not the engineering cause—even though any one of them could have been a root cause. Engineers often refer to engineering failure cause as root causes. I call them proximate causes. The confusion is due to the multitude of different definitions in various standards.

EO
Equipment operator: one level up from a tender. The roving operator who starts and stops most heavy equipment requiring local monitoring. Typically a senior experienced operator with 10 or more years of experience.

EPA
Environmental Protection Agency

EPRI
(The) Electric Research Power Institute: a voluntary, industry sponsored research organization that performs most research for the generating industry.

EQ
Environmentally qualified: a special class of (essential) nuclear equipment required to perform shutdown and monitoring activity following a hypothetical design basis accident. Typically, equipment with organic and elastomeric materials that is susceptible to temperature aging. These require special scheduled maintenance programs by law.

Event tree
A logic tree that shows the pathways from a primary event upward to a final outcome. Used to identify contributors to overall outcome risk.

Evident
Evident failure: a failure in which the failed item should be evident to a qualified operator performing their normal duties. Contrast with hidden failure: one that no one would be aware of except by monitoring or performing special checks and tests.

FAA
Federal Aviation Administration

Fail safe
Fails in the safe direction or position, *e.g.*, a fail-safe air-operated control valve would fail shut if that were the "safe" direction. This is how the feedwater-regulating valve in some fossil plants fails since it avoids flooding the steam drum. Fail-safe valve positions include FAI—fails as is, FC—fails closed, and FO—fails open, for example.

Failure Mode
Repetitive manner of failure intrinsic to design and application.

Failure finding
Testing to identify a hidden failure. Startup test of a redundant train,

channel checks, and alarm checks are examples of standard (on-condition) failure-finding tasks.

Failure

(Broad) An unsatisfactory condition that fails to meet expectation. (Specific) A condition that fails to meet a performance standard.

Failure maintenance

Failure-based maintenance: maintenance based on a failed condition. A type of corrective maintenance.

Failure mechanism

Failure mode and a cause.

Failure substitution

Substitution of a major failure with a minor one. Redesign to lower the consequences of failure.

Fault tree

Logic tree of outcomes traced to primary events through system logic modeling that enables the calculation of failure probabilities for certain events.

Feeder

Equipment that feeds a continuous, controllable stream of product into a process. A coal feeder, for example.

FERC

Federal Energy Regulatory Commission: a federal commission charged with the regulation of interstate power sales.

Fishbone

Fishbone diagram: Ishikawa diagram.

Flash

For computer applications, memory set permanently in EPROM. Also short for "flashover."

Flashover

Electric plasma arc due to failed equipment or protective device. A high-energy arc that can cause injury or fire. Commonly occurs in switchyards, on switchgear, or in large motors and controllers, which use medium voltage (2000-8000 V) electric equipment. Requires high voltage due to the electric resistance of air. Once initiated, however, requires another circuit interruption to terminate.

FMEA

Failure modes and effects analysis: systematic review of the ways that equipment is expected to fail, and consequences of the failure. A qualitative enumeration of failure modes. Developed as a discipline in the early 1960s as a technique to improve reliability. Made into a Mil spec standard, and required as a part of some defense design proposals

FMECA

Failure modes and effects *criticality* analysis: FMEA with criticality calculation based on standard published component reliability figures. FMECA extends FMEA to a numerical basis. This in turn allows techniques such as reliability allocation to be used as a design tool.

FOR

Forced outage rate: forced outage hours(forced outage and service hours) expressed as a percentage. A NERC reliability measure.

Fossil

Fossil-fired boiler or combustion turbine fueled with fossil fuel.

Function

Output(s) provided by a system.

Functional failure

Loss of one or more system outputs. Loss of a system purpose.

GADS

Generation availability data system: a statistical reporting system operated by NERC that summarizes broad categories of electric generating plant performance.

Gaitronix

Plant public address and personal communication system.(A trade name.)

GE

General Electric. Supplier of power plants, turbines, and electrical equipment.

Graybeard

Experienced personnel who know the ropes. Usually with 20 or more years of experience.

Gun Deck

Perform superficially but completing the paperwork to perfection.

GUI

Graphical user interface: a screen interface that uses a mouse to isolate and execute commands

Gyrol

(slang) A soft-start single speed transmission between a motor and a load. Used on large coal belts. Named for a manufacturer.

Hard time

Time-based—not condition-based. Sometimes used for emphasis to indicate activity that could be worked as OCM, but is left at hard time due to program maturity, or intent.

Hard wired

Unchangeable, unmodifiable. (Slang) Impossible to mess up. Contrast with "jumpers."

Heat rate

Heat required to generate a MWhr of load. Literally, the efficiency of the plant to convert fuel to generation. The inverse of efficiency. Typically fossil range is 6000-12000 BTU/MWhr

HEU

Hydraulic equipment units: a hydraulic equipment skid.

Hidden

Not evident to operators performing their normal routine duties.

Hidden failure
Opposite of evident. Describes a failure not normally evident to the operator without special instrument and test.

House
Main plant.

HRSG
Heat recovery steam generator: steam generator used for combustion turbine heat recovery.

HTGR
High temperature gas reactor: a nuclear reactor cooled by helium. The retired Peach Bottom 1 and Fort St. Vrain plants were these types of plants.

Hydro
Hydroelectric

I&C
Instrument & control

IEEE
Institute of Electrical and Electronic Engineers

Ignitor
Device used to ignite a fuel stream in a boiler, usually oil or pulverized coal. An electric ignition source (spark plug) and fuel—usually gas or oil that assures a flame for combustion.

Important
Worthy for consideration of scheduled maintenance. A classification used by EPRI for streamlined RCM.

In service
A device during its useful life. Typically used to describe equipment, train, or plant that is operating, or capable of being operated.

Infant mortality
Failures shortly after entry into service due to quality, defect, and other latent causes that drop as service life increases. For electronics, this used to be the basis for "burn-in" of equipment.

Inherent capability
The equipment's intrinsic service capability related to design.

Inherent reliability
The reliability level supported by the intrinsic design. An upper limit to the performance reliability of a plant or equipment.

In-op
Inoperable

Inoperable
(Nuclear terminology) Not capable of performing design-basis functions. A system can be operating, yet be inoperable because certain accident or other scenario design assumptions can't be met. Based upon technical specifications, inoperable status can force a plant to shut down until operability is reviewed and assured. (Secondary) Unevaluated for operability; standing by until a qualified person (usually an engineer) evaluates and declares the equipment capable of performing its design function. A real or virtual equipment status category.

INPO
Institute of Nuclear Plant Operations: a self-regulated nuclear industry oversight body originated by the ANI after TMI. A quasi-regulatory nuclear body performing many self-oversight functions.

Instruments
Devices that make a transducer conversion of condition to human-readable format. Commonly visual, but occasionally acoustic and other formats.

Interlock
A device that prevents one device from operating until other requirements are met. These are usually based on personnel or, machine protection, or both. Sometimes the protection of the general public is a factor. For example, mid-1970s vintage cars had seat belt ignition interlocks. The engine wouldn't crank until the seat belt was fastened. These were eventually eliminated based upon public outcry.

Interlocks
Devices that prevent undesirable equipment operations. For instance,

automobile hood releases are interlocked with a manual release latch requiring the vehicle be stopped to open the hood.

IPP
Independent power producer. A producer outside the generating and transmission company's jurisdiction but who has the right to sell power to the generating company's grid on an economic dispatch model.

ISO 9000
A European common market standard requiring process and mapping certification. A certification standard that assures basic process controls are in place for production manufacturing.

Jumpers
Temporary power or control cables that defeat the purpose of a control device, including interlock. The purpose of a jumper is to temporarily defeat a control or interlock to facilitate maintenance, or bypass a failure.

KISS
Keep it simple stupid: a military term.

LAN
Local area network: a shared computer or microprocessor network.

LCM
Life-cycle maintenance

LCO
Limiting condition for operation: the technical specification limit that requires shutdown or entry into a grace period when exceeded. A grace period, when expired, must be followed by appropriate action-often shutdown, if the condition can't (or hasn't) been corrected. A risk management tool for nuclear power plants.

Learning
A process that reduces times and cost to perform an activity. Used in manufacturing to represent the general improvement in design and cost as products enter and proceed through a production life cycle.

"Legacy" systems
Existing, company-developed systems.

Life-cycle
The progression of a product from introduction through production and into obsolescence. It occurs over many years.

Life-cycle maintenance
A maintenance plan with the overall product life-cycle strategy in mind.

Life-cycle cost
Total cost throughout the product life cycle including disposal costs. Often the initial cost is the driving factor in a purchase decision. Like the owner of a new European sports car, the owner may find that the total operating costs far outweigh the initial cost.

Like-for-like
Like-for-like replacement: exact replacement. In contrast with replacement by an improved or superior part. A replacement that minimally maintains performance.

Living
Ongoing, changing, and evolving.

Loaded cost
Costs including overhead charges that may not be applied at the plant level. Overhead charges pay for staff and corporate services not ordinarily charged directly at the plant level.

LTA
Logic tree analysis: an RCM decision analysis for classifying failures by type. This in turn influences the maintenance strategy selected. One of the more confusing aspects of traditional RCM for new users.

Lubrication
A process of replenishment of aging lubricants that are a part of the equipment.

LWR
Light water reactor: the types of commercial nuclear plants licensed in the United States. They in contrast with a heavy water reactor, such as the Canadian Candu reactors.

Maintain
To preserve in an orderly state.

Maintainability
The capacity to maintain equipment. The design consideration of maintenance to provide access, turnaround, tools, and other support requirements to facilitate maintenance.

Maintenance
(as defined by 10CFR50.65) The aggregate of those functions required to preserve or restore safety, reliability, and availability of plant structures, systems, and components.

Maintenance rule
(10CFR50.65) An NRC (federal) rule that requires nuclear plants to perform maintenance monitoring for in-scope structures, systems, and components, and their safety functions and take corrective action appropriately. Informally called the "maintenance rule."

Maintenance strategy
A plan for the maintenance of a component or equipment in an RCM format using a combination of CNM, OCM, TBM, OCMFF, and the resulting CDM. NSM is the null strategy.

Markov analysis
A type of conditional probability analysis used for the prediction of successful events with preconditions. One use is the likelihood of starting emergency diesel generators after faults.

Mechanism
See: Failure mechanism.

Metal clad
Metallic cladding applied to some nuclear fuel types for protection. For many fuels, the primary fission product barrier that restrains fissionable gases from release to the environment.

MIL spec
Military specification. A military standard derived from U.S. DOD equipment procurement specifications. MIL-STD-2173 (AS), *i.e.*, addresses provision for FMECA reliability analysis for procurement.

Mill
Coal mill. A machine that pulverizes coal to a fine combustible dust, mixing it with air in the process.

MIS
Maintenance information system (*See*: CMMS).

Mixed waste
Waste that includes both radioactive material under the jurisdiction of the NRC and hazardous material under the jurisdiction of the EPA.

Mod
Design modification: a change to a plant's fundamental design. Sometimes as simple as a part upgrade, or as complex as replacing a precipitator with a bag house. Most fall somewhere between these extremes, and many times won't be recognized as a design change (in non-nuclear applications).

Morpheline
A volatile organic chemical used for treating feedwater where inorganic chemicals aren't acceptable.

MORT
Management oversight risk tree

MPFF
Maintenance preventable functional failure

mREM
One thousandth of a REM: the common practical measure of exposure. Typical radiation jobs incur several mREM of exposure. Big jobs—tens of mREM. Bigger jobs correspondingly more. Typical limits are 40 mREM/week.

MSG-3
Maintenance steering group standard-3: commercial aviation maintenance standard for RCM-based maintenance programs maintained by the Airline Transport Association (ATA).

MTBF
Mean time between failure: the average period between failures for a failure mode.

MTTR
Mean time to repair: the average time to restore a failed component to service for a given failure mode.

Multiple failure
More than one concurrent failure. In contrast with single failure.

MW
Megawatt: 1,000 Kilowatts. A city of one million people has an electric demand of about 1,000 megawatts, or 1,000 watts per capita. This is the output of a relatively large two-unit generating station of 500 MW each. This is a common standard.

NDE
Non-destructive examination: evaluation of a material condition such as welds without destructive examination. Uses methods such as radiograph, ultrasonic inspection, and replication to assess condition.

Near miss
An event which breaches several levels of protection—usually leaving one remaining fault barrier.

NEC
National Electric Code

NERC
North American Electric Reliability Council. A voluntary organization, that sets standards and rules for interconnected transmission system generating station requirements to assure reliability of the transmission system. The country is divided into 10 contiguous interconnected regions. NERC regional committee members take responsibility for meeting requirements that assure transmission system reliability, such as establishing and maintaining "rotating reserve" and standby reserve requirements. These are generating units immediately available or available on short notice to come online to meet contingencies. NERC also measures the overall reliability of member units. Whenever a unit is brought online or removed from service, the plant records the nature of the status change and cause.

NFPA
National Fire Protection Association

NRC
Nuclear Regulatory Commission

NSM
No scheduled maintenance. A plan of using condition monitoring to wait for a maintenance requirement to become evident. Legitimacy is based on theoretical and actuarial studies that form the basis for reliability centered maintenance.

O&M
Operations and maintenance

OCM
On-condition maintenance. The first check/inspect part of an on-condition/condition-directed maintenance pair. A combination of task, limit, and performance interval.

OEM
Original equipment manufacturer: the original supplier. Contrasts with the "secondary market" or "after-market" supplier.

Old hat
Graybeard. A very experienced person.

On-condition
A scheduled maintenance activity with a specific monitoring method and failure resistance limit identified that experts have agreed detects resistance to failure. Upon exceeding this limit, resistance to failure has declined so that failure will occur. Equipment is removed from service for maintenance, at this point. It can be as simple as measuring the thickness of remaining tread on a tire or as complex as modal analysis for vibrations. The key concepts are resistance to failure and explicit, repeatable limits that trigger condition-directed maintenance.

On-condition/condition-directed pair
A two-part maintenance activity that is unique to RCM.

OOS
Out of service.

Operate
To run, maintain, and dispatch production from a facility. To exercise disgression in the asset to generate income.

Operationalize
Make useable in an operating environment. For example, a standard must always be operationalized. This process develops the infrastructure and insures that the utility can make an activity work in a production environment.

Operator
The owner-operator entity. More commonly, a person charged to monitor, configure, and report plant conditions, working on shift.

OSHA
Occupational Safety and Health Administration

OTF
Operate to failure. A "characteristic" of RCM over-emphasized in aftermarket books. "No scheduled maintenance" or maintenance required on an interval exceeding the asset's useful life is a more appropriate term in the author's opinion. *See also* NSM.

Outage
A scheduled production down period to facilitate maintenance. Outage maintenance is comprised of on-condition, time-based, and condition-based maintenance. For nuclear units, this also provides the window to refuel the reactor for American BWRs and PWRs.

Overhaul
To rebuild by teardown, reassembling with new consumable parts, and reworking all parts and components to a "like new" condition. Terminology applied to any large complex piece of equipment, from diesel engine to turbine disassemble/reassemble work.

Pareto
Vinceto Pareto, Italian mathematician and statistician. Pareto demonstrated the statistical presentation of data in block chart format by fre-

quency and expressed an early version of the now trite 80/20 rule that summarizes "skewness" often present in statistical data.

Pareto chart
Data presentation in block chart format ordered by frequency—most frequent to least.

PC
Primary containment

PCRV
Pre-stressed concrete reactor vessel

PdM
Predictive maintenance

Peak load
Load added only as demand requires. Since demand varies by hour, day, week, and season, some units will be started and stopped with demand according to "economic dispatch" rules. These units are loosely termed "peakers" and largely comprise hydro, combustion turbine, gas-fired, and a few coal-fired boilers. Nuclear units are not peakers.

"Peaker"
A plant used to supply system peak load periods. Gas turbines, gas-fired boilers (sometimes re-powered coal-fired units), pumped storage, and diesel may comprise units in peaking service. Usually not economically dispatched until all base load is available due to high fuel cost.

Permissive
Logic permissive: a control scheme that must be completed for an action to be permitted.

P&ID
Process and instrumentation drawings: design drawings supplied with the plant (along with vendor O&M manuals) by the A-E to aid in performance plant maintenance and modification.

Pillow block
A bearing housing shaped literally like a pillow. Often installed as a separate assembly for large equipment like coal belts.

Planned maintenance
Prepared maintenance plans for equipment that requires repetitive maintenance. May include standard clearance points, parts, tools , and resources such as labor and contractors. Planned maintenance is made up of scheduled, on-condition, condition-directed, and some condition-based maintenance.

PM
Preventive maintenance: planned scheduled maintenance activity. Also, slang for PM work orders. Also, a scheduled maintenance program. Sometimes used to refer to the discretionary part of the scheduled maintenance program in a regulatory enviroment.

PMO
Preventive maintenance optimization: a maintenance optimization process that streamlines RCM. Primary advantage is simplification of paperwork and coding based on the LTA of TRCM.

"Poke yoke"
Make user-friendly and simple: a Japanese term summarizing a technique that stresses task simplification to remove or diminish the possibility of error. Poke yoke devices are devices which serve the same purpose.

Population
Statistical population

Pot
Potentiometer: a variable resistor often installed in instrument loops to facilitate calibration.

Potential failure
A failure that is imminent based upon exceedance of a failure resistance standard. Examples: High-pressure boiler tube wall thickness less than 0.025 inches, machine vibration amplitude in excess of that machine's specified limit (say 5 mils at 1800 RPM).

PRB
Powder River Basin (coal): a distinctive Western low-sulfur coal characterized by low heat rate, high volatility, and dust. Widely used based on low sulfur content and price.

Precursor
Precursor event. An event that predicts susceptibility for ("precurses") a more severe event. An event that predicts future events of the same nature, but more severe in consequence. A leading risk indicator.

Predictive maintenance
Maintenance to diagnose conditions and predict future maintenance requirements. Gradually being supplanted by condition-directed and on-condition terminology.

Premature failure
Failure prior to planned end-of-life.

Premature removal
Removal from service ahead of schedule due to unsatisfactory service, or selection as part of an age-exploration sample for new equipment.

Present value
Present value of money: total cost adjusted by discount factor and time.

Preventive maintenance
Traditional term for scheduled maintenance (10CFR50.65). Predictive, periodic, and planned maintenance actions taken prior to failure to maintain SSC within design operating conditions by controlling degradation or failure.

Preventive maintenance program
PM program: a process that maintains equipment in a state of readiness to support production requirements. All the supporting elements required to support the PM process, including work identification, part support, training, scheduling, and work planning.

Primary failure
The immediate failure. The first failure. A tire blowout causing an accident is the primary failure. *See also* secondary failure.

Process
A defined way in which something is transformed. A process must have specified inputs/outputs and processing technology.

Profound knowledge
Intrinsic, hard-to-replicate knowledge of a process. Almost always proprietary, whether by formal intent or functional cost to extract. Often the basis for competitive advantage. Term coined by W. E. Deming.

Proximate cause
The immediate, local cause. Not necessarily a root cause (based on RCFA), but known in engineering circles as root cause. The apparent cause of failure. The cause evident at the failure location.

PUC
Public Utility Commission: the government entity that regulates the traditional utility environment in the public interest. Also known as PRC, RC, and other acronyms.

Puff
Low-pressure "explosion," large enough to damage large low-pressure boiler walls, ducts, and mills. Overpressure on the order of several inches of water. Because of the limited extent of over pressurization, commonly called a "puff." to designate minor nature

Pulverizer
Coal mill

PWR
Pressurized water reactor

RAM
Reliability availability and maintainability. A type of analysis that looks at the total reliability of a system and factors in maintenance turnaround time. Developed for aerospace and other high-cost applications such as weapons programs to establish theoretical baselines for performance expectations.

RCFA
Root cause failure analysis. Analysis to uncover root causes of problems. There are approximately 10 different root cause techniques. They can be further delineated into stochastic and statistical groups. It's important to understand the RCFA context. Nuclear units, for example, don't use statistical RCFA.

Redundant
(Webster's) "More than enough. Excess." Important safety, operational, instrumentation, and other features are provided with redundancy in engineering designs to assure their availability. Anyone who's ever gazed into the cockpit of a commercial airliner has seen the four-fold redundancy of critical instruments such as altimeter, direction, and roll. Redundant equipment is provided in duplicated and triplicate precisely because the function is critically required.

Regulate
Broadly speaking in a process control perspective—"control." For transmission and distribution control, "regulate" describes the remote operation of a generator to provide instantaneous load following. Some power plants are reserved for base loading—principally nuclear units and very large fossil units which are hard to start up and shut down, or which don't follow load well. Typically, hydro, small fossil, and combustion turbine "peakers" are used for regulation. They follow the load through the course of the day, adjusting for instantaneous load changes.

Reliability
The ratio of successful missions to total trials. The degree to which an operating unit meets the expectations of the operating entity (usually owner) between scheduled down periods. The expectation that the SSC will perform its function upon demand at any future instant (10CFR50.65).

REM
Roentgen equivalent man: the primary measure of radiation dose exposure used in the nuclear industry in the 1980s. Superseded by dose equivalents in Sievelts (Sv), where 10 0 REM = 1 Sv.

Repair
Restore to specifications using welding or other processes. More than "rework/replace." Typically involves certified personnel and testing to assure specification.

Replace
A rework task where a "like for like" part replacement is performed. Commonly applied to filters, lubricating fluids, greases, and other "consumables" that require repetitious service during equipment operation.

Technically, exact replacement. An upgrade to a superior part—like an improved filter design—involves a technical specification change. This distinction is important in controlled aerospace and nuclear applications.

Restore
Return to exact specification. Most PM work is technically of a restoration nature.

Rework
Unneccesarily perform work again. A job may require rework due to infant mortality failure (relatively common in power applications), unserviceable parts, failed performance tests, or (occasionally) lack of appropriate documents and certifications (nuclear and aerospace work). For fossil, bad welds are typical of a rework requirement.

Risk
What can happen (scenario), its likelihood (probability) and its level of damage (consequences).

RO
Reverse osmosis: a common type of water makeup train purifier.

Root cause
The basic cause. In traditional RCFA, a cause that, once removed, prevents recurrence (a stochastic perspective). This context is different from a root cause on an Ishikawa diagram, which takes a probabilistic perspective.

Round
A scheduled activity that checks a large part of a facility. Generally comprised of a series of area checks with intermittent on-condition checks. Also, a scheduled review of screens on DCS systems.

Round sheet
A round logsheet, updated in real time by round logging devices. A sequence of readings currently being superceded by round logging devices.

Run to failure
A misnomer, a term intended to summarize the "no scheduled maintenance" aspect of many planned maintenance tasks. Misleading because virtually none of these tasks result in functional failure. *See also* NSM.

Safe life limit
A part lifetime based on a known aging characteristic for a part with direct safety impact. The part lifetime is limited based on manufacturer tests, conservative design factors, and standards or codes in the absence of service aging experience.

Schedule
Enter an activity into a scheduling system.

Scheduled maintenance program
The series of scheduled planned maintenance activities. The program that schedules the planned maintenance program Commonly, PM program..

Secondary failure
Indirect failure. Failure that results from a primary failure. A boiler tube leak caused from steam cutting from another tube leak is a secondary failure.

Service
Maintain.

Shifter
Shift supervisor, called operating engineer at some plants.

Significant
Equipment that has either a safety or economic impact, thereby warranting review for potential PM benefits.

Simple item
An item characterized by very few failure modes. A relative term. Review of the failure history of a simple item results in very few repetitive failure modes recurring with great frequency. A filter and a journal bearing are two example of simple items. The building block for complex items. Contrast with complex item.

Single Failure
One concurrent failure. A simple failure to diagnose and correct, in contrast with a multiple failure.

Six Sigma
A quality goal based on the reduction of failure frequency to less than one in two million events.

Smoke
To destroy something. Occasionally from overload, continued use in failure, or abuse. Smoking a motor, breaker, or starter are examples.

SNAFU
Situation normal all fouled up: a fiasco on a large scale. An organizational mix-up.

SP
Surveillance program: a planned functional test program made up largely of on-condition and on-condition failure finding tasks used at nuclear power plans to verify the functional capability of standby, protective, and alarm equipment

SPC
Statistical Process Control. A statistical study of processes that provides measures of process capability and control. Widely used in manufacturing of high-quality products. Advocated for floor-up quality control.

SRCM
Streamlined RCM: an abbreviated version of RCM that simplifies RCM using a two-path critical/non-critical approach. Developed by the EPRI.

SSC
(10CFR50.65) Structures, systems, and components.

Standby system or train
One that is not operating and only performs its intended function when initiated by automatic or manual demand signal.

Startup
Plant startup. A special period lasting from several minutes (for fast-

start combustion turbines) to several days (for large baseload coal and nuclear plants) that requires manual intervention, reconfiguration, and direct support to place the plant in an operating phase.

Stroke
Operate, or test operation of, as "stroke" a valve.

Substitution
Replacement of an OEM part with an aftermarket one.

Super session
A super-session resembles MS Windows in which multiple applications can be kept running so the user can jump between applications without need to formally shutdown and restart applications. Early applications could only run one per terminal—like some DOS-based PCs even today. This is a significant productivity tool.

SWOT analysis
Strength-weakness opportunity threat: a type of subjective risk analysis.

Synthetic
Synthetic oil: chemically constructed lubricants, in contrast with distilled column fractionated oil common in traditional lubricants. Such lubricants generally possesss superior qualities, but at a price. Synthetics cost four or more times more than specified traditional lubricants.

System
A defined equipment group that performs a specific set of functions. Usually, the A-E's plant documents provide a list of all plant systems, their major equipment, functions, and expected operating conditions. Very often all the related CIC lists and vendor O&M manuals are provided in binders (1970-1980s vintage units) that organize all information about a plant. They are retained along with A-E design drawings in document centers or shops for reference doing maintenance.

Tagno
Tag number; same as CIC.

Tag outs

An equipment control technique that facilitates work on equipment in an operating plant. Also known as clearance. A controlled technique to seperate energy from equipment to perform work.

Task

A single activity with a failure prevention aspect. The basic building block of a PMWO. Usually, a PM consists of enough tasks to make effective use of operator trip time to and from the work location. Many simple tasks require 10 to 15 minutes to perform and are listed in vendor manuals. Invariably, tasks must be organized into larger work activities or rounds to be done cost effectively.

TBM

Time-based maintenance. Roughly equivalent to hard time maintenance with one slight distinction: The "on-condition" part of a two-part on-condition/condition-directed maintenance pair can be considered as time-based. It's scheduled off the same software system as the TBM task activities and looks virtually the same from a scheduler perspective.

Tech spec

Technical specifications. All equipment has technical specifications used for reference in performance testing for deterioration. Nuclear plants also have technical specifications that provide a basis for operating licenses. They must operate within these specifications or shut down.

Tender

Job title for the lower seniority operators who roam the plant's service and outside areas, monitoring, servicing, and configuring equipment.

Time card

A charge for time, typically made against an activity or account. In some CMMS systems, a time card and a work order are combined for maintenance workers.

TMI

Three Mile Island. The Pennsylvania nuclear plant whose trip and shutdown in March 1979 set the nuclear industry spinning from the adverse publicity and cost. The most serious commercial nuclear power plant event in North America.

Total cost
Total life cycle cost. The total cost of operations, as distinct from operating and maintenance (O&M) cost, or startup or installation cost.

TPM
Total productive maintenance

TQM
Total quality management: a field of quality management that received a great deal of promotion in the 1980s as traditional manufacturing faced competitive pressure form overseas suppliers.

TRCM
Traditional RCM

Trip
Automatic or manual shut down of a piece of equipment. An operator can manually trip a turbine or a breaker could trip on a ground fault protection relay.

Tripper
Tripper belt. The coal unloading belt, the last of a series of belts that moves coal from a railroad unloading point (often a rotary dumper) to the "house"—the plant.

UAL
United Airlines

Unit
Generating unit. One increment of generating capacity at a plant

Useful life
Economically useful life. The period of time when an item can be expected to operate with predictable cost and performance.

VAR
Volt amp reactive: in power flows, this portion of power provides voltage, but does no work. It's necessary to support the voltage in transmission and distribution systems

Violation
A citation for violation of an article of law. Common jargon used in the nuclear industry for citations under 10CFR50 and related parts of the federal registrar. Very much subject to interpretation and established precedent. Becoming common usage at fossil plants as EPA, OSHA, DOT, and other agencies spread their wings.

VOM
Volt-ohm meter

VWO
Valves wide open: for a turbine, the maximum practical load that can be placed on a machine.

Walk around
Area check: A tour looking for general failure evidence or environmental factors. Sometimes performed by management or non-routine performers at plants.

Wear
To impair, consume or diminish by constant use, handling or friction; to tire or exhaust.

Wearout
Fail gradually, with degrading performance allowing a long period to evaluate alternatives and options. Similar to "non-failures" except that performance specifications and expectations aren't met. Gradual performance deterioration until further service is no longer cost-effective.

Weibull
A specific mathematical distribution named after Lauritz Weibull, who first used it extensively to model failure with age, infant mortality, and randomness characteristics.

Weibull analysis
An analysis of failure data to fit the observed measurements to a Weibull distribution. This can be done with specific Weibull analysis paper or using software.

WO
Work order: a work authorization that conveys not only information

about a job, but job accounting and performance information. Very similar across all industries from auto repair to power plants.

WSCC
Western States Coordinating Council: the region of NERC covering the Western states. One of eight NERC regions.

Xerox-style benchmarking
Process benchmarking. More complex than traditional benchmarking because the process is also examined and compared.

Zerks
Grease fittings. After the manufacturer trade name.

Zero time
To reset the component aging clock to zero after an overhaul. To make the item statistically indistinguishable from "new" based on mission goals, performance, and failure criteria.

Zonal inspection
Inspection of an area or "zone." Typically includes environmental conditions, leakage, and other non-specific conditions that an experienced person is expected to know. A pre-flight walk around aircraft check is a zonal inspection.

Further Readings

No Silver Bullets

RCM is not a silver bullet. Ultimately, improved performance comes from better maintenance selection, timing, and performance. RCM helps with selection timing and provides tools to raise awareness. Both timing and performance benefit from heightened equipment awareness. Timing improves first, maintenance performance, later.

As maintenance programs improve, two things become evident. Crises decrease, but maintenance costs run higher. As crises decrease, overtime, low productivity work, material parts expense, and service expenses fall. After a year—long enough to capture secondary cost factors—production unit costs start to fall. More "mega-wiggles" are produced, so unit costs drop. This decrease in unit production cost due to increased availability is a major benefit.

Long term effects are an increase in worker productivity and a decrease in maintenance costs. These changes take approximately 2-5 years. This time period allows the benefits of reduced overtime, better quality work, and improved productivity to build up. The requirement for measurement to roll-up, as well as the time to allow fundamental changes to influence machine performance causes the delay. Expect 1-2 years to see an improvement in maintenance unit costs. For highly reactive environments with high maintenance costs—such as those solely focused on in-service failures—improvement can take place more quickly—as short as six months! Improvement is seen as a decline in unbudgeted maintenance expense and a decrease in both total hours and cost by system. In order to see the decrease, you must first have system-level performance measurement.

Aggressive RCM implementation can increase short-term costs. Implementation makes the staff aware of the weaknesses in measurement and administrative systems. A measurement infrastructure must then be developed. The need for increased infrastructure begins to grow in other areas. Requests for productivity tools and cost-effective modifications rise. Training needs are recognized and their requests increase. It takes time to recover these up-front investments. To see RCM benefits accumulate, measurement capacity is required. Some facilities are only just gaining the capability to conduct performance and cost measurements on systems and equipment by drilling down. New CMMS systems are creating these exciting measurement possibilities.

"Tough Love" Maintenance Work Screening and Prioritization

Condition monitoring programs generate work orders when no failure has occurred. This work request typically goes into a planning backlog, where it will be reviewed, discussed, worked, and eventually closed. Low-priority work arises because there is no obvious "failure." Very often this work resides in backlogs where it languishes until it is canceled. These backlogs can be large and visible. Regulated nuclear plants have agencies that base "effectiveness" on backlogged counts (approximately 500 triggers "escalated" maintenance program action).

Few people are willing to cancel work requests, at any plant. Thus a backlog becomes a Catch-22 barrier. Some environments backlog WO's "to work," where they become lost. Operators do not like cancellations. They perceive them as a slight unless there is personal follow-up and contact. Cancellation is distinct and "cancelers" are distinctly accountable. I reviewed the backlog of the service water system work orders at a decommissioned, re-powered nuclear plant in preparation for its restart as a fossil combined-cycle plant. In more than 1,000 of the MWRs still active at the time of nuclear shutdown, fewer than 10% could be classified as representing failures in the most general context. In fact the mechanics had already identified these failures and had completed (or were completing) them. Most "failures" were hypothetical conjectures written up by anyone (and everyone) in the plant's final operational phase documenting what they *felt* were problems.

I was asked to help clean mess this up. With the regulatory veil lifted, we quickly canceled most of them. Many were written by QA auditors, outside inspectors, and contractors who had no basis for identifying anything as a "failure." At that time, this plant lacked an engineering group capable of WO screening. Screening was beyond the authority of the scheduler/planners. Few had the gumption to do what was needed during plant operations—disapprove the groundless requests. We did not know of a theory called RCM at that time and if we did, let me assure you, we would have implemented it!

The lesson is "garbage in, garbage resides." An important thing any plant can do is to reduce this cholestric CMMS traffic. Setting up specific "on-condition" failure criteria helps. Any work order request that fails to meet these criteria should go onto a clock "suspense" file to have work specified or "time out." If no resolution can be found within a reasonable period, such as ten days, it should be canceled. Obviously you cannot fix a problem that cannot be specified. Maintenance, as with medicine, is patently expensive exploratory surgery. If, after you request it, a WO does not exceed system or equipment performance specifications, then it is a design change request. Maintenance is very ineffective at performing design changes, yet many WO's strive to do just that. The work initiation process should ask originators to specify the exact nature of the failure. When it does, the justification monkey is on their backs.

Plants that take this approach see a dramatic reduction in hypothetical and non-failure WO tasks. This clears the air to focus on things that have truly failed. The on-condition/condition-directed maintenance pair is also a powerful focusing tool. For failed CDM "tags" there should be little, if any, waiting. Prioritization is unnecessary. The equipment basically needs to be fixed.

This is "tough-love" maintenance. It is hard at first, but a world of benefit follows.

Missed PM

A nuclear plant declared a reactor core isolation cooling (RCIC) pump "inoperable" based on a low-lube oil level. On his rounds, an operator dipped the sump and the level came up just below the low-level mark. The obvious and simple thing to do was to add oil. However, the plant was operating, and HP (Health Physics) was leery about exposure risk for a simple PM. They held up the work order, putting Operations in the awkward position of having to perform an operability assessment. Asked for my opinion, my initial reaction was, "You have got to be kidding." I compared the exercise to driving my car about 600 miles to the station in a winter storm, low on oil. In that situation a running engine was certainly a personal safety issue. Their equipment was only in standby. This answer was all too simple. Several hours later, the system engineer (armed with my assurances) put pen to paper and with a few documents, laid the issue to bed. Everyone breathed a sigh of relief, the clock ended, and the crisis was over. This was a bread and butter system engineer routine. About 10 similar events happened per week. Some points are worth discussion. If operability is really a concern, why not simply add oil? The dose was several millirem and oil addition only takes two-hours (with all the paperwork) to perform. This would have required assessing the PM program value, vis-à-vis the "ALARA" (As Low As Reasonably Achievable) method.

I have made enough decisions to burn up my share of some equipment, including bearings, on various operational problems. We never lost a piece of equipment with low oil that we were monitoring. The RCIC level was being monitored, in addition to being in standby service. It is likely that correct oil level was never initially established since the equipment was in mint condition and had no leaks. Although oil sumps vary, wetted roller bearing levels can get well under nominal range and still perform nicely. This equipment had synthetic oil. The "supply oil" function failure level falls well below the low-level limit on typical lubricated equipment. It is a simple matter to confirm this from the equipment's oil sump drawing. Failure modes show up as pumping air (from a low oil level) and foaming (from a high oil level), but you need equipment in service to see this.

Clearly, we prefer to keep the oil level in range, but my experience has been that this equipment is insensitive to small oil level variations. There are many occasions when performing oil addition at inopportune times can be prevented, keeping this allowable variation in mind. This was what I had alluded to with the car metaphor. I will operate with low oil on a 4.5-quart system (*e.g.*, down to three quarts), and never have problems, even driving hundreds of miles in winter weather. Equipment margins are available to be used, and should be used to support ALARA and HP, but not to the degree that Operations must escalate to a plant shutdown level.

About 10% of the checks I do on the road indicate I am low on oil. I drive high mileage cars, odometered at 140,000 miles or more, and these use a bit of oil. The car I previously mentioned used about a quart per thousand miles. I also add oil at favorable times and weather (daylight, wind less that 30 mph, no precipitation), except in emergencies. I have driven upwards of a thousand miles with oil below the dipstick on a couple of select occasions. I have yet to burn up an engine. Likewise, the RCIC pump could have used some oil, but consumption was well known and trendable. Oil could have quickly been added in an emergency.

A broader question routinely asked in a nuclear plant is "Is a missed PM cause to declare equipment inoperable?" Based on statistical reviews of "missed PM" operability assessments the answer would generally be no. I cannot recall one that declares equipment inoperable! In an RCM context, the answer could be yes. This is particularly true if the task is the condition-directed maintenance follow-up to an on-condition measurement. In this case, the clock is running. Limits are based on failure resistance. The beauty of engineering (specification) failure is its slow, progressive nature. There is a period of time before the failure progresses to rendering the equipment inoperable.

This is the dilemma of PM programs. Intervals are created for a reason, but invariably some are missed in even the best of programs. An effective program performs a high percentage of PMs on schedule; an ineffective program does not. Intervals are set with great conservatism and are widely spaced in most programs. When completion percentages of PMs "on the books" are in the 5-10% range, it is hard to claim that there is an actual PM program. Often, these hopelessly over-conservative paper programs are set up in the hope that a few PM's will be accomplished. My observation is that many facilities fall into this latter category. This is partly due to over-conservative intervals and absence of life cycle maintenance program design. Few people run their cars this way while most are generally successful. Few are engineers or mechanics. The real tribute is the robust nature of the equipment, and the capacity of operators to perform condition monitoring—they keep these facilities viable. How much cleaner, efficient, and simple is the specified program that is largely completed! Where these program types are in place, costs and failure rates are substantially lower!

ALARA

An HTGR nuclear plant had intermittent control rod problems. As the reactor engineer, I proposed PM to restore several of the nine inoperable spares to service level. We did not have any spare control rod assemblies. Unfortunately, control rods are not only contaminated, but activated, and potentially cause substantial exposure. Working on the lower activated areas resulted in exposures of up to 10 millirem per hour. The neutron absorbers were far too hot to work with directly. Fortunately, assembly work was many feet away. For three years, Health Physics (HP) held the PM work orders based upon ALARA ("as low as reasonably achievable" radiation exposure). We could not perform control rod assembly maintenance. Then an event occurred. The plant scrammed during startup, and six control rod drives failed to insert. The alert went up the corporate ladder to the president, since the plant was under shutdown order. Suddenly, it was an all-out sprint to develop and implement control rod-drive PM plans. Now HP was receptive.

We knew what needed to be done, but were woefully short of spare parts. We puttered around developing work plans and failure information. It was months before we could start work. When we did, the exposures from control rod overhauls were between 20 and 500 millirem per drive. The total overhaul project, as I recall, required a grand total of around a 100 man-REM over the course of one and a half years. HP learned to tolerate control rod PM-related exposures. But HP was still a maintenance work barrier. All work orders in this plant went through HP—even work orders in the switchyard! On a good day, walking a WO around for sign-off's took four hours. The fact that the plant was the radioactively cleanest in the country and less than 1% of work hours involved contamination or radiation spaces could not influence this turnaround.

At one critical point, while rebuilding control rods, we reattached the highly-radioactive neutron absorbers to the drive assembly. We would figuratively "burn up" mechanics on their weekly administrative radiation exposure limits. A few subtle points help explain why the plant was eventually forced to shut down. First, HP and ALARA fundamentally did not recognize PM in their plan. They discounted any work that was not a crisis. HP was much more receptive to broken equipment maintenance directly supported by the plant manager. This reflected the prevailing culture at the plant.

Second, most of the control rod drive failures were secondary failures. The absence of a startup maintenance strategy on control rod drives (and a host of other equipment) necessitated earlier overhaul performance. For radioactive equipment, the longer an overhaul interval can be stretched the lower the man-REM exposure. HP ALARA, as practiced, ultimately increased the life-cycle exposure for the plant. Theoretically, PM warrants ALARA recognition. Most HP administrators and technicians know little about maintenance. They do not trust the maintenance supervisors and workers—after all, they cause 95% of the HP workload, including contamination events! Work groups reflect prevailing culture. This HP department approved work based on a call from the plant manager. This was ultimately not competitive for a commercial nuclear plant.

Optimizing radiation exposure and maintenance costs remains a challenge at nuclear units today. Recently, HP concerns were again a barrier to full-scope PM plan implementation. The nuclear world has not improved in 20 years. One solution is to put the life-cycle maintenance strategy on an RCM-based foundation and pre-approve planned work. Condition-monitoring programs, on-condition maintenance, fixed time maintenance, and condition-directed maintenance activities can be reviewed and pre-approved by HP. ALARA should not be a barrier to PM. ALARA is, in so many ways, simply another cost of doing nuclear maintenance that has to be optimized in an overall plant context. As with safety, ALARA concerns must be placed on a common playing field. An activity avoided today that will incur a future exposure ten times as great does not implement ALARA.

One Use of OTF

Some plants are reluctant to drop activities that have no PM value. Some tasks are "applicable" in identifying a failure, but are not cost effective. OTF can document an uneconomic plant activity. When there is no identifiable failure that can be related to the task, the activity is recommended to "drop." Many drops are scheduled outage tasks.

For example, at one plant the weld liners on a continuous blowdown tank were identified as low value. The liners protected the tank from flushing blowdown water. In another case, welding of fatigue cracks from thermal expansion was a "drop". The functionality required was compressive, not tensile strength. Yet welding cracks is a major outage work activity.

Some other non-beneficial PM examples include rebuilding valves that were like new upon disassembly; time-based solenoid valve part replacements for low-importance equipment; and overhauling redundant boiler feedpump lube oil system exhausters. Inexpensive equipment with a redundant installation is always a candidate for condition monitoring. A host of literature provides guidelines as to when equipment can be effectively run to failure (remembering that this case still benefits from operator monitoring). Standby and spare equipment run times are often only a fraction of manufacturer's calendar-time rework recommendations. All too often these recommendations become the basic interval for equipment PMs. This partly reflects traditional CMMS scheduling limitations. Reducing these activities is a substantial saving for HVAC and other support systems.

Typically, work performers know over-performed PM's well. They visually examine filters, read the differential pressures during replacement, or notice "as found" calibrations consistently "dead on" specified calibration values. Operators can also identify ineffective PM candidates. They know partially loaded equipment, well-maintained environments, and other favorable factors that lead to low stresses and longer-than-average service intervals. At some plants the PM list itself is an ideal starting point to screen ineffective PM. PMs which have not been done as scheduled or have only been sporadically performed are candidates for interval extension or outright elimination.

"Standards" can document the activities that the plant selects to perform as time-based or on-condition PMs and those they reserve for condition monitoring. They can also control factors such as intended parts, lubricants, and services that have a substantial impact on service life. A premium lubricant, contrasted with a discount one, can improve lubrication time intervals by a factor of five. A superior part supports longer service intervals than an inferior one. It is common to see lubricants and other "consumables vendors" replaced without consideration for product performance or service interval appropriateness. Standards can help avoid discount purchases that backfire. For your vehicle, you might purchase a superior lubricant on the basis that it would reduce service intervals. Would you also increase service frequency when you dropped back to a low-quality substitute? If you did not, you would greatly accelerate deterioration and reduce the overall life of your vehicle.

Maintenance mythology exists everywhere. Some is founded on facts, but much of it is culturally based. OTF can help manage the things people want to do, but which have no value. As with revving an auto engine before shutting it down—which is supposed to make it easier to start—there is no factual basis for many activities. To trim ineffective tasks, someone has to document that they see no value in a task and put that in the organization's history. Leave it to the work advocates to explain how it adds value. Many times a credible story simply cannot be found.

For an organization moving toward condition monitoring (from conservative time-based maintenance or fixed time overhauls) an important caution is to assure that the organization builds an appropriate maintenance performance response. A condition-directed maintenance task has no benefit when the maintenance organization cannot turn it around in time to avoid final failure. Indeterminate work that takes preference over the time-based and condition-directed tasks will sidetrack the program. Maintenance credibility is the key to overall success. A credible program works simply, understandably, and with commitment.

Risk Management: FEAR

Why do plants perform low-value maintenance work? In a nutshell, the answer is fear. When you do not understand what things do and how they interrelate, you fall back to doing what has always been done. When you are uncertain about failure causes, a defensible action is to fix something. I have never heard a performer, manager, or utility cited for performing unnecessary maintenance, even where a long history of infant mortality problems was evident. Much "exploratory surgery" maintenance has gone awry, particularly in outages. Medium-sized equipment such as compressors, pumps, and motors, which are not highly visible equipment, result in a great deal of spontaneous work.

Many engineers and managers have college degrees, but limited hands-on experience for how things actually work, age, and fail. Many maintenance workers have wonderful insight, skill, and intuition, but are at a loss to challenge "official" work scopes and descriptions formalized in plant CMMS systems. Maintenance workers ignore written instructions, or take them as general guidance based on this knowledge. Maintenance is tradition-oriented. Change happens slowly. Blame is common. Any change creates a target for blame. In one case, a rotary-dumper motor-brake setup was extended from two-month intervals (inconsistently performed) to four months, assuming maintenance could perform it to the schedule. The total work scope declined by 50% in the maintenance streamlining effort. The interval was missed, however, putting the scheduling and the PM team on the firing line.

Assessment revealed:

- The plant missed the two previous scheduled quarterly PM performances (on the same equipment).
- The scheduling system "lost" this PM hardcopy for two months.
- The nominal quarterly PM was used as "fill-in" by this shop.
- The electrician performer had physical limitations.
- The electrician performer had never done the work under training.
- The dumper used a complex master-slave control that was sensitive to control adjustment.
- The work done had not been documented.

Peers questioned the assigned work performer's selection for the job. The physical location required light build and this man was big. The unresolved questions pointed towards plant maintenance administration, not RCM intervals. Unfortunately, the plant had no process to perform a root-cause assessment, so few things changed. With little documentary evidence, and only speculation as to what actually happened, it is hardly more than hearsay to guess at the work performed or its problems. When companies work without documentation and work plans, tracing mix-ups is difficult. This basic administered scheduled maintenance program could not *deliver* TDM or OCM/CDM. We had pushed the performance interval to the limit without considering the plant's capacity to perform and deliver PM! We thought that simplifying the work significantly would be adequate to assure task performance; it didn't.

What is the solution? When more people have access to information that supports performing maintenance (or not), better decision-making will result. The development of maintenance strategy is helpful. Consistent compliance with approved maintenance plans, expected from licensed airline mechanics and maintenance, reduces the "seat of the pants" decisions. These decisions may lack insight into the task performance basis, but remain essential to keep informal programs running. More dialogue and development of trust between and among workers, maintenance schedulers, and analysts is necessary. The adoption of an RCM-based PM format takes the guessing out of all TBM/CDM tasks. "Just do it" becomes the mantra. Maintenance can be objectively measured based upon the completion degree of the resulting plan. Uncertainty and judgment will still be present, but in the development and diagnostics of CM-type work request where problems must be defined. As Franklin Delano Roosevelt once said, "The only thing we have to fear is fear itself." The same applies today in maintenance.

A Case for Overhauls

Nolan and Heap make a strong case against traditional "overhauls." They address the commercial aircraft industry, focusing on jet turbine engines. This creates the case for *conditional* overhauls. Conditional overhauls correct primary and secondary failures but do not exhaustively replace non-aging parts and components. They should be used more in traditional utility maintenance to manage costs. For turbines and other large rotating machines, however, there is still a case for "overhaul", based on grouping many multiple independent failure mode PM tasks and the extensive disassembly required for large machines. For a turbine, many inspection tasks need to be performed based on time and risk. Some include:

- Instruments

  - Penetration weldment inspection
  - Failed thermocouple replacement
  - Failed pressure sensing line replacement
  - Calibrations
  - Control connector inspections

- Stages

  - Blade deposits removal
  - LP stage tie wrap inspections
  - Blade root tip crack inspection
  - Bearing dimensional checks
  - Steam cut checks
    - Across gaskets
    - Along the split casings
    - At penetrations
  - Bolt crack inspection
  - Rotor crack inspection

Overhaul activity requires turbine stage disassembly. Disassembly/ reassembly for a large 500 MW machine alone is a 4,000-work-hour task. Even with a three-shift coverage working six days per week, it takes a 20-man/shift crew several weeks to complete! Most hours are required in performing disassembly/reassembly. For this basic time investment, it is absolutely critical to achieve maximum leverage and minimize risk. Large grouped activities, driven by disassembly time, create "overhauls." I believe the term does have utility.

Effective overhauls require using both time-based and on-condition maintenance risk management. For example, consider the rotor bore crack inspection. Rotor bore cracks are low probability events, but ones with great potential consequences. What is the utility's risk profile for such events? Are cracks minor cost factors, or major generation risk factors? Manufacturers recommend performing inspections on every overhaul. Our experience has been that doubling those intervals was suitable. This adjustment provides risk-management, but also substantially reduces overhaul costs. A decision such as this can only be made unit by unit based on specific inspection results. This requires maintaining a history. Unit loading influences crack propagation. Double-shifted units might not support this decision. Changing a unit's service would require re-evaluating this on-condition task. Overhaul tasks can be time-based or on-condition. For example, performance efficiency, loading behavior, and main bearing vibration-level trends are on-condition indicators. Time based age mechanisms include blade root tip cracks, tie wrap cracks, and control valve deposits. Instrumentation can convert time-based tasks to on-condition tasks. Blade deposits can be monitored by careful stage efficiency tests. That necessitates instrumentation maintenance such as calibration.

Feedwater chemistry influences the rate of metal transport as well as blading deposits. Feedwater heater leakage or boiler operation can influence oxygen levels in transient periods, as well as the need to perform an overhaul. Condenser leaks influence feedwater chemistry especially at facilities that lack full-flow demineralizers. These secondary factors can be primary failure causes. Overhaul timing can be improved using a combination of:

(1) known aging performance

(2) history since last overhaul

(3) condition-monitoring as an ongoing risk control practice

What are the on-condition (condition-directed) tasks? Typically, they include:

- bearing temperatures (thermocouple replacement)
- bearing vibration (bearing inspection and rework)
- performance (specific problem identification and correction— like blade deposits and erosion)
  - especially stage performance
  - load capability
  - response
- control valve position trends (valve stem and seat rework)
- valve stroke tests (valve packing and operators)
- turbine protection tests (protective devices rework)

A turbine in baseload service needs to operate smoothly at steady load. A load-following machine, on the other hand, must "ramp" (move to a new load level) smoothly. Failure to perform either well could indicate control valve seat erosion or stem bending.

Owners are extending turbine overhaul intervals. Five-year nominal overhaul intervals are being pushed upwards of 12 years, based on condition monitoring. The nominal overhaul period is less important than what the unit performance and experience support. One unit could not achieve five-year overhaul intervals due to copper transport from feedwater heaters. Copper turbine-stage blading plate-out severely limited loads. Overhauls should be extended using combined information. As generation gets more competitive, achieving maximum production with no schedule interruptions is more important that ever.

Conditional Overhauls

Conditional overhauls specifically correct equipment failure, its cause, any secondary failures, and nothing more. A conditional overhaul is an opportunity for traditional workshops. The basic idea of only fixing the obvious primary and secondary damage from a failure is widely used in the commercial world. Aircraft turbines receive conditional overhauls. A conditional overhaul extends to automobiles, diesels, and other large equipment such as power turbines. We conditionally overhaul equipment at home. When an automobile engine fails due to a main bearing, we evaluate the remaining life on the engine. Then we either perform a selective complete bearing replacement (on a relatively new engine), or bearings and cylinders on an older engine. If there is any ring or valve damage we fix that too, based on the equipment's age and our inspection. To apply conditional overhauls, one must understand the time-based needs for the equipment. Then, when a failure occurs, they must evaluate the failures occurring in the context of the equipment's age. Effort shifts to fixing the failed equipment. This contrasts with a traditional shop practice of tearing failed equipment down to be rebuilt up from the frame.

We lost a large compressor at a plant. We knew we had finishing stage problems, but we lacked the staff adequate to follow up the job. Although the nominal overhaul interval had been four years, only 18 months had elapsed. The assigned mechanic did a complete compressor teardown.

When it was complete, the only problem noted had been the premature failure of the fifth stage compressor wheel. It was an expensive overhaul at $250,000; on the other hand, fifth stage replacement ran around $30,000.

In another example, a gearbox was lost due to a failed retainer. A grinding noise had resulted in the early gearset shutdown. The mechanic tore into the gearbox with no specific instructions other than to correct the problems. We found a missing retainer immediately after opening the gearbox, 10 minutes into a six-hour job. We proceeded through the entire disassembly from one end to the other, although we suspected the missing retainer was the sole problem. On completion, we confirmed this. Guide bearing misalignment from the missing retainer was the sole problem. It could have been replaced, the cover installed, and the entire assembly run successfully even before we had finished our exploratory surgery. RCM states that this partial rebuild strategy is sound and that we should use it. This approach should be built into a facility work maintenance process. Jet engine actuarial studies showed that there is no statistically different performance between a conditionally overhauled machine and the completely overhauled one.

This lesson is counter-intuitive to most shop thinking. The feeling is that it cannot be good until you have looked the entire machine over. Given that many shops implicitly direct mechanics to perform work as they see necessary, conditional overhauls are a prime opportunity to reduce low-value work. To practically implement conditional overhauls, however, mechanics must recognize the difference between a conditional and a full overhaul, committing to perform conditional overhauls when appropriate. Many mechanics enjoy equipment work, especially teardown maintenance. This is why they are mechanics; they excel in their jobs. Everyone in maintenance has to remember that performance of the maintenance, in itself, is not the purpose. It is to keep the equipment functional with as few resources as necessary.

Part Aging Dispersion

In maintenance, we often lack specific information on *when* components and parts entered service. We may also lack failure mode statistics—which modes are dominant, their likely causes, age dispersion at failure, and related factors. We need statistically representative samples, but these are hard to develop other than at plants experiencing very high failure rates. Even here, samples vary due to different classes of equipment, vendors, operating modes, and maintenance plans. We rarely have more than a few detailed samples upon which to base conclusions. Multiple parts suppliers, alternate mechanics and technicians, and rework all increase failure modes complexity and variability.

Process standardization reduces dispersion and is common in manufacturing. Standardization can simplify maintenance failure analysis. We can considerably improve equipment by using maintenance performance information to identify dominant failure modes and benchmark these to manufacturer standards and industry data. With specific failure resistance criteria, we can sample and project comparable equipment failure rates for similar environments. Specific failure resistance criteria are one implementation key. Applying similarity analysis to identify similar applications for specific failure modes of concern is another. These should reflect operating and environmental factors such as the plant, external operating factors, service, and use. This approach is similar to the design of experiments. We make inferences when there are many factors at play. The analyst's skill enables us to sift through history, find meaningful data for the failures in question, and develop the results into simple conservative programs. This is as much an art as it is a science. It must be based firmly on what facts are known. Techniques to firm up and manage risk improve the result and lead to risk management strategies that are insensitive to exact timing. Actual experience and performance in service matters most. Teams perform better than individuals. Experienced equipment reviewers help. Exact analysis, documentation, and peer reviews can base an assessment on objective opinion. This can also backfire when analysts are too conservative. Conservatism results when analysts aren't trained in statistics.

Another way to manage risk is to assure an effective condition monitoring program is in place. With this, analysts can be comfortable selecting meaningful intervals. This is particularly true for economically based failures.

Economic-based failure prevention constitutes the lion's share of a PM program by task numbers and work-hours. For economic-based failure prevention tasks, the lesson is that intervals are extended too conservatively. The use of condition monitoring, age exploration, and other hedges can reduce the tendency to incrementally extend equipment inspection intervals. A database of equipment components and their failure modes is also helpful, as are benchmark intervals. A characteristic of modern equipment is the combination of one or several dominant age-based failure modes and an underlying complexity. The composite exhibits mixed characteristics. A strategy of managing the known aging failures with on-condition or time-based maintenance, as appropriate based on certainty of aging and organizational capability, combined with condition monitoring, maintains this equipment very well. The challenge for operating organizations is to develop simple standard applications of this strategy.

People lacking confidence and experience are uncertain. "Sunday" analysts are squeamish as opposed to hands-on staff or experienced analysts. Databases and experience provide confidence. For economic failures, rapidly getting intervals to the age-based failure range is the only way to learn what aging equipment failure modes are present, how long they take to develop, and what applications they can support.

Maintenance Budgeting

To focus on managing maintenance costs a plant or company must have a motivating driver. The traditional utility environment has not provided this focus. Some companies do not invest in productivity growth. Operating budgets can influence how productivity grows.

In my experience, changes are made in a previous year's budget adjusted for corporate goals (minus 5-10%, whatever the corporate accounting office desires). Take existing staff salaries and add overtime percentage (about 5%), services, and historical material costs. Budget for non-routine events such as major scheduled outages. Adjust for historical trends such as inflation. The resulting budget is last year's plus a percentage. Then hope for the best!

In my years as manager, we wound up chronically over our maintenance budget. A catastrophic year, such as one involving a turbine failure, could double budget expenses. How could budget overruns be sustained year in and out? How could they be tolerated with no corporate response? Companies have "bought the farm" on maintenance. They accept budgets and expenditures as they occur since no one ever figured out an alternative. Corporate offices presume that historical performance adjusted for predictable events is the most reliable budget performance predicator. Corporate staffs manage catastrophic plant outage risk by spreading them over all of the plants in the system, effectively self-insuring. If a company owns 20 major generating units and one major unbudgeted failure occurs per year (such as a $10-million generator rewinding), it is buried in the $150 million of budgeted maintenance expenses for generation. At the fleet level, costs, including catastrophes, are relatively predictable.

Corporate accounting departments can project likely expenditures with certainty beyond formal budget submittals. This approach provides a simple cost-plus expense budgeting plan. It has no bias to reduce costs. The approach presents barriers to promising new maintenance processes. It is averse to changes and to risks. It does not provide a return to plants for better production or maintenance performance. Emphasis is on existing staff and contract expenses, not innovation. It fails to allocate money for long term improvements. In an environment like this one, struggles over money for RCM, or any non-routine activity, will always exist. What are the ways to overcome these mindset barriers?

- innovation
- standard measures
- value focus
- contingency budget reallocation

Maintenance and plant managers have been rewarded for being conservative. This incentive is powerful. To achieve change, incentives must change. Incentive should encourage innovation. To see what innovative approaches work requires measurement. Simple measurement and feedback has shown its value repeatedly. Executive management should lead the way in promoting risk acceptance. An outstanding approach would be for executives to literally raid the pantry—allocate part of the plant loss contingency budgets for developing cost-effective, innovating technology. They should also fairly address any work loss from productivity changes with affected workers. Success with RCM means this problem will be encountered. The competitive generation market will stimulate change. How much and how quickly is a matter of conjecture. The beauty of RCM is the profound maintenance process knowledge it provides. For those willing to master the subject, maintenance need not be a mysterious, budget-busting free agent.

Bigger Opportunities

Most plants adequately manage turbines and boilers. RCM benefits come from "in-between" systems that are high in cost and potential production impact, but without chronic outage causes. These systems often include:

- sootblowers
- sootblowing air
- flue gas and overfire air
- ash Handling
- coal handling
- coal Milling

Savings are found at the knee of the Pareto system cost curve.(See fig. A-1 and A-2. These are discussed earlier in the book.)

These system losses provide high-value opportunities. They influence the generation of secondary failures (sootblowers for example) or redundancy loss (coal handling). Their elimination pays double dividends providing cost reductions with production increase!

Success in one area eventually spills over into others. Leadership, persistence, and commitment are necessary. Pilot programs that were not nurtured withered on the vine. As RCM is implemented there are some specifics to look for, including:
- condition monitoring
- PM's defined as tasks
- total existing PM count

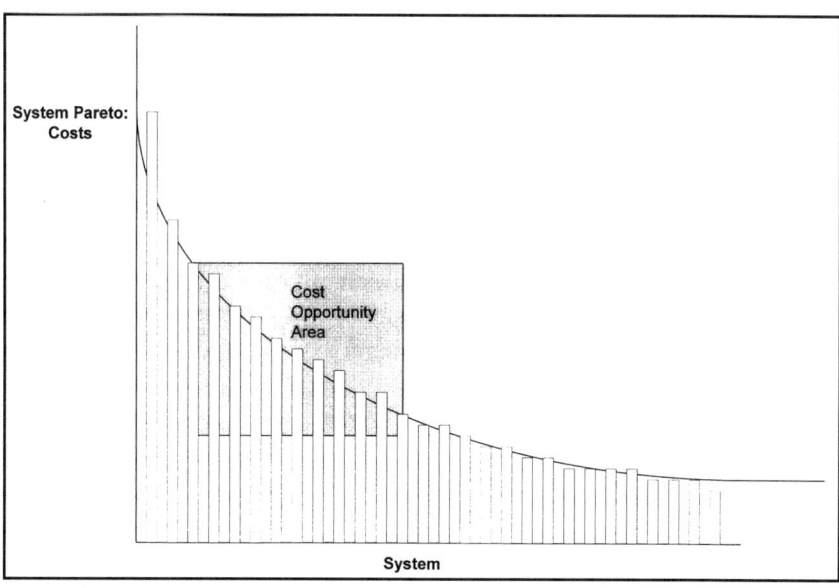

Figure A-1 Pareto System Cost Curve : Costs

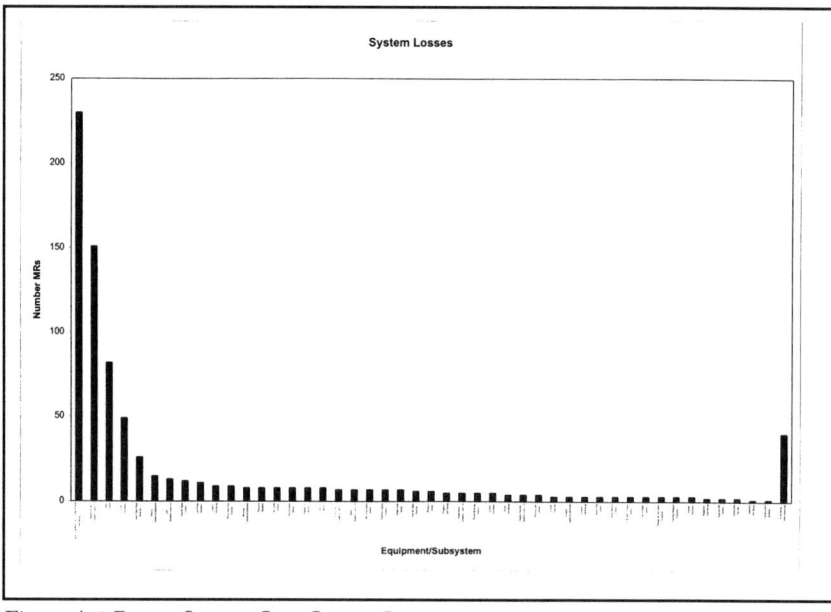

Figure A-2 Pareto System Cost Curve : Losses

- formal on-condition/condition-directed P maintenance pairs
- failure limits explicitly identified
- "must-do" CDM task identified

These increase numerically.

- Prioritization rules
- Change
- Prioritization simplified

These are more defined.

For high-cost, failure-prone systems, materials, services, and work-hour costs are high. Costs decrease, first as work-hours, then in services and material costs. For example:

- "E" MWRs
- Overtime

...are leading indicators, and decline. Systems that lack redundancy will see reliability and availability increase. For redundant systems high cost activities drop.

For existing PM programs:

- hours
- CNM
- work hours
- shift to planned work
- percentage of work tasks completed as preplanned OCM/CDM pairs
- preplanned CDM work tasks
...increase.

Checklist: State of Maintenance Health

Review List

Purpose: *This checklist helps identify the availability of an effective time-based maintenance scheduling system. This scheduling system supports the basic PM program foundation.*

PM Health

- scheduled maintenance work list
- active PM Scheduler (usually residing on the CMMS system)
- PM completion reporting
- operationalized PM Tasks
- PM grouping
- system level performance measurement
  - Cost
  - Hours
  - Reliability/availability

RCM Health Sub-list

- unit operations goals
- systems ranking for importance
- system goals
- PM bases and identification as TBM/OCM(CDM)/CM
- OCM/CDM limits
- OCM/CDM pairing
- CDM performance measures
- system functional failures measured
- failure analysis

Maintenance Process

Traditional CMMS system request work is based on noted problems. This is the corrective maintenance model. Maintenance begins with a problem. A second model, scheduled maintenance, supplements and extends the fundamental model. Combined, these two processes comprise the basic CMMS software areas. Identification of problems, *a posteriori*, is how traditional maintenance works. Proactive maintenance requires shifting to an *a priori* model. This is what RCM provides.

Operators understand equipment problems. You cannot understand the problems until you become familiar with the system, equipment, its capabilities, and what you need to do with it. Response-based maintenance is the first improvement over disposable equipment. It has a significant capacity to reduce cost. This was the motivation for developing the space shuttle. A $500 million satellite that can be recovered and reused at the expense of a single $100 million shuttle flight represents substantial savings.

Response-based maintenance is very cost-effective compared to the alternative, which is nothing. It is the first basic step in any maintenance program. The next is an intuitively harder one—scheduled maintenance. Scheduled, or preventive, maintenance represents the second maintenance program improvement level. Scheduled maintenance is much harder to implement than failure-based maintenance. Many companies satisfy themselves with response-based maintenance implementation. To see what scheduled maintenance can do, consider first what it cannot.

For any component, there is some residual failure rate tied to random failures that are inherent in the component based on design and production processes. This represents a minimum failure floor below which we cannot drop. No minimum limit represents perfection (an impossibility), but through highly evolved, mature designs are not far removed from perfection. Equipment is often economically retired before aging is evident. Factors establishing the floor include:

- design
- materials
- construction
- environment
- operation

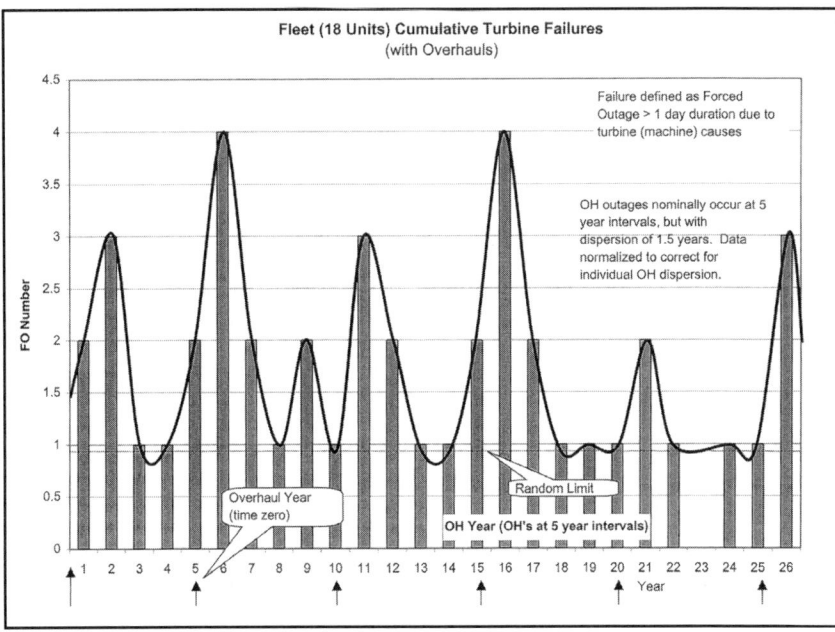

Figure A-3: Cumulative Turbine Failures.

Failure occurs when stress exceeds capability. Design, materials, and fabrication provide equipment with capability. Operating stresses in a perfect, variation-free world would never exceed design. In the real world they do. Designers must anticipate field loads and conditions. Suitable materials, manufacturing, and dimensions assure products perform adequately, with a factor of safety. Systems designers build systems from components and equipment. They are not exact. They stretch design envelopes with operating and environmental assumptions. Some application stresses exceed design expectations. Designers work based upon experience. A residual failure rate is always present in efficient design. A perfect maintenance program would achieve the residual inherent capacity of the design, which is the minimum floor in our example above. Discovering this floor with scheduled maintenance and extending it through design is the focus of applied RCM. That 90% of component failure modes do in fact realize this inherent capacity with virtually no maintenance is the discovery of RCM. For this reason we must use PM tools with great care!

So, where does PM fit in? Condition-directed maintenance lies somewhere in between "absolutely no maintenance" and "inherent reliability limit." Response-driven maintenance works well as a first step— and this is where many organizations find themselves. Further strategies move closer to design-limited reliability. Scheduled maintenance effectiveness has been validated by long-term measurement turbine overhauls. (Fig. A-3)

Peaks are limited by suitable PM tasks that lower failure rates for some period. Performing "maintenance" establishes an intermediate failure level reliability curve. The scheduled maintenance plan drives the failure rate towards the inherent reliability floor.

For some failures, operational changes or re-design are necessary. Failures caused by external environmental factors require review of environmental controls. The characteristic opportunity of each failure varies from one application to another. Some are minor, others huge. Many other factors require a consideration of their added value. Infant mortality or quality control issues can influence the "run-in" period with a high failure rate. After some period of time the failure rate drops to an inherent baseline level.

Condition Monitoring

Most "condition monitoring" (CNM) maintenance is initiated through operations. "Condition monitoring" is monitoring without specific failure criteria. This is a double-edged sword. It can be hard to rank, prioritize, and perform condition monitoring due to its generality. In the absence of time-based and on-condition work order categories an organization can measure its CNM-originated work. This is based on the work percentage coming from Operations. If Operations originates 70% of the work orders, then about 70% are "no scheduled maintenance". Scheduling, planning, and engineering initiate most of the balance of outage, PM, and modification work.

Time-based maintenance comprises the traditional on-condition, failure finding, and time-based rework/replace planned maintenance. If a plant can tag originated work from time-based work orders, then they can measure the RCM maintenance workload as follows:

Summarizing,

PM (time-based):	rework/replace	TBM
	check/inspect/test	OCM (including OCMFF)
CM (corrective):	operations	NSM
	engineering/S&P & design changes	

A fraction of the CM reflects functional failures. Measuring the fraction involves reading WO's or checking logs. Few CMMS systems have fields to record "functional failures." Few operators discriminate between functional and other failures. Logs typically provide functional failures, where they are maintained.

A quick way to re-align CMMS systems to measure RCM-based work strategy is to relate condition-directed maintenance (CDM) to OCM work orders. Or simply perform all "on-condition" directed work as part of the original OCM work order. There are now three basic work order classes:

PM (time-based)	(1) TBM
	(2) OCM/CDM (including OCMFF/CDM)
CM (corrective)	(3) NSM/OTF
	(Failed)

This approach provides a quick way to measure existing processes. Now, what do the numbers mean?

Different equipment and systems have different profiles. Redundancy shifts the profile towards OCM/CDM, and ultimately to condition-monitoring (NSM-OTF). Defined aging with high-production impact biases towards TBM. This profile can be helpful to benchmark existing maintenance plans quickly. An irregular profile suggests a more in-depth review.

Benchmark profiles are only being developed in the power industry. Grouping can show absolute work order numbers any way desired, but "hours worked" is a common benchmark comparison quantity. PM work hours are inherently low. Most non-outage PM jobs are simple tasks. Organizations generally need to increase the OCM/CDM work fraction. This work involves an explicit failure resistance measure and when exceeded that measure initiates work. This requires:

- explicit failure limits
- work performance focus and priority on failure limit exceeded WOS

This combination is often seen in traditional instrumentation programs. A significant amount of out-of-calibration and failed instrumentation work is identified. When restored, immediate operational reliability improvement results—a quick payback.

Redundancy and Functional Failure Measurement

Redundant equipment presents two important failure modes. The first is redundant train failure, such as with two-out-of-three in a standby feedwater pump train. The second is protective device failure. Protective device functions are redundant in one sense; they alert operators to a primary function failure. In the absence of this failure mode they serve no useful purpose. Protective devices *should not act* until an event occurs. Their lifetimes are spent in standby. Like redundant backup equipment, no backup is needed until a primary fails. For protective devices, inadvertent component failure can create an unplanned event. In this case, a functional failure results. An unintended, incorrect feedwater level trip constitutes a protective action failure. A spurious, high-vibration alarm trip on a steam-driven boiler feedpump, with no actual elevated vibration, is another alarm control failure.

Functional tests, near misses, or demand events often reveal protective device failure. Because safety devices have multiple redundant backup devices, demands rarely utilize all backup devices and their trains and infrequently fail to actuate protection. Precursor events are often near misses revealing lost redundancy and risk exposure. Nuclear power plants monitor hidden functionality under required technical specifications and general public safety considerations. Fossil units are not as structured. System functional failures (FFs) provide a performance measure. Identifying FFs is subjective, particularly where functional requirements have not been formally identified. "Total work orders" as a system performance measure is arbitrary; "system work hours" and costs are not. Work hours relate directly to labor costs, and indirectly to total cost. A measure is needed to track failure performance. Maintenance preventable functional failures (MPFFs) are measured for nuclear plants. Some fossil plants track large unbudgeted maintenance expenses, but few track system level availability. Vague, subjective measures are hard to track and have less value for improvement measures.

Two measures generally correlate with failures and are simple to use. First, system emergency work orders are useful as an absolute number and percentage of the unit's total WO's. The second is system overtime. Reactive maintenance strategy for a system reflects a high overtime rate and emergency work order percentage. Both measures are widely available. For plants that measure hours and work orders down to the system level, these provide a ready indicator of performance.

Condition Monitoring or
Condition-Directed Maintenance?

Morpheline has a pungent smell. Once you smell morpheline, as with smoldering coal, you remember it. It is used as a volatile feedwater treatment at some steam plants. At a fossil plant, on a Main Steam maintenance optimization project, we were working on steam leak detection tasks. Valve packing, turbine steam seals, and pipe cracks cause steam leaks. The question is "What is an appropriate maintenance task to identify steam leaks?"

Large leaks, noise, steam release, and increased makeup signal that there is a problem needing to be identified. Noise usually accompanies steam leaks. Saturated steam leaks exhibit vapor. Inability to maintain makeup is a sign that a system is not secure. Changes in make-up trends are one clue to the presence of a small leak. Leaks in inaccessible areas of the boiler must be inferred. For accessible leaks local inspection, vapor tests, and ultrasonic tests are the best identifiers. Valve packing is the most common source of steam leaks. Checking a valve's lantern ring compression is a good time-based packing measure. Most operators and mechanics learn this on the job. Steam piping is all lagged so minute hairline cracks are not evident. Some condensation dripping out from covered lagging is all that is seen. Visual inspection for condensation is a highly effective leak check.

The operators at a morpheline-treated plant shared one detection method they use, which is the sense of smell. Your sense of smell is acutely sensitive. They identified steam leaks by morpheline odor. When you have a sensitive diagnostic like this, there is no point in considering test equipment. Our sense of smell is adequate to monitor a whole boiler unit in an enclosed building. It can be effective for detection of other serious conditions, such as coal fires, too. Of course, knowing that there is a leak and finding it are two different things. Finding leaks can require sensitive equipment.

What is an appropriate strategy for steam leaks? Unless you have piping fatigue, piping leaks are random. Welds and stress risers experience aging and NDE exams for cracks are appropriate in aging-stress areas. Where erosion is a concern, wall thickness measurement is valuable. Inspection with radiography can be effective. The assumption that cracks will be detectable before catastrophic failure supports on-condition inspections and condition monitoring. Plant lighting, cleanliness, and accessibility are equally important. Thus for steam leaks we have:

Source	Action	Strategy Type
Packing	Check for leaks	CNM
	Take-up lantern ring	CDM (Take-up)
Seals	Check for leaks	CNM
	Performance test	CDM
Cracks	Check for leaks	CNM

What is the difference between CDM and CNM? CDM has explicit thresholds and is scheduled. CNM is informal, although the two are very close in performance. For operating tasks, it becomes somewhat arbitrary as to the category in which an activity fits. CNM generally requires more experience and skill to apply. Interpretation is subject to opinion. Some operators note everything, while others see very little. Experienced, skilled operating staffs use CNM with high degrees of success. If CNM is effective, and the organization can support on-demand CDM effectively, then CNM should be used (if only) because scheduling is simplified. CNM depends on environmental factors. When suitable environments are absent, monitoring sensitivity drops. As with finding oil leaks on an oil-soaked floor, CNM evidence can be missed when:

- background noise is high
- randomness prevails
- lighting is poor
- housekeeping is poor

Operator Training and Life Cycle Maintenance Cost

After WWII, my dad was stationed in Japan as a part of the occupation army. My mother followed him over. Lacking much to do and wealthy (compared to the devastated Japanese), she looked for ways to increase her mobility by traveling, shopping, and entertaining herself. Unfortunately, she had never learned to drive, and "Pop" was not interested in teaching her how. He cited her innate physical inability, certain loss of the car in a wreck, insurance costs ... obviously groundless, irrelevant reasons why she should not drive. She had a friend, Cookie, of a similar independent persuasion. Cookie, having a car, offered, encouraged, and even insisted to teach her how to drive. The point of this story is to point out the value of training.

Cookie taught my Mom to drive—in a manner of speaking. She taught my mom to drive with *two feet*—one on the gas pedal, the other on the brake, at all times. My mother never was a "good" driver in the sense of being consistent on the accelerator or brake. (On the other hand, she never had a serious accident.) And Cookie was the reason.

And she was hard on brakes. If she was lucky, she would get 10,000 miles on brake linings. Typically it was around 5,000. In high school my brother and I kidded about how, when mom drove away, her brake lights remained lit until she dissapeared over the hill. That was about a half-mile up a long grade from our house!

Average drivers get around 40,000 miles to a set of brake pads or linings, and many get much more. I get around 80,000 to 100,000 reguarly. (Of course, this depends on where you drive, how you drive, quality of linings you purchase, and other factors.)

Think a moment about the life-cycle costs. She put at least 300,000 miles on various cars over the years. (She worked and commuted 50-100 miles each workday over most of her life.) At around $200 per brake job—a competitive 1990's rate ($300 is probably more like it) we have 20 (100,000/5,000) jobs per 100,000 miles, or around 60 total jobs. In today's dollars these added up to 60 ($200) or $12,000, conservatively. Probably more like $18,000 considering secondary damage when the brake lining work got missed. Then throw in the present value costs over the years and you are up around $20,000. Then consider we haven't started to value anyone's time! What would the training cost have been—a few hours? At perhaps, $50/hour for a skilled driver (using today's rates). Cost benefit ("benefit to cost") is 10,000/$100 (round terms), or well over 100/1 conservatively.

Missing this kind of opportunity on a personal level is expensive; in business it is uneconomic. Unfortunately, for all the same reasons, businesses regularly miss the opportunity to train employees, especially operators, in the optimum use of equipment to manage costs. I hesitate to use "correct" because that presumes there is a "correct" way, and there is not one. There are only costs.

Strategically, one distinction between excellent (low-cost, high-reliability) companies and "also-rans" is the ability to train people cost-effectively. Why are there so many also-ran companies in business? One reason is protected markets, such as the traditional utility industry. Another is market inefficiency. Many American companies see their product benchmark costs as unfavorable overall, while their training costs are negligable, and they cannot make the connection. They lack the profound business knowledge to relate training costs to final product costs. A former boss of mine, R. O. Williams, used to jokingly ask "Who's the most expensive person on the payroll?" At the time he was the highest-compensated executive in the company. His answer, "The worker who is not trained."

Failure Complexity

Failures can be classified as simple or complex. Simple failures involve single faults and modes without interactions or secondary failures. Aging failures—in which a specification is exceeded, such as pump wear—demonstrate simple failure. Complex failures involve hidden failures and secondary failures. Multiple failures may be complex. We might have known that we had a steam-packing leak, but had been unaware of secondary instrumentation damage that developed afterwards. "Hidden" secondary failures mask the extent of the failure. Instrumentation failures are often hidden. This suggests a strategy to diagnosis complex failures. Complex failures can result from common causes such as environments. For example, a complex economic failure in a prototype nuclear plant resulted from a failure to act on the helium interspace gas moisture monitor alarms. These monitors forewarned of moisture in purge helium used in keeping instrument penetrations dry and clean. The moisture-related condensation increased the instrument failure rate. By the time the scope of moisture contamination was identified, we had a major instrumentation rework effort on our hands. This type of failure is protected by instrumentation. These failures are categorized as "common mode failures" when equipment independence assumptions are invalidated. This is of particular concern in the nuclear and aerospace industries.

Overlooking environment maintenance is a common mistake. Environments or process inputs can fluctuate from normal specifications causing complex failures. Examples include:

• Building HVAC for a PRB coal-fired unit.

The building's louvers and dampers were an integral part of the HVAC. They froze due to moisture in the wintertime. They were hard to get to, so gradually they went out-of-service. With about 50% of the total cooling capacity out, the building was extremely hot in the summer—around 120F in areas of many ignitor logic circuits. Mechanics, electricians, and instrument technicians had to cope with higher rates of ignitor failures under such adverse circumstances. They understandably did not want to work in the hot areas during these high-failure periods.

In winter, with many louvers stuck open, the cold air blasts and moisture coming in froze more equipment. Although not as cold as it was outside of the building, the cold blasts from the prevailing Arctic winds caused temperature excursion failures.

High flame scanner, ignitor failures and control drift costs were the secondary failures; they had a common root cause in the absence of design environmental conditions.

• Instrument cables in a nuclear power plant.

A steam leak occurred on a complex, high-pressure, steam pipe-articulated thermal expansion joint. Although the immediate area of leakage was shielded with a heat-resistant blanket, the increased temperature and moisture resulted in condensation along many of the connector pins for the plant's primary coolant flow instrumentation. About 20 large control cables terminated with 100 pins each. When these started to fail by intermittent grounds, the plant went into technical specification grace periods while the connectors were removed and cleaned. Not all connectors could be reworked at load. The scope of the problem was not evident until an unscheduled shutdown. The joint was repaired at this time, but the cable pin problems persisted for months despite extensive rework.

• Water chemistry for a coal unit.

The unit suffered a condenser tube leak. The contamination of the condensate quickly dropped the feedwater pH to the "under 7.0 range" with no subsequent boiler trips. Dispatchers were pleading with the unit's crew to not trip the boiler. The unit went down four hours later on a boiler tube leak, but not before extensive scaling occurred in the boiler's waterwalls. Subsequent tube leaks and heat transfer imbalance required special chemical cleaning. Cost of the chemical cleaning alone amounted to nearly one million dollars. The cleaning added over a week to the outage scope of a 350 MWe unit. Lost generation cost amounted to $500,000 at the company's generation value added rate (difference between purchase cost and generation cost).

• Fire monitoring for a coal unit.

The unit's methane detection and fire protection deteriorated due to wet environmental conditions causing a high rate of control failures. Acid runoff from coal-dust water-suppression sprays caused a high incidence of instrumentation faults. Water spray was required because the plant's original dry dust suppression was inadequate for the dusty coal supplied. Water sprays were added when the dry dust system proved incapable of managing the dust and being maintained at the same time.

The dry dust suppression system was rebuilt after it was pointed out that the station's license certificate explicitly demanded it. The replacement system was installed under adverse winter conditions at over three times the budgeted cost under the pressure of a regulatory compliance consent agreement. The replacement system, in spite of installation troubles, was highly successful and returned the unit to license compliance Local papers groused afterwards that the company was inappropriately excused from fines for voluntary compliance.

- Instrumentation and controls for a coal handling area.

Coal handling equipment that monitored the tramp iron, belts, and alarms went out of service for a variety of reasons. Coal handling did not warrant resources beyond the "emergency" level. At the unit age of 15 years coal handling system costs had taken a number three position behind the boiler and the turbine. It appeared a matter of time before direct coal handling outages impacted production. Coal handling equipment:

- *failed to crush coal to specified size*
- *failed to remove iron or metal*
- *required heavy washing to manage fire risk*
- *suffered repetitive spills that required manual cleanup and heavy washing*

- Hydraulic fire at nuclear unit.

A nuclear unit had hydraulically actuated bypass valves. The hydraulic fluid, a commercial, stabilized synthetic, was thought to be immune to fire, based upon supplier promotional literature. Hydraulic leaks from the valves occurred chronically. Pans were installed to catch and funnel leaking fluid into catch cans of approximately ten gallons capacity.

One of the drains plugged up, the tray overflowed, and the leaky fluid dripped down onto exposed reheat steam safety valve hardware two levels below. These started smoldering. Because of the non-conventional plant design, the exposed parts of the safety valves were slightly above the flash point of the "fireproof" fluid. The smoldering fluid ignited, and the little fire eventually triggered the fire detection system. An operator responded and extinguished the fire. Subsequent flashover after the flame was extinguished extensively damaged an area of intense cable, instrumentation, and control equipment adjacent to a cable spreading room. Damage repair took a focused effort of nearly 90 days and a special release from the NRC to restart the unit.

The area of the fire was congested, dirty, poorly lit, and the facility had historical problems of the hydraulic valves, especially oil leakage, that were a root cause of the blaze. Direct costs of repairs were between $10 million and $20 million dollars.

Common to each of these events was the failure to maintain a specified environment or conditions followed by the inability to control subsequent multiple equipment failures. In some cases, they led to a major event. At the time, plant operators thought they had no alternatives; their focus was on generation. After the event, or when the cost and reliability impact of the secondary failures was evident, the primary failure problems were corrected at great expense.

Some events are humorous in hindsight but for those that suffered through the crises, or participated in the front-end decisions that later lead to problems, they were disheartening. The residual problem corrections required extensive efforts that detracted from the strategic needs of the plants. The lessons here include the importance of maintaining environments and the relative ease of diagnosing and correcting simple failures in contrast with complex ones. This lesson, in fact, is so important it needs emphasis.

A primary benefit of a comprehensive (implemented) RCM-based PM program is the focus on identification and correction of failures while they are simple and less expensive to correct from either a production or cost perspective.

In conclusion, after events such as these, the availability of a comprehensive hidden-failure test program is very helpful to assure all controls and instruments of consequence are restored. Nuclear plants have these programs implemented as surveillance plans. Fossil units do not have an equivalent.

Interval Extension with Age Exploration

Age exploration—a fundamental design tool—is formalized by RCM. Design engineers have always seen product improvement as their design role and goal, but widespread TQM application has shown wrench turners can improve designs, too, including quickly extending maintenance intervals.

My experience has been that PM interval extension is done with great fear and caution. Whether it is the oil in your car or the turbine overhaul for a power plant, everyone gets queasy extending intervals, particularly if his name is on the extension request! Consequently, we extend intervals incrementally. Five to 10% increases are typical, and we do not see the accelerated effect of any major *faux pas*. Increases in minor intervals dilute the benefits of major technical, parts, or materials upgrades. This is true when a valve diaphragm polymer is improved for temperature resistance or the lube oil in equipment is upgraded to synthetic. These superior products age much better in service, and we need that knowledge. To see the difference in performance, we must either perform a painfully detailed condition-monitoring assessment, or select a few candidates for accelerated aging and see how their materials perform. The "safe" way is to do a material aging characteristic study. For example, fatigue aging relates to the stress energy levels of the metal, and there are a number of ways to test this in the lab. A lubricant's replacement interval is based upon successful machine performance up to a given level of oil deterioration. We can measure the end-of-life characteristics of the oil (or other material), and then compare the superior product's performance.

At "end of life" a traditional distillate-based lubricant shows physical and chemical property changes. Viscosity, dissolved metals, particulate count, total acid number (TAN), and other indicators demonstrate aging. If we can correlate any one of these to the aging of the equipment, such as total acid number, then we can estimate lubricant aging for other lubricant comparison, such as a synthetic. We may find that the other lasts twice as long in service. This approach is exact, engineering-based, and controlled.

Complex equipment provides a different challenge. It may not exhibit a dominant failure mode. We may not have enough experience to see how it performs in service. However, we need a maintenance plan. If we use the OEM's recommended interval and observe no failures over the service period, how should we go about extending the interval? Here is where RCM provides a useful tool. First, we need to quantify the failure mode in question. It must not have a "safe-life" limit. We must be assured that the failure will not create a personnel, public, or other hazard. If it does, we should have, through the supplier and other agencies, a great deal of information on which to fall back. If it does not—as is the case in 90% of the PM activity in a typical plant—our next task is to reasonably extend the interval. The RCM approach says that actuarially we have a solid basis to extend the interval a substantial amount—about 50%! This is usually a shock. For me, this is still like leaping off a cliff. Based upon studies performed for "no predominant failure mode" complex equipment, we will do very well with large service interval extensions by fitting a "no experience" template. These large extensions are exactly what we need to identify dominant failure mode characteristics in complex equipment.

This type of extension either very quickly extends parts out to where a lifetime can be identified, or very quickly achieves substantial reductions in PM hours performed and associated cost. In the context of an ARCM-based approach, it can be done with very little economic risk. In this way, we greatly accelerate the rate at which we learn the dominant failure modes and their appropriate PM intervals.

Before RCM, few would perform substantial part life extensions. Now we can extend intervals with some comfort. Not only are large extensions possible, but they are statistically justified. In fact, there is very little statistical justification for initial intervals for most equipment. Typically, the first few failures are assumed to approximate mean life. We wind up with greatly conservative service intervals from the onset. A corollary concerns cases where PM's have been missed with no adverse failures developing. This experience justifies extending intervals to the discovery limit. These add legitimacy to interval extension. With work performers included in age exploration of parts performance, we can advance quickly to more accurate realizations of potential equipment lifetimes.

Operate-to-Failure Is Not What It Seems

Operate to failure (OTF): *Definition: A planned-maintenance choice to perform no scheduled maintenance activities prior to equipment "failure." Equipment "failure" identifies a service requirement; e.g., maintenance is performed when the equipment self-identifies a maintenance need. OTF selection can be based upon (1) absence of a functional failure impact on a system, its safety, or environmental performance; or (2) the absence of an applicable, effective maintenance task.*

Operators monitor OTF equipment for condition and performance. OTF is more accurately described as "no scheduled maintenance" or NSM. NSM better conveys the meaning, for the plan does not allow OTF equipment to proceed to functional failure. NSM is a legitimate strategy because failure actuarial studies show that 90% of component types will not fail during service in a preventable way. They are "inherently reliable" and will not benefit from PM over their service lifetimes. OTF is a misleading term because it implies that components will eventually fail in service, when in fact, they will not. This is not appreciated practically by many operating and maintenance staff.

Key Points

Part of a system: An item selected for OTF is part of a system. Item failure should have no impact on the system functionality to be an OTF candidate. This could be due to:

- redundancy
- low failure impact
- acceptable risk (for random failures, for example)
- inherent reliability

No Direct System Impact: The item must not impact any essential system functions.

- functional failures absence

- production impact absence
- safety impact absence
- environmental impact absence

Engineering-Specified Failure

- Engineering specification-defined failure is based on a gradual deterioration towards "out-of-specified" condition. When a specification has been developed it provides a measure of failure resistance. In many instances failure measures are implicit; an exact failure limit has never been identified. Frequently, this is because the component is inherently reliable and very little failure experience is available. (Fig. A-4)
- "Failure" deterioration is based upon a specification; the item retains residual performance capability. The "failure" in this case is a continuous process and is tolerable for a brief period beyond the specified limit as performance deteriorates.
- "Spec"-based failures are proactive; they are based on design limits, not catastrophic events. Design limits incorporate margins.
- There are no known or specified applicable/effective tasks or corresponding failure resistance limits that have been identified or agreed upon.

Random Failure Nature

- Random failure characteristics often cannot be eliminated; design has reduced the failure risk to an acceptable level by redundancy or inherent reliability characteristic

Cost

- The cost of replacing the failed equipment is lower than that of preventing the failure.

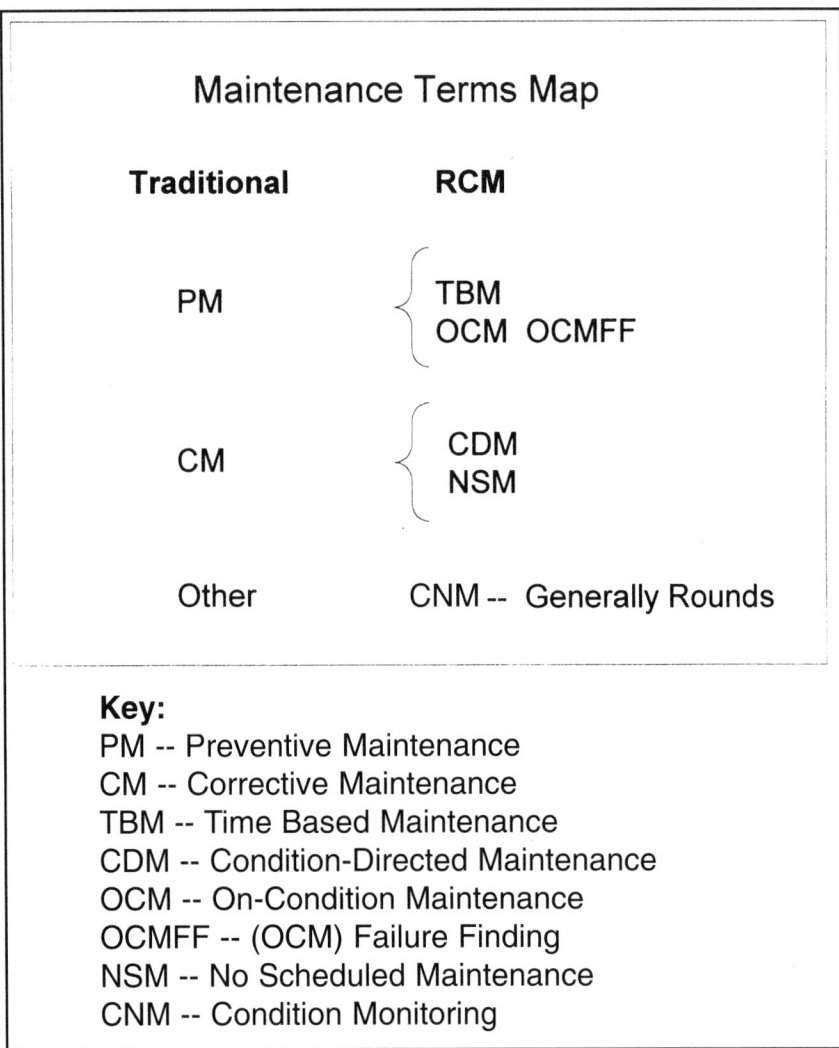

Figure A-4: Maintenance Terms Map

Identification

- The failures are typically identified by operators on area checks, walk-arounds, or in use.
- Failures may be identified through outage work.

Typical Candidates

- any function where the failure resistance is ambiguous, not explicitly identified, or not yet worthwhile
- instrumentation (non-critical)
- small items with local failure effects and no secondary damage failures
- "engineered" equipment with wear-out limits for gradual deterioration
- small tools
- minor/hidden function items
  * not in service
  * event driven
  * risk-acceptable
  * joint probability of failure vanishingly small; risk acceptable

Largest Typical Plant Application

- I/C program instrumentation

Very often, substantial amounts of non-critical instrumentation can be effectively "run-to-failure" for calibrations & other maintenance. These savings can be large.

Examples:

- Home

1. light bulbs
2. small TVs, other consumer electronics
3. small appliances
4. watches

- Plant

- inherently reliable components
- passive components
- service components

1. small motors, pumps, valves
2. corrosive service pipe
3. tank liners for acids, caustics
4. flooring
5. structurally redundant steel
6. small piping
7. out buildings (pump houses, etc.)
8. cable
9. conduit
10. pipe

Maintenance Discipline

When maintenance programs struggle with PM, it may reflect a problem with discipline. Developing and following a work plan reflects maintenance discipline. Discipline means the ability to comply with standards, no matter what their source. Correctly initiating work orders, working to procedures, working to schedules, meeting deadlines, writing work summaries on completed work orders, signing completed work—all can be reduced to basic work habits that demonstrate commitment to standards. Work habits are hard to learn and easy to compromise.

The Navy relaxed standards in the early 1970's. Candy, food, and beverages increased the food residue in sleep areas. In short order, on some ships, shipboard spaces looked like dumps. Cockroaches became shipmates.

Discipline requires standards, training, and reinforcement. Unfortunately, reinforcing behaviors is not the strength of traditional maintenance. Unaccountability can prevail. Lawsuits have been filed against companies stemming from the most trivial attempts to exercise standards and authority. Submitting signed, accurate time cards, keeping tools stored, cleaning work areas, even wearing shoes to work were all cause for complaint. Companies that lack discipline coincidentally struggle with PM. No amount of paperwork performs PM. It takes someone who knows and cares about equipment and abouty his facility. If discipline is absent, companies may need to add it to their strategic goals.

I am not an advocate of authoritarianism. I do, however, believe there are fundamental standards and processes. Everyone working at his own pace will not cut it competitively. Companies must identify and adopt their own basic standards. Without standards,there is no maintenance process—no foundation to build upon. Focus on political and social goals at some utilities left them fundamentally without discipline. Companies without standards will not be competitive.

Pre-stressing Tendon Buttonheads

Fifteen years ago, a nuclear plant had a corrosion problem with its containment concrete pre-stressing tendons. Water droplets had accumulated in some of the tendons, and around 40 buttonheads (144 per tendon) on the anchor hardware had popped, up to five on a single tendon. These indicated wire failures inside the tendon conduit tube. Without going into too much detail, accumulated moisture had initiated a corrosion cell on about eight of the 500 tendons and the result was an "operability" concern of the vessel containment pre-stressing system. As it was an essential system, nuclear safety was involved. The episode became a plant startup issue. Analysis and review showed that the tendons that experienced failures were a small population of the longitudinal and bottom-circumferential tendons. The common factor was a tendency for the tendon tube grease to drain away from the tendon head and button hardware where it had been applied, after exposure to the heat. Later, moisture in the tube—hypothetically from original construction and with the temperature differential—formed mass transport cells and condensed as water onto susceptible bare anchor wire, buttonheads, and buttonhead wire extensions.

The few popped heads had given a warning. The heads were nominally inspected every five years, and several had been found popped on the first tendon inspection. Functional performance could be demonstrated by measuring prestressing tendon lift-off force—a lift-off test.

This constituted on-condition maintenance. After the second event, we embarked upon a mad rush to develop diagnostic techniques for the button head corrosion monitoring that did not require "liftoff" of the prestressing tendon. Liftoff of a single tendon required the location and installation of a short stroke "pancake" jack, shims, and techniques to unload the tendon and hardware. Once unloaded, they could be inspected. Performing one liftoff occupied a three-man crew an average time of two shifts. We were performing many lift-offs.

Corrosion developed in the area of wire within 10 inches of the anchor hardware. Most of it was located just under a heavy anchor plate, which acted as a cold trap. One diagnostic technique was to use NDE ultrasonic exams to monitor the acoustic sound reflection from the buttonhead. It could detect whether or not there were button heads in good condition. Tuned to look only for those heads that exhibited corrosion within five inches of the button, the instrument would detect buttons with complete corrosion nearly 100 % of the time. The corrosion process was incomplete, so many wires had partial or slowly developing corrosion. They would give weak reflections—sometimes none at all—with a wire that exhibited surface corrosion. Ergo, the test was not perfect, but it would detect complete failure with certainty if that failure occurred within five inches of the end hardware. As one who did much of the testing, I was confident that we had a method that would detect failed tendons with high accuracy. I felt we had a test that would save us thousands of dollars in liftoff tests. Liftoff testing of the accessible tendons had been our only recourse. Unfortunately, this was a nuclear plant. Our Quality Control and Quality Management voided our test on the basis that it was not perfect. It could not *prove* that buttons were in good condition; it could only detect failed buttons. Therefore, we continued to perform liftoffs at a rate of two to three per week. This kept a team of three mechanics busy doing liftoffs for a year.

Did this test meet applicability criteria? Effectiveness was not an issue, if it could be used! The time to remove and inspect one tendon head was about two man-hours; one to remove the cover, another to perform ultrasonic inspection of 144 buttonheads. In contrast, it took 48 hours to lift-off and visually inspect. Therefore, we had an imperfect test in a nuclear environment and a charter that sought perfection. We continued doing liftoffs until the plant was permanently shut down for high cost. Developing effective on-condition/condition directed maintenance pairs is challenging work. It can depend greatly on the plant regulatory and cultural environment for its success. Static, regulated environments will not be conducive to any new techniques or methods. They are far too demanding!

Statistical Process Control (SPC) Systems

Process control is a sensitive indicator of overall system health. A process that is "in control" can be verified by performing a series of quick, statistically-based, running-average checks. "Statistical process control" (SPC) provides a wide range of literature addressing techniques to identify and measure process control. Most plant people can look at controller output signals and make momentary judgments as to whether or not a process is "controlling." This is an intuitive check. Knowledge of SPC can make monitoring more intelligent and useful. Other processes become candidates for control assessment. We may not realize that different processes exhibit controls, and that a problem originates from a secondary process or system input.

Failure of a process to be "in control" is statistically easy to determine. A process that is "out of control" has other problems that need identification and correction. Determining when a process is out of control requires control limits. For example, it is common to find that operators adjust controller setpoints for personnel preference. This results in processes that are statistically "out-of-control." While this subject is of paramount importance in manufacturing, it also applies in power generation.

For example, different consequences resulted when operators preferentially loaded along the generator excitation-loading "D" curve. Adjustments eventually percolated throughout the unit, establishing a new equilibrium. In another instance, we found that different operators had different control schemes for boiler sootblowing. Some were more effective than others based on blowing times, boiler differential, and temperatures at a given loading and firing rate. Operator preference was an expensive choice, for this boiler became plugged when slag and ash buildups ran away. The station's philosophy though was that the twenty blowing schemes assured a clean boiler. Therefore, this operating practice was preferred, despite the evidence of SPC.

Typically operators are unaware of the broader ramifications of their operating styles and the impacts this can have on equipment reliability and aging. Some may be beneficial, while most are not. SPC provides a technique to view specific controlled system outputs and evaluate their condition. SPC identifies systems that are becoming unstable well before they fluctuate out of specification, asserting its usefulness as a powerful predictive tool. As with all predictive tools, it is only cost-effective when applied judiciously. Broad-based SPC application without cost-effectiveness consideration is inappropriate.

RCM Software Applications

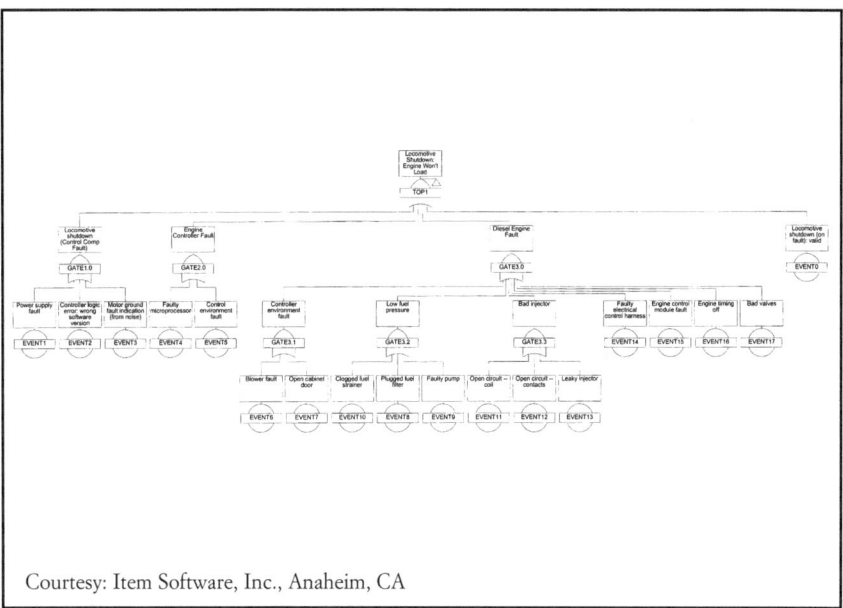

Courtesy: Item Software, Inc., Anaheim, CA

Fault Tree Analysis: FaultTree+ provides an example of a Windows-based engineering reliability software tool. In the past, engineers avoided exact reliability analysis partly due to the tedious nature of developing manual fault trees. Software like FaultTree+ simplifies analysis so that relatively large fault trees — 6000 elements and up—can be developed and analyzed with relative ease. Lookup tables of standard component failure rates are combined with analytical simplifying techniques. These make fault tree analysis a real option for the shop-floor engineer. The fault trees themselves focus efforts on fault areas of concern — from the minimum "cut sets" of interest (and non-interest), to the overall contributions that various faults have on overall reliability. (A cut set is one way the top event can occur. The top event is typically an undesired outcome — a failure.) The locomotive example here is typical; the fault tree illustrates design areas not previously considered in the overall unit reliability. Sensitivity analysis can test the numerical values used for the event probabilities, as well as the model logic itself.

Often overall "top" event reliability is known, but the individual reliabilities are not. Individual reliabilities may be taken from generic tables. In any event, the Fault Tree identifies the fault paths of interest, and their logic, providing the opportunity to focus on the critical few that matter. In the example, we build the fault tree on FaultTree+, assigning failure data as we go. We have several selections of failure models and information to select from. Once complete, we can run the analysis and see whether the frequency of occurrence of the top event — here, engine failure to load—fits our experience. Very often it doesn't, but we now have specific guidance on where to look for additional data. The fault tree thus supports the continuous evolution of a maintenance and operating plan based on facts. For users who have never used fault trees, they help understand the complexity of multiple failure data and help focus efforts in selective areas for maximum results. It's common for a fault tree model of a problem to draw out risk areas not previously appreciated.

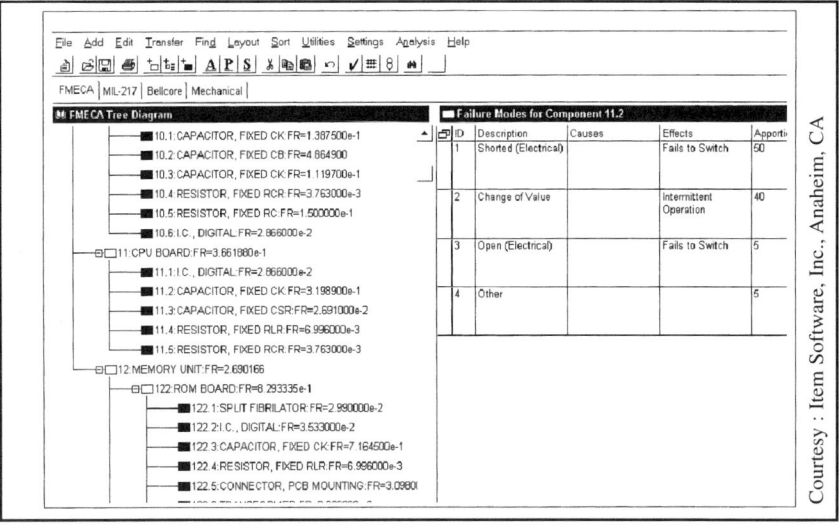

Failure Modes and Effects Criticality Analysis: FMECA provides a failure mode and criticality analysis tool. The difference between Failure Modes and Effects Analysis (FMEA) and FMECA is the calculation of "criticality" — a numerical calculation of the combination of probability and consequences ("criticality") that allows the ranking and

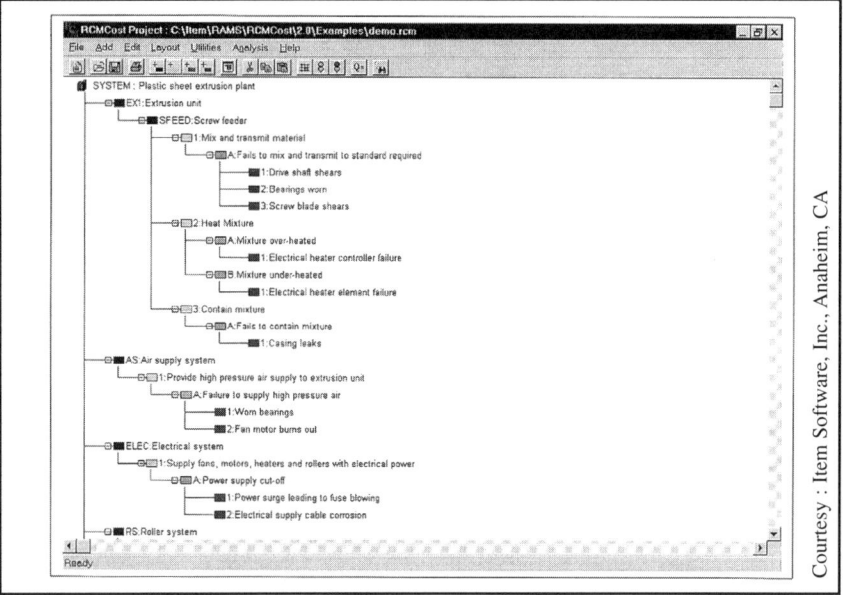

Courtesy : Item Software, Inc., Anaheim, CA

focus on failure categories of interest. The traditional weakness of FMEA's is the inclusion of "the trivial few," often over-ranked, to the point where analysis isn't cost-effective. The addition of criticality can avoid this problem. Practically, the engineer must learn by rote memorization what typical probabilities are and apply these with speed and impact. Contrary to what some say, numbers should always be based on data, or validated. Those available or derived from seat-of-the pants estimates are almost always exceptionally conservative in plant applications. A failure modes and effects analysis documents the modes of interest; the criticality hangs a number on the mode that allows relative comparison. FailMode is an easy to use FMECA tool that also supports later analysis in related ways. The equipment hierarchy, in particular, can be reused many times for different analysis.

RCM Analysis: RCM can serve as a product design optimization tool. Product optimization (as used in manufacturing) can view products from a life-cycle cost perspective, optimizing the combination of initial, operating and maintenance costs. "RCMCost" provides a design tool for the evaluation of products and development of suitable maintenance programs. Starting with an hierarchical model, FMEA is developed from which costs (like criticality) can be developed. The common file structure in Item Software's RCM Cost and related products (Fail Mode, Fault Tree+...) means that designers can develop a FMECA, perform fault tree analysis, and review "critical" failure modes together during design. Alternative maintenance strategies can then be developed, explored, simulated, and optimized. Any combination can help to develop a product manufacturer's initial installation and recommended scheduled maintenance program. Clearly, the same software can be used by the end users (operating organizations) to review and evaluate their maintenance practices, costs, and risks, and optimize their maintenance programs. There are at least seven RCM software products available. Some support one or more applications more easily than others. Software users must understand their own needs and then explore the market alternatives.

CMMS Example: Power FM
Courtesy: Asset Works, San Antonio, TX
and New Century Energies, Denver, CO

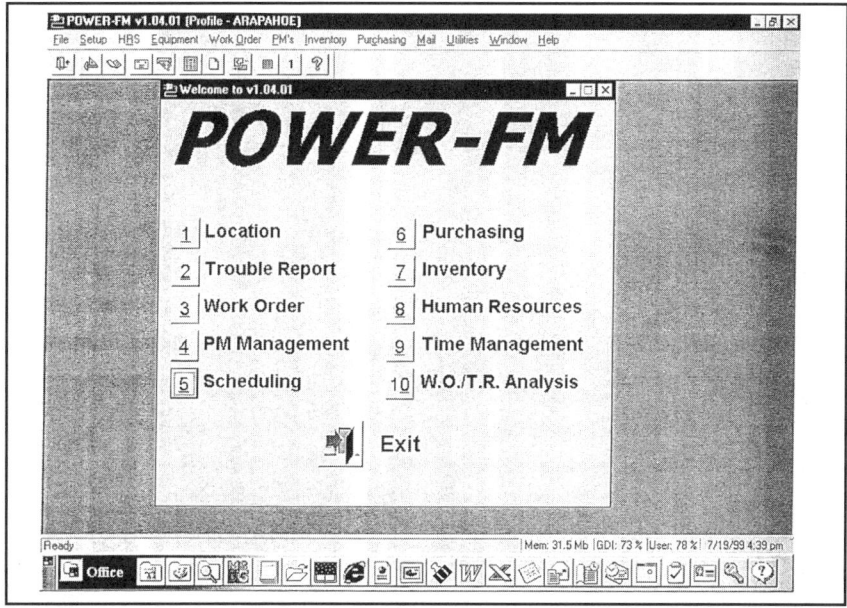

Splash Menu

Splash Menu. The startup or splash menu shows the major functions offered by the CMMS and graphically suggests use of a mouse — the trademark of a GUI interface. Since many non-routine CMMS users are not typists the GUI interface is essential for speed and convenience. Note that most CMMS systems even today use traditional terminology since this is what users know. (We could equally call PM Management "Scheduled Maintenance.") Users can view and update different areas by controlled authorization. Most systems offer generous "view only" data privileges but restrict updates to specific work areas.

Equipment Hierarchy. The hierarchy provides a convenient way for plant workers and staff to quickly locate any equipment of interest for the purpose of identifying, selecting, or reviewing work and related failures, resources, and costs. The hierarchy (in a GUI environment) dou-

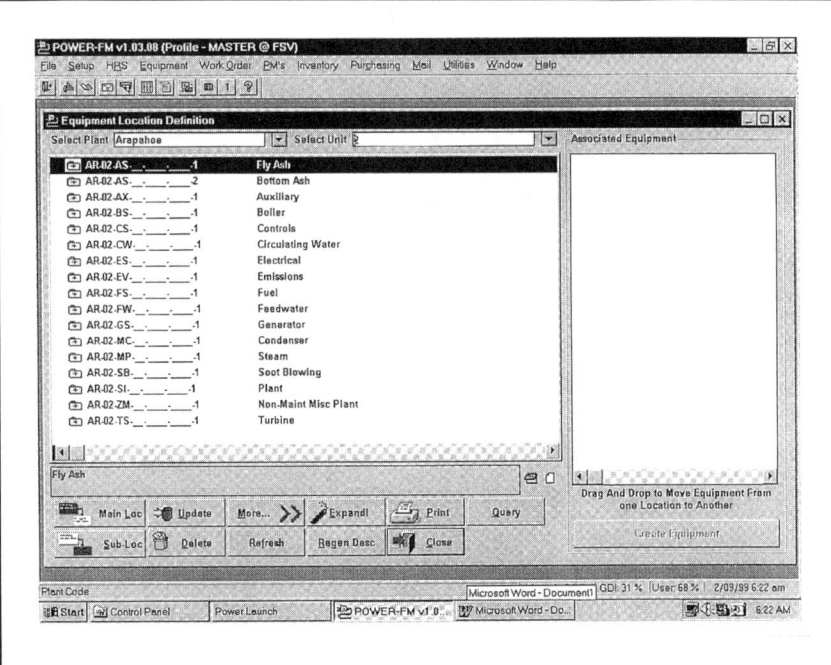

Plant Registar (System Level)

bles as what was once called the plant register. It relationally lists all equipment that the plant anticipates may require work over the facility's life. The hierarchy should uniquely identify components down to the level just above that where you expect parts replacement. For example, if the tubes in a compressor aftercooler heat exchanger tube bundle are the last items replaced, then the aftercooler heat exchanger should be identified as a unique part in the compressor assembly.

Developing an appropriate level of detail is a fine art form that only comes with experience. Too little and work can't be traced to failed components; too much and the hierarchy becomes complex and difficult to query. A "good" compromise is that a moderate-sized single fossil-fired generating unit or process facility should have between 500 - 2500 "coded" components. Over 10,000 should raise a concern for excessive detail; under 250 raises concern for two little. The purpose for coding any equipment (of course) is to identify and perform applicable and effective maintenance. This hierarchy was selectively opened

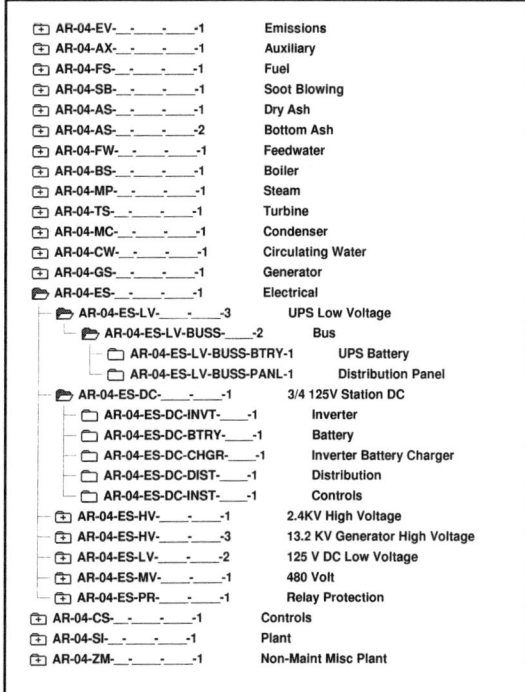

⊞ AR-04-EV-__-___-___-1	Emissions
⊞ AR-04-AX-__-___-___-1	Auxiliary
⊞ AR-04-FS-__-___-___-1	Fuel
⊞ AR-04-SB-__-___-___-1	Soot Blowing
⊞ AR-04-AS-__-___-___-1	Dry Ash
⊞ AR-04-AS-__-___-___-2	Bottom Ash
⊞ AR-04-FW-__-___-___-1	Feedwater
⊞ AR-04-BS-__-___-___-1	Boiler
⊞ AR-04-MP-__-___-___-1	Steam
⊞ AR-04-TS-__-___-___-1	Turbine
⊞ AR-04-MC-__-___-___-1	Condenser
⊞ AR-04-CW-__-___-___-1	Circulating Water
⊞ AR-04-GS-__-___-___-1	Generator
🗁 AR-04-ES-__-___-___-1	Electrical
└ 🗁 AR-04-ES-LV-___-___-3	UPS Low Voltage
└ 🗁 AR-04-ES-LV-BUSS-___-2	Bus
└ 🗀 AR-04-ES-LV-BUSS-BTRY-1	UPS Battery
└ 🗀 AR-04-ES-LV-BUSS-PANL-1	Distribution Panel
└ 🗁 AR-04-ES-DC-___-___-1	3/4 125V Station DC
└ 🗀 AR-04-ES-DC-INVT-___-1	Inverter
└ 🗀 AR-04-ES-DC-BTRY-___-1	Battery
└ 🗀 AR-04-ES-DC-CHGR-___-1	Inverter Battery Charger
└ 🗀 AR-04-ES-DC-DIST-___-1	Distribution
└ 🗀 AR-04-ES-DC-INST-___-1	Controls
└ ⊞ AR-04-ES-HV-___-___-1	2.4KV High Voltage
└ ⊞ AR-04-ES-HV-___-___-3	13.2 KV Generator High Voltage
└ ⊞ AR-04-ES-LV-___-___-2	125 V DC Low Voltage
└ ⊞ AR-04-ES-MV-___-___-1	480 Volt
└ ⊞ AR-04-ES-PR-___-___-1	Relay Protection
⊞ AR-04-CS-__-___-___-1	Controls
⊞ AR-04-SI-__-___-___-1	Plant
⊞ AR-04-ZM-__-___-___-1	Non-Maint Misc Plant

Equipment Hierarchy (to find WO's)

by "drilling down" with the mouse to the area of interest. Folders with a "+" on the folder icon contain embedded "children", so the user knows where to find the details.

Trouble Reporting. Condition-Monitoring begins when trouble is identified. Initiating Trouble Reports (TR's) has traditionally been a primary means of initiating maintenance. In an RCM context, the vast majority of failures will fall under the No Scheduled Maintenance category and will be initi-

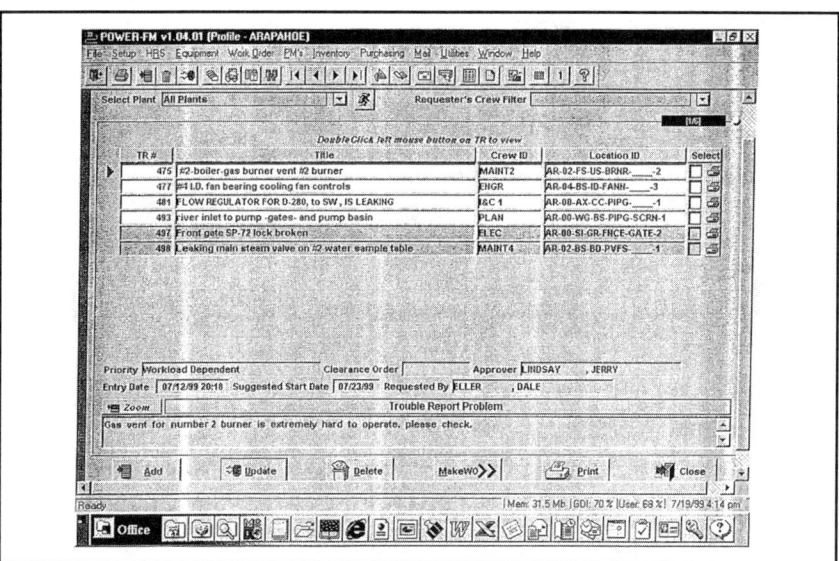

Trouble Reporting: TR-WO relational list

ated as TR's. Success with TR's hinges on ranking the nature and importance of the failures. Documented failures and planned actions for important equipment — equipment in the plant's CMMS register— facilitate sorting through TR's quickly to (1) extract high-impact failures that warrant high-priority, and (2) allow the option to pre-plan work, and work pre-planned work on many NSM-type failures. A TR should clearly identify a problem in the title — not just a piece of equipment. Since the TR title converts into supplemental documents like a work order, the TR's title, initiator, plant impact and priority, reported date, and work start target date need identification. Obviously, a planner and scheduler will have to review and adjust the initiator's request with overall plant schedule and resources.

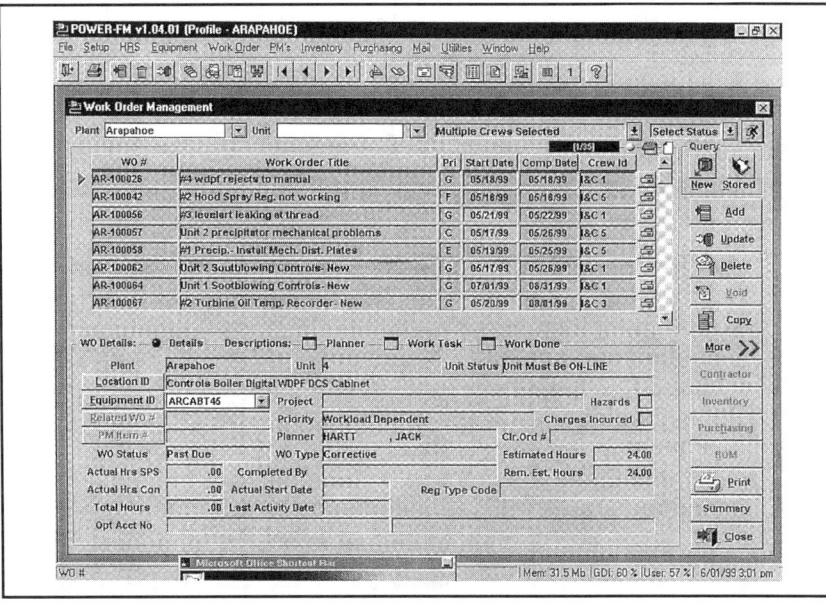

Work Orders: Sheduler's Active WO List

Work Orders. After screening, planning, and approval, TR's become Work Orders (WO's). PM masters issue as Work Orders. (They are prescreened and preplanned.) The CMMS scheduler — an internal clock—checks PM scheduling information against time and, when scheduled, converts a PM master into 'real' WO's. A WO on a CMMS is an electronic document. By itself, a WO can do nothing. But

		7/21/99
Work Order # : AR-100938	Related work Order # :	Page 1 of 3
Title : 7th stage heater, #3unit vent steam leak	PM Item # :	

Plant : Arapahoe   Unit : 3   Unit Status : Unit Must Be at REDUCED L(

Location : Feedwater High Pressure 7 Heater

Equipment : AR-HETR-20   Heater

Tag Number:

Crew : Welders - Moore   MOORE ,CINDY   Clearance Order :

Planned Start Date : 7/21/99   Planned Completion Date : 7/21/99

Problem / Work Description :
[TR# 474 Requested by DAVID WELCH   on 07/12/99 14:36:00 - vent valve on east side of heater
to d.a. 4th level. valve blowing steam and leaking condensate   could either be bonnet
gasket or piping behind somewhere. need to remove insulation to investigate.Insulation was removed
by Roland Bolduc ,found 1/8 inch hole in valve body. Valve will have to be replaced. Took steam off of
heaters long enough to find leak, steam is back on heaters until replacement valve is available ? Area
is barricaded off for safety. Complete high pressure heater clearance will be needed to replace valve.
THis will restrict load 6 nms. This should be done as soon as possible.]

Work Task :
Welder Stamp_____Name_____

Pipe: Size & Schedule:1" sch 40__ Material: A106 Gr.B/ 316ss__Stores Code:____

Valves: Size & Type: 1" Angle Globe SW__ Material: A105____Stores Code: 7726790____

Manufacturer: Rockwell/edwards__Figure No._Fig. 849Y_

Fittings: Size & Type;1" 3000# SW___ Material: A105___Stores Code_____

Weld Rod: GTAW_____ SMAW_____GMAW_____

Procedure_____1A-1__1A2__3A-1__4A-1__5A-1__8A-1__Other_____

WPFP 1A-1 GTAW: ER70S-2   SMAW: E7018   Preheat: 60F____Other_____
WPFP 1A-2 GTAW: None   SMAW: 6010/7018 Preheat: 60F
WPFP 5A-1 GTAW: ER90S5-B3 SMAW: E9018-B3 Preheat: 400F

Process SMAW___GTAW___ GMAW____ Other_____
Schedule/Thickness_____ Material_____
Position_____ Post Weld Heat Treat:_____
Weld Identification (if required)_____

----------------Hold/Witness Date Inspector   Accept/ Reject   Comments

Fit-up					
Position-					
Preheat					

Work Orders: CM Work Order

it is an electronic authorization to use resources and perform work.
This may be printed as hardcopy or left as an electronic authorization
to allow work to be performed and reported entirely through PC interfaces. A WO number generally authorizes time charges, parts usage,
and contractor support. Thus, a CMMS generally ties electronically

445

```
                              ≡📠                            7/21/99
                                                           Page 2 of 3
  Work Order # :  AR-100938        Related work Order # :
  Title :   7th stage heater, #3unit vent steam leak        PM Item # :

Root-_____  |   |   |   |   |
Interpass-_____  |   |   |   |   |
PWHT_____  |   |   |   |   |
Visual_____  |   |   |   |   |
UT_____  |   |   |   |   |
RT_____  |   |   |   |   |
PT_____  |   |   |   |   |
MT-_____  |   |   |   |   |

Other-_____  |   |   |   |   |


Accepted by_____ Date_____
```

```
                              ≡📠                            7/21/99
                                                           Page 3 of 3
  Work Order # :  AR-100938        Related work Order # :
  Title :   7th stage heater, #3unit vent steam leak        PM Item # :

  Work Done : _____
              _____
              _____
              _____
              _____
              _____
              _____
              _____
              _____
              _____
              _____
```

Page 2 CM Work Order

into time reporting and payroll, purchasing, and inventory systems. On our splash menu we see these as Time Management, Purchasing, and Inventory. (Human Resources provides a register of employees that supports Time Management.) WO's originate as (1) TR's, (2) PM masters, and (3) work originating from the Planning and Scheduling Department. The last category includes design changes and in some cases, informally scheduled outage PM work and condition-directed work that isn't formally tied to a TR or PM. From a measurement perspective, the goal is to drive all WO's into their appropriate categories for measurement & accounting purposes.

PM #	PM Title	Type	Crew	Curr PM Date
AR-210	1 Attemperator Controls Calibration	Timed	I&C 1	
AR-161	1 Attemperator Roll Tube Crack Inspection	Overhaul	CON	
AR-122	1 Bailey Drive Positioners Check: Mill Fdrs& Fans	Overhaul	I&C 1	
AR-225	1 BFP Recirc Calibration	Overhaul	I&C 1	
AR-51	1 Blowdown ID Fan Draft Taps (Yearly)	Fixed	I&C 1	
AR-211	1 Boiler Camera Checks	Timed	ELEC	
AR-232	1 Boiler Drum Level Xmtr Calibration	On Demand	I&C 1	
AR-100	1 Boiler Feedpumps LO Sampling	On Demand	MAINT3	
AR-338	1 Boiler ID/FD Fans Inspections (18 month)	Overhaul	PLAN	
AR-162	1 Boiler Inspection	Overhaul	CON	
AR-266	1 Boiler Wet Bottom Inspection	Overhaul	PLAN	
AR-334	1 Bottom Ash Clinker Grinders (Common)	Timed	MAINT3	7/15/99
AR-322	1 Bowser ROTO Fill (Common)	Timed	MAINT3	
AR-206	1 BTG Boiler Master Controls Check	Timed	I&C 1	
AR-260	1 Check Boiler Steam Drum High/Low Level Trips	Overhaul	I&C 1	
AR-340	1 Chemical Addition Pumps Lube (Common)	Timed	MAINT3	
AR-227	1 Circ Water Cond WB Header Press/T Taps Cal	Overhaul	I&C 1	
AR-262	1 Circ Water Condenser Waterboxes Check	Overhaul	OPR5	
AR-164	1 Coal Mill -- Check Tensioners	Timed	OPR5	
AR-113	1 Coal Mill Checks	On Demand	MAINT2	
AR-359	1 Coal Mill Feeder Gearbox LO Replacement	Timed	MAINT3	
AR-130	1 Coal Mill Overhaul -- Tonnage/Performance	Overhaul	ENGR	
AR-358	1 Coal Mill PA Fan Cplgs	Timed	MAINT3	
AR-124	1 Coal Mill Primary Air Flow Testing	Timed	ENGR	6/30/99
AR-356	1 Coal Mill Springs Fdr Dmpr EPGSE	Timed	MAINT3	
AR-203	1 Coal Mill Temperature Recorder	Overhaul	I&C 1	
AR-355	1 Coal Mills Gearbox	Timed	MAINT3	
AR-99	1 Coal Mills LO Sampling	On Demand	MAINT3	
AR-357	1 Coal Mills PA Fan Bearings	Timed	MAINT3	
AR-80	1 Cond Pump Motor Bearings EPGSE	On Demand	MAINT3	
AR-222	1 Condensate Pump Capacity Test	Timed	ENGR	
AR-218	1 Condenser Air Ejtr & Cond Pump Safeties Liftoff	Overhaul	CON	
AR-263	1 Condenser Tube Leak Inspection (Waivable)	Overhaul	PLAN	
AR-212	1 Condenser Vacuum Alarm/Trip Test	Overhaul	I&C 1	
AR-144	1 Extraction Non-Return Air Dump Check	Fixed	OPR5	7/7/99
AR-143	1 Extraction Non-Return Valve Pkg Takeup	Timed	PLAN	6/22/99
AR-142	1 Extraction Valve Checks	Timed	OPR5	
AR-120	1 Feeder Calibration	Timed	ENGR	6/30/99
AR-223	1 Feedwater Control Calibration	Overhaul	I&C 1	
AR-97	1 Feedwater Pump/Htr Limitorques	Timed	MAINT3	
AR-217	1 Feedwater Safety Valve Liftoff Tests	Overhaul	CON	
AR-33	1 Furnace Draft Taps- Ciran	Fixed	I&C 3	7/2/99
AR-224	1 FW Heater Level Controls Calibration	Overhaul	I&C 1	
AR-221	1 FW Pump Capacity Test	Timed	ENGR	
AR-141	1 Gas Fuel Regulation Cal/Chk	Overhaul	I&C 1	
AR-92	1 Gen Seal Oil Pump Skids	Timed	MAINT3	
AR-286	1 Generator Hydrogen Analyzer Cal/Check	Fixed	I&C 3	
AR-264	1 Generator Hydrogen Coolers Cleaning/Inspection	Overhaul	PLAN	
AR-157	1 Generator Seal Oil Samples	On Demand	MAINT3	
AR-335	1 Generator, Trans, & Elect Bus Protection Relays	Timed	PLAN	
AR-268	1 High/Low Side Switchyd Breaker Tests	Overhaul	PLAN	
AR-145	1 Hotwell Level Controls Chk Cal	Overhaul	I&C 1	

WO Lists PM: WO Masters

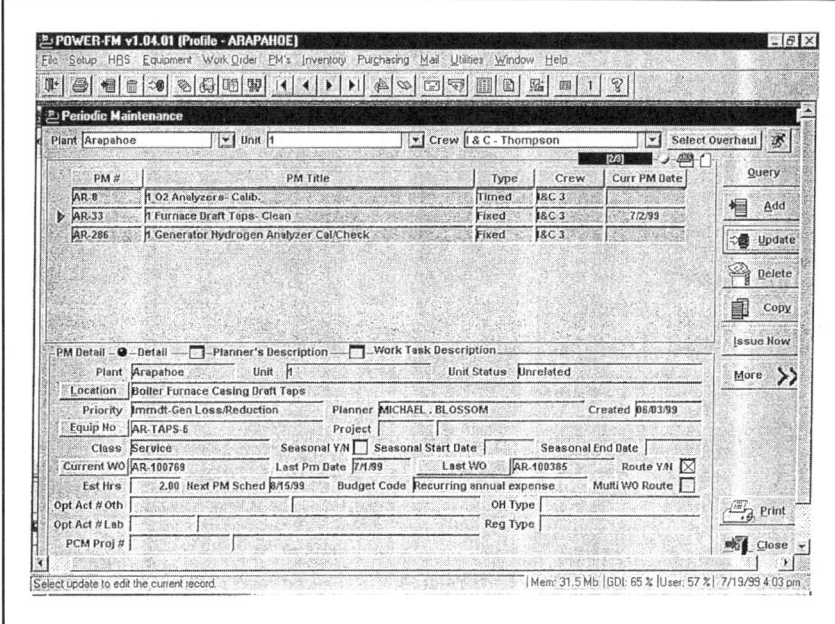

PM: Periodic Maintenenance Crew WO List

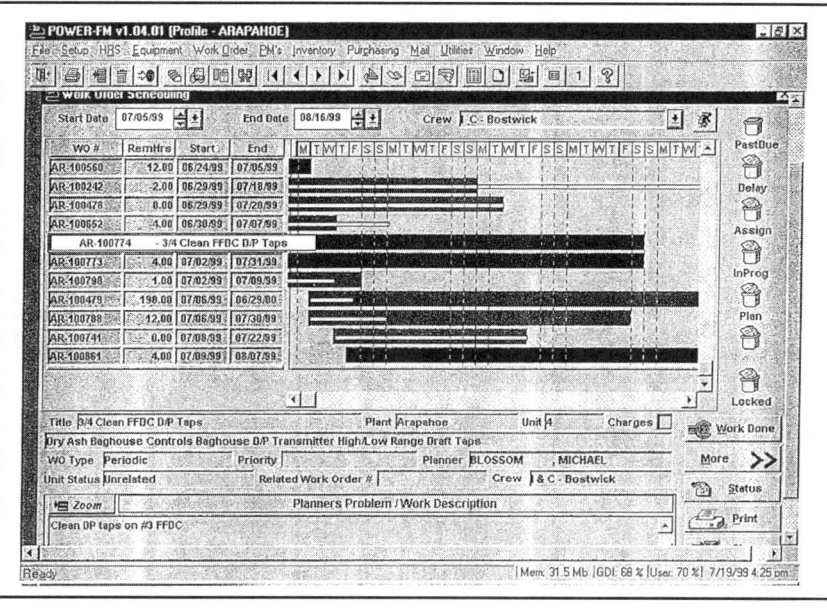

Scheduling: WO Scheduler

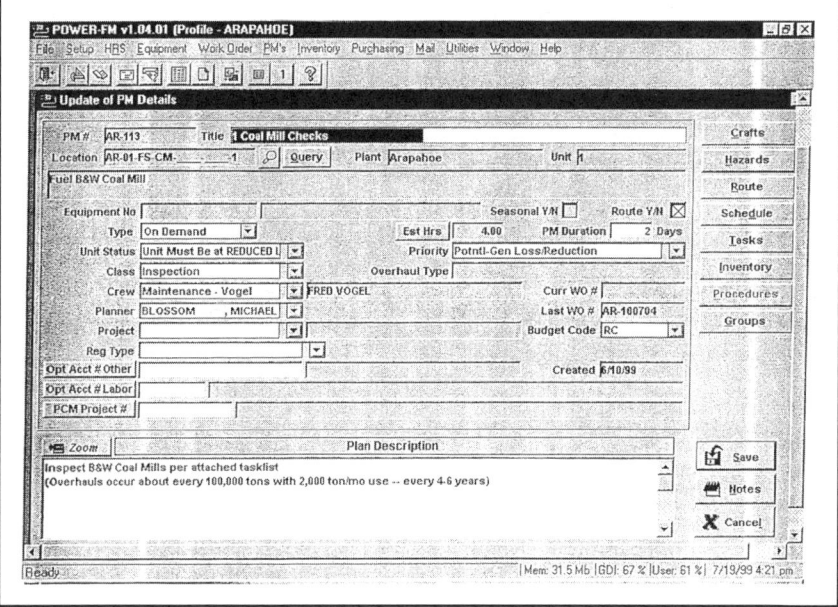

PM WO - Coal Mills WO

PM Periodic Maintenance. A PM master (or standard) must be available "on file" as the source for all scheduled maintenance work orders. This may also provide a convenient repository of preplanned work that is initiated on-demand in response to TR's or Condition-Directed Maintenance. The PM attributes includes the work description, plan, resources, parts, and scheduling information. A PM master is extracted and converted electronically into a work order by the CMMS's subroutine scheduler or an on-demand "issue now" action, much as a person used to make copies of a master form for data entry using a copier in former times.

Scheduling. Work Scheduling reconciles the reality of resources and time (to perform work) against the desires of the work requesters (to have problems corrected). Scheduling is the hallmark of maintenance effectiveness, so a convenient and simple scheduling system is critical to overall work scheduling success. Power FM offers here a modified Gantt Chart format that visually displays key scheduling information.

WO numbers and title can be displayed by GUI mouse features,

```
6/8/99                              ▤🏛                     Page 1 of 2
                         PREVENTIVE MAINTENANCE
   PM #     AR-94          Title  Feedwater Pump/Htr Limitorques 4
   Type     Timed          Unit Status  Unrelated         Est.Hrs.  2.00      Budget Code  RC
   Location  AR-04-FW-__-___-___-1                         Equipment ID


   Crew   Maintenance - Vogel                              PM Duration    2  Days
   Seasonal   Non-Seasonal  Seasonal Start Date            Seasonal End date        Created  6/8/99
   Task                                                    Planner  BLOSSOM MICHAEL .
   Capital Acct No.                                        Class  Lubrication
   Project                                                 Priority  Workload Dependent
   Route ( Y/N )  Has routes        ● Single  ○ Multi      Overhaul Type

                 ESTIMATED HOURS FOR THE PM GIVING CRAFT DETAILS AND HOURS
                    Craft Name                             Estimated Time (Hours)
   Mechanic                                                         2.00

                              ROUTES FOR THE PM
                 Location                                 Active Y/N   Curr WO #   Last WO #
   Feedwater Extraction HP6 3rd Stage Piping Extraction MO Isolation Valve

      Equipment   ARVALV761       Valve                     Active      -            -
   Feedwater Extraction HP4 10th Stage Piping Extraction MO Isolation Valve

      Equipment   ARVALV765       Valve                     Active      -            -
   Feedwater Extraction HP3 13th Stage Piping Extraction MO Isolation Valve

      Equipment   ARVALV766       Valve                     Active      -            -
   Feedwater Extraction HP5 7th Stage Piping Extraction MO Isolation Valve

      Equipment   ARVALV771       Valve                     Active      -            -
   Feedwater High Pressure Feed Piping FW Reg Control MO Block Valve

      Equipment   ARVALV790       Valve                     Active      -            -
   Feedwater High Pressure Feed Piping FW MO Reg Station Bypass Valve

      Equipment   AHVALV791       Valve                     Active      -            -
   Feedwater High Pressure Feed Piping FW "A" Htr Bank MO Block Valve

      Equipment   ARVALV793       Valve                     Active      -            -
   Feedwater High Pressure Feed Piping FW "B" Htr Bank MO Block Valve

      Equipment   ARVALV794       Valve                     Active      -            -
   Feedwater High Pressure Feed Piping FW "A" Htrs Outlet MO Iso Valve

      Equipment   ARVALV795       Valve                     Active      -            -
   Feedwater High Pressure Feed Piping FW "B" Htrs Outlet MO Iso Valve

      Equipment   ARVALV796       Valve                     Active      -            -
   Feedwater Boiler Feed 4C Pump Piping & Valves Discharge Isolation Valve

      Equipment   ARVALV811       Valve                     Active      -            -
   Feedwater Boiler Feed 4A Pump Piping & Valves Discharge Isolation Valve

      Equipment   ARVALV814       Valve                     Active      -            -
```

PM: Periodic Maintenance PW WO with Route

WO status is displayed visually by color (Past Due, Delayed, Assigned, In Progress, Completed, and Locked). In Progress and degree of completion are visually confirmed by the white completion line that displays hours charged as a fraction of estimated workhours. Little features like

450

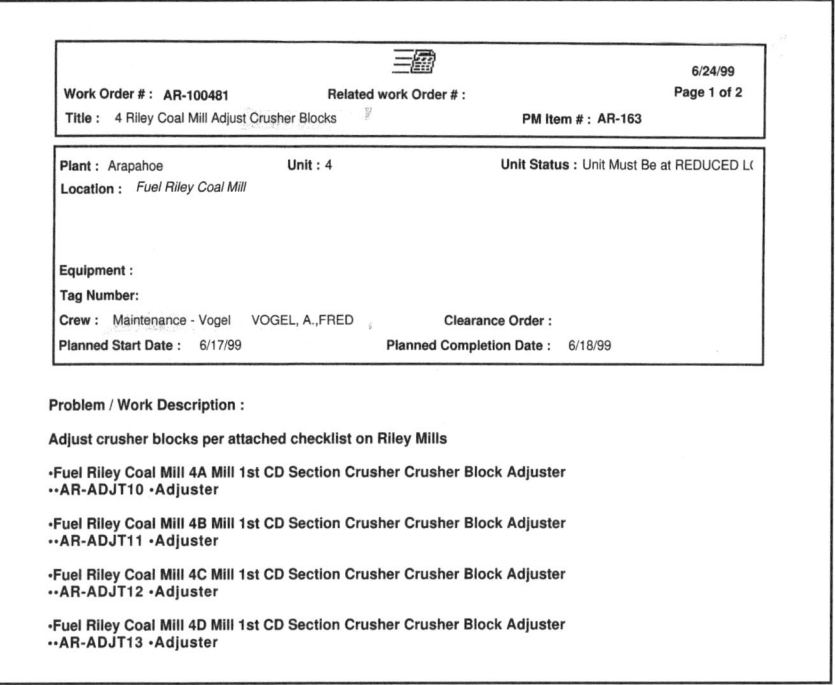

Work Order # : AR-100481   Related work Order # :

Title : 4 Riley Coal Mill Adjust Crusher Blocks   PM Item # : AR-163

Plant : Arapahoe   Unit : 4   Unit Status : Unit Must Be at REDUCED L(

Location : *Fuel Riley Coal Mill*

Equipment :

Tag Number:

Crew : Maintenance - Vogel   VOGEL, A.,FRED   Clearance Order :

Planned Start Date :   6/17/99   Planned Completion Date :   6/18/99

Problem / Work Description :

Adjust crusher blocks per attached checklist on Riley Mills

•Fuel Riley Coal Mill 4A Mill 1st CD Section Crusher Crusher Block Adjuster
••AR-ADJT10 •Adjuster

•Fuel Riley Coal Mill 4B Mill 1st CD Section Crusher Crusher Block Adjuster
••AR-ADJT11 •Adjuster

•Fuel Riley Coal Mill 4C Mill 1st CD Section Crusher Crusher Block Adjuster
••AR-ADJT12 •Adjuster

•Fuel Riley Coal Mill 4D Mill 1st CD Section Crusher Crusher Block Adjuster
••AR-ADJT13 •Adjuster

PM: Periodic Maintenance WO

this allow anyone to quickly confirm (1) that work is in fact in progress, and (2) that it's had so many hours of time charged — and presumably worked. Work that is stalled or parked is also visually clear. Key information for selected WO's is displayed on the same page without jumping around.

WO/PM Lists. Lists can be generated by many sort orders to support any of a variety of standard plant work review activity: daily work, outage work, scheduled work, skill category or department work. These lists must provide the key information — WO number, title, priority importance, crew ID, and scheduled completion date—in any sort order. For PM masters the priority is supplemented by WO type that should roughly translate into RCM scheduling options — Hard Time: TBM, Scheduled Tests & Checks: OCM, On Demand and No Scheduled Maintenance (preplanned): CDM. Other categories such as Overhaul allow convenient grouping into scheduled work categories for

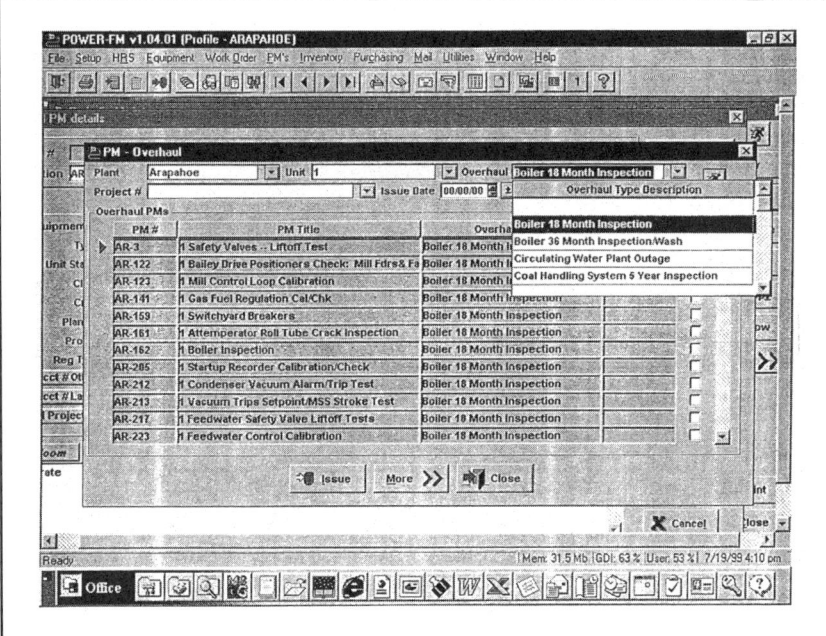

Overhauls: Boiler 18-month Outage WO List

outage. Double clicking the top field of any column resorts the list by that column, and a second time will reverse the sort order (top-to-bottom goes to bottom-to-top).

Overhaul. Overhauls are special groups of work activity that are issued as large groups of activity at the same time. Overhauls such as six-year turbine tear-down and inspections can be developed as many separate individual WO's for the many activities that must separately be performed. These can then be tagged as a specific outage — such as an "18-month boiler inspection"—and issued as one single group (or be issued based on selected items from the group). The "overhaul" group then gets issued as a single clump of work. While the benefits to this are not immediately apparent, let me personally attest that many hours were spent in former days issuing many of the individual WO's that made up large power plant outages — as many as 1,000! The time savings and simplicity of this feature are tremendous. Typically, these activities are the same ones the plant wants to download for their Project Management Software, to be able to schedule the outage in rote detail. The

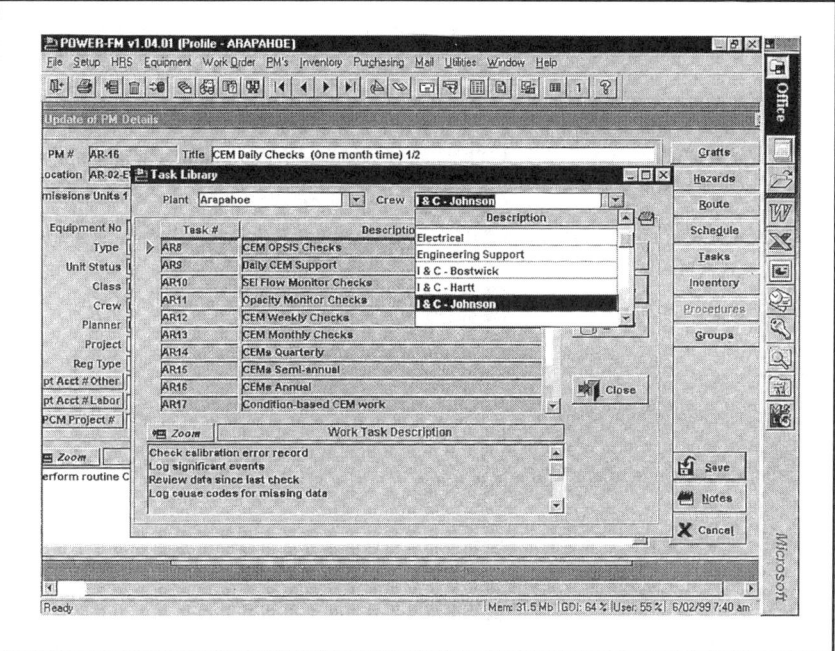

PM Periodic Maintenance: CEM PM Work Orders (by Crew)

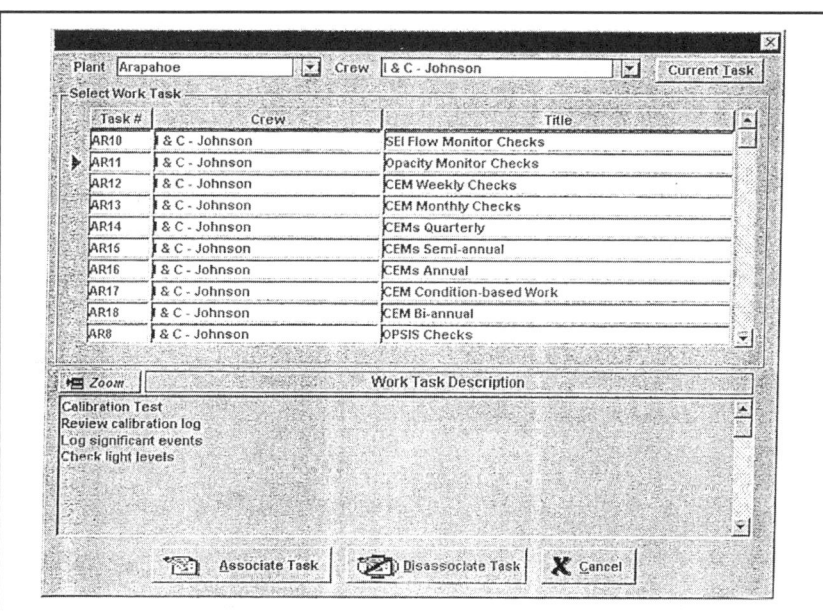

PM Periodic Mainrenance: Crew, CEM, PM, Wo's (All Selected)

		6/24/99
Work Order # : AR-100390	**Related work Order # :**	**Page 1 of 3**
Title : Coal Feeder/Mills EPGEAR Lube 4 (Replace)	**PM Item # : AR-48**	

Plant : Arapahoe   **Unit :** 4   **Unit Status :** Unrelated
Location : *Fuel Riley Coal Mill*

Equipment :
Tag Number:
Crew : Maintenance - Vogel   VOGEL, A.,FRED   **Clearance Order :**
Planned Start Date : 6/15/99   **Planned Completion Date :** 6/16/99

Problem / Work Description :

Lube Feeder/Mills -- per spec:
Mill: EPGEAR: AMOGEAR 150 or EQUIV
Motor: ROTO: AMER IND 68 or EQUIV
Mill Oil System Filters: FRAM PH8A
Based on oil analysis extend one interval (six to twelve months)

•Fuel Riley Coal Mill 4C Mill Outboard Bearing
••AR-BRNG460•Bearing

•Fuel Riley Coal Mill 4C Mill Motor Inboard Bearing
••AR-BRNG461•Bearing

•Fuel Riley Coal Mill 4C Mill Motor Outboard Bearing
••AR-BRNG462•Bearing

•Fuel Riley Coal Mill 4D Mill Inboard Bearing
••AR-BRNG463•Bearing

•Fuel Riley Coal Mill 4D Mill Outboard Bearing
••AR-BRNG464•Bearing

•Fuel Riley Coal Mill 4D Mill Motor Inboard Bearing
••AR-BRNG465•Bearing

•Fuel Riley Coal Mill 4A Mill Motor Outboard Bearing
••AR-BRNG466•Bearing

•Fuel Riley Coal Mill 4B Mill Motor Outboard Bearing
••AR-BRNG467•Bearing

•Fuel Riley Coal Mill 4D Mill Motor Outboard Bearing
••AR-BRNG468•Bearing

•Fuel Riley Coal Mill 4A Mill Motor Inboard Bearing
••AR-BRNG469•Bearing

Routes: PM WO Route

software facilitates download for detailed outage planning.

Work Tasks. Work Descriptions can be added in text format to provide master preplanned jobs that can be pasted into a Scheduled PM master or appended to a WO. Work Tasks are unique based upon work crews. An IC technician work task would have little value for a

```
                              ≡▦                              6/24/99
   Work Order # :  AR-100390        Related work Order # :    Page 2 of 3
   Title :   Coal Feeder/Mills EPGEAR Lube 4 (Replace)        PM Item # :  AR-48
```

•Fuel Riley Coal Mill 4A Mill Inboard Bearing
••AR-BRNG480•Bearing

•Fuel Riley Coal Mill 4B Mill Oil Pump Filter
••AR-FLTR71•Filter

•Fuel Riley Coal Mill 4C Mill Oil Pump Filter
••AR-FLTR72•Filter

•Fuel Riley Coal Mill 4D Mill Oil Pump Filter
••AR-FLTR73•Filter

•Fuel Riley Coal Mill 4A Mill Outboard Bearing
••AR-BRNG481•Bearing

•Fuel Riley Coal Mill 4B Mill Inboard Bearing
••AR-BRNG482•Bearing

•Fuel Riley Coal Mill 4B Mill Outboard Bearing
••AR-BRNG483•Bearing

•Fuel Riley Coal Mill 4B Mill Motor Inboard Bearing
••AR-BRNG484•Bearing

•Fuel Riley Coal Mill 4C Mill Inboard Bearing
••AR-BRNG485•Bearing

•Fuel Riley Coal Mill 4A Mill Oil Pump Filter
••AR-FLTR70•Filter

```
                              ≡▦                              6/24/99
   Work Order # :  AR-100390        Related work Order # :    Page 3 of 3
   Title :   Coal Feeder/Mills EPGEAR Lube 4 (Replace)        PM Item # :  AR-48
```

Work Done : _____
            _____
            _____
            _____
            _____
            _____
            _____
            _____
            _____
            _____
            _____

Routes: PM WO Route (Continued)

mechanic. A task could be viewed as a simple procedure that can be "pasted" onto a WO to provide guidance, rather than typing in the plan on that specific WO. The tag is electronic, however, so the transfer only occurs when the work is issued. When you statistically analyze the work

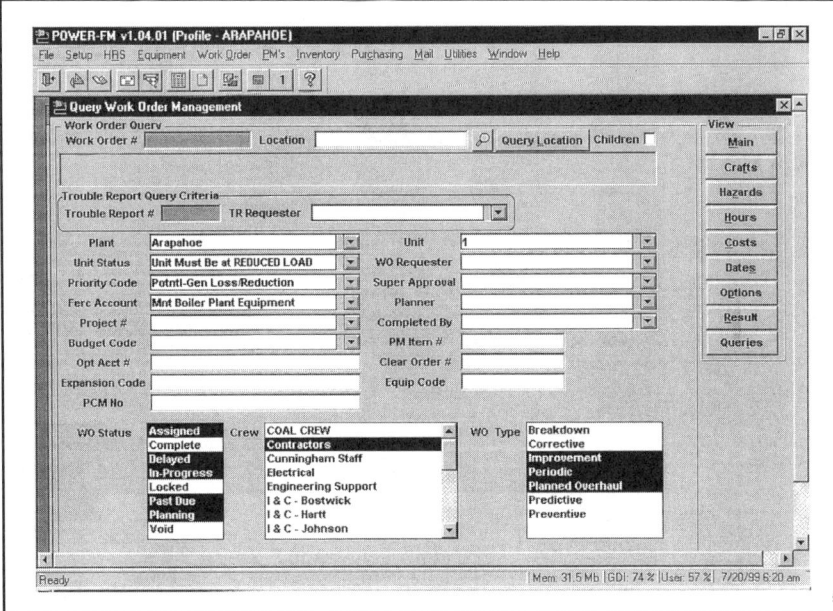

Query: WO "Query" Selection

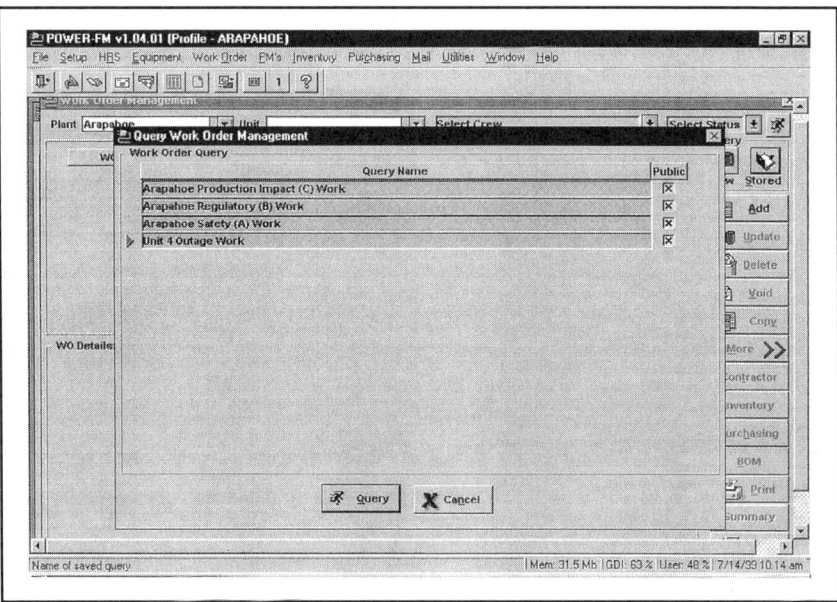

Query: Completed WO "Query" List

PREVENTIVE MAINTENANCE

PM # AR-98	**Title** Feedwater Block Bypass Limits		
Type Timed	**Unit Status** Unrelated	**Est.Hrs.** 2.00	**Budget Code** RC
Location AR-04-FW-__-____-____-1		**Equipment ID**	

Crew Maintenance - Vogel		**PM Duration** 2 **Days**
Seasonal Non-Seasonal **Seasonal Start Date**	**Seasonal End date**	**Created** 6/8/99
Task	**Planner** BLOSSOM MICHAEL .	
Capital Acct No.	**Class** Lubrication	
Project	**Priority** Workload Dependent	
Route (Y/N) Has routes	● Single ○ Multi **Overhaul Type**	

ESTIMATED HOURS FOR THE PM GIVING CRAFT DETAILS AND HOURS

Craft Name	Estimated Time (Hours)
Mechanic	2.00

ROUTES FOR THE PM

Location	Active Y/N	Curr WO #	Last WO #
Feedwater Extraction HP6 3rd Stage Piping Extraction MO Isolation Valve			
Equipment ARVALV761 Valve	Active	-	-
Feedwater Extraction HP4 10th Stage Piping Extraction MO Isolation Valve			
Equipment ARVALV765 Valve	Active	-	-
Feedwater Extraction HP3 13th Stage Piping Extraction MO Isolation Valve			
Equipment ARVALV766 Valve	Active	-	-
Feedwater Extraction HP5 7th Stage Piping Non Return AO Check Valve			
Equipment ARVALV767 Valve	Active	-	-
Feedwater Extraction HP5 7th Stage Piping Extraction MO Isolation Valve			
Equipment ARVALV771 Valve	Active	-	-
Feedwater Extraction HP4 10th Stage Piping Non Return AO Check Valve			
Equipment ARVALV772 Valve	Active	-	-
Feedwater Extraction HP3 13th Stage Piping Non Return AO Check Valve			
Equipment ARVALV773 Valve	Active	-	-
Feedwater High Pressure Feed Piping FW Reg Control MO Block Valve			
Equipment ARVALV790 Valve	Active	-	-
Feedwater High Pressure Feed Piping FW MO Reg Station Bypass Valve			
Equipment ARVALV791 Valve	Active	-	-
Feedwater High Pressure Feed Piping FW Reg Control Valve			
Equipment ARVALV792 Valve	Active	-	-
Feedwater High Pressure Feed Piping FW "A" Htr Bank MO Block Valve			
Equipment ARVALV793 Valve	Active	-	-
Feedwater High Pressure Feed Piping FW "B" Htr Bank MO Block Valve			
Equipment ARVALV794 Valve	Active	-	-

Route: PM WO Route (for MOV's)

done in a large facility, you find it is highly repetitious (*e.g.*, a few failure modes dominate), and the advantages of this feature to pre-plan even "on demand" CDM and NSM-type failure work are tremendous! The beauty of this feature is the capacity to standardize planned work plans and revise the work plan for tens or even hundreds of equipment PM's and WO's with one standard change.

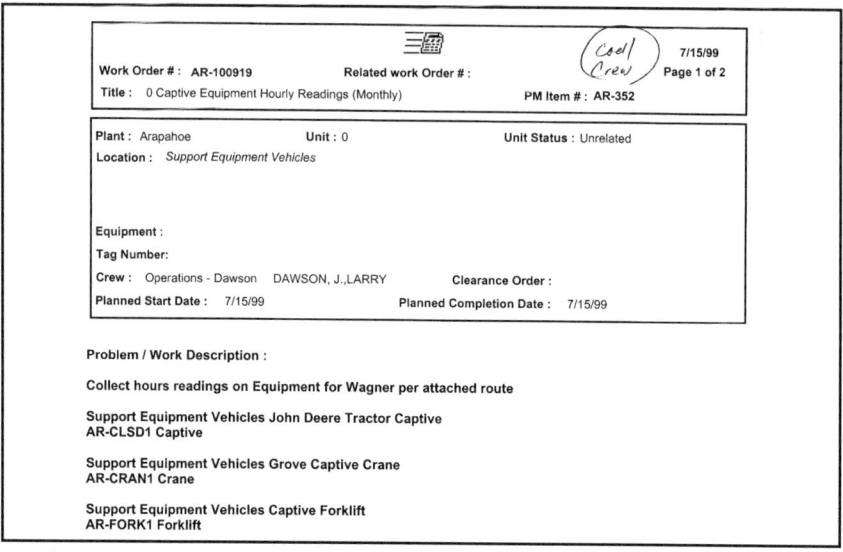

| Work Order # : AR-100919 | Related work Order # : | | 7/15/99 |
| Title : 0 Captive Equipment Hourly Readings (Monthly) | | PM Item # : AR-352 | Page 1 of 2 |

Plant : Arapahoe   Unit : 0   Unit Status : Unrelated

Location : *Support Equipment Vehicles*

Equipment :

Tag Number:

Crew : Operations - Dawson   DAWSON, J.,LARRY   Clearance Order :

Planned Start Date :   7/15/99   Planned Completion Date :   7/15/99

Problem / Work Description :

Collect hours readings on Equipment for Wagner per attached route

Support Equipment Vehicles John Deere Tractor Captive
AR-CLSD1 Captive

Support Equipment Vehicles Grove Captive Crane
AR-CRAN1 Crane

Support Equipment Vehicles Captive Forklift
AR-FORK1 Forklift

Hardcopy: Documents Site Vehicle PM WO

Routes. Routes are groups of repetitive PM activity performed together on one WO. A lubrication route, for example, lists the lubrication locations for one type of grease or lube oil in a given plant area. It conveniently packages activity for streamlined performance. A calibration route does likewise. A group of pressure transmitters with the same calibration interval is calibrated with the same standard job plan in repetitive sequence. The key to successful route performance is the ability to group and perform many brief PM's together in sequence as a group — reducing trip time.

Query. Many query features are available to organize and sort information. Pre-programmed queries (built into the basic software) provide several options for standard information displays. Special queries can be created "real time", used, and saved into libraries (where desired). Information selection and display

Hardcopy Documents. When printed, hardcopy should clearly illustrate the work scope and performance information easily, to facilitate work. Once complete, the required work performance summary and time charges may be returned to a plant clerk for data entry, or (as ever more common) be entered directly into the CMMS by the worker as a job completion step.

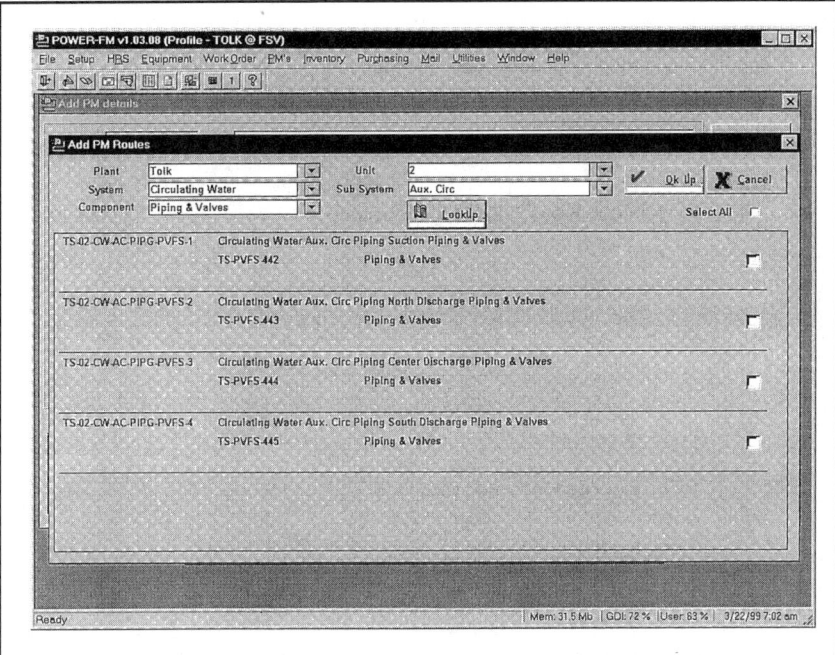

Routes: PM Route Developemnet

WO #	Work Order Title	Pri	Start Date	Comp Date	Crew Id	
AR-100922	1 Gen Seal Oil Pump Skids	C	07/16/99	07/17/99	MAINT3	
AR-100737	1-3 Sootblower A2E Wallblower Lubricator	C	07/01/99	07/01/99	OPR5	
AR-100769	1 Clean Furnace Draft Taps	C	07/02/99	07/23/99	I&C 3	
AR-100709	1 Mill Fineness Check	D	08/02/99	08/03/99	ENGR	
AR-101036	1 Boiler O2 Analyzers Calibration	D	07/30/99	07/30/99	I&C 3	
AR-100708	1 Feeder Calibration	F	07/30/99	07/31/99	ENGR	
AR-100340	Sootblower Retract Checks	F	06/24/99	06/26/99	OPR5	
AR-100341	Sootblower Retract Checks	F	07/10/99	07/12/99	OPR5	
AR-100923	1 Lube Oil Transfer Pumps Skid Lube	F	07/16/99	07/16/99	MAINT3	
AR-100953	1 LO Checks ID/FD Fans	F	07/21/99	07/21/99	OPR5	
AR-100711	1 Coal Mill Primary Air Flow Testing	F	08/04/99	08/05/99	ENGR	
AR-101046	1 Generator, Trans, & Elect Bus Protection Relays	F	07/31/99	08/09/99	PLAN	
AR-100906	1 Sootblower Retract/SL Gearbox EPGEAR	F	07/15/99	07/17/99	MAINT3	
AR-100343	Sootblower Retract Checks	F	06/24/99	07/08/99	OPR5	
AR-100342	Sootblower Retract Checks	F	06/25/99	06/27/99	OPR5	
AR-101050	1 Precipitator Rappers EPGSE (Common)	G	07/31/99	07/31/99	MAINT3	
AR-100595	1 Extraction Non-Return Valve Pkg Takeup	G	06/22/99	07/01/99	I&C 5	
AR-100915	1 Bottom Ash Clinker Grinders (Common)	G	07/15/99	07/15/99	MAINT3	
AR-100985	1 Bottom Ash Clinker Grinders (Common)	G	07/24/99	07/24/99	OPR5	

WO/PM Lists: Prioritized List of Combined CM/PM WO's

RCM Software Examples: Rcmtrim ™
Courtesy: ERE, Inc., Arvada, CO

Startup Menu. The startup menu defines the major RCM software functions. Since users are often not typists, a GUI interface improves speed and convenience. Different users can view or update different areas. Since use is restricted to a small group of engineers and analysts, control requirements are simpler than for a CMMS. "View only" use includes interrogating the database for known hardware failures and failure data, the failure "bases" (*e.g.*, failure basis, plural), and strategies addressing known failures. Systems should be at the highest level in the database and be identifiable by general category such as fossil, nuclear, or chemical process, and/or other general classification. Broad systems classes such as control, service and power conversion should be available for later analysis and sorting.

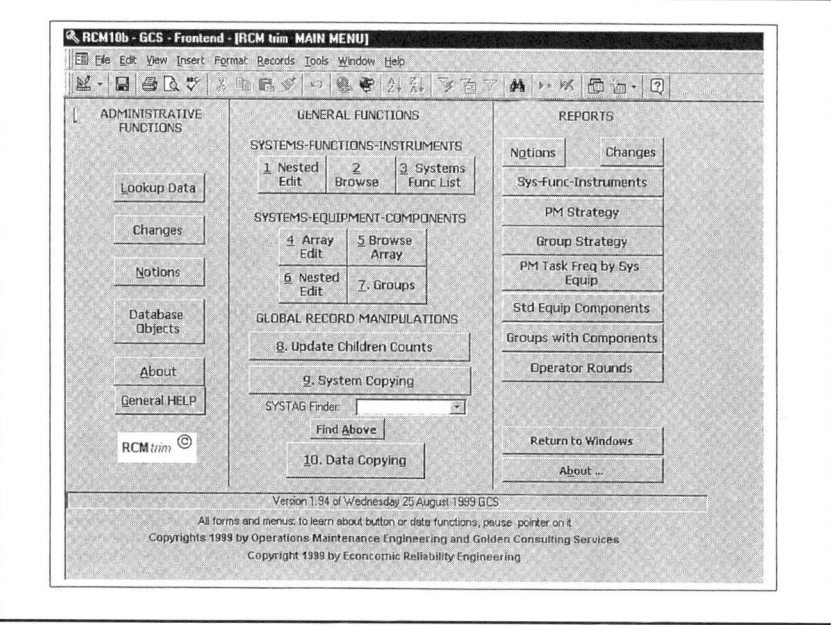

Startup Menu

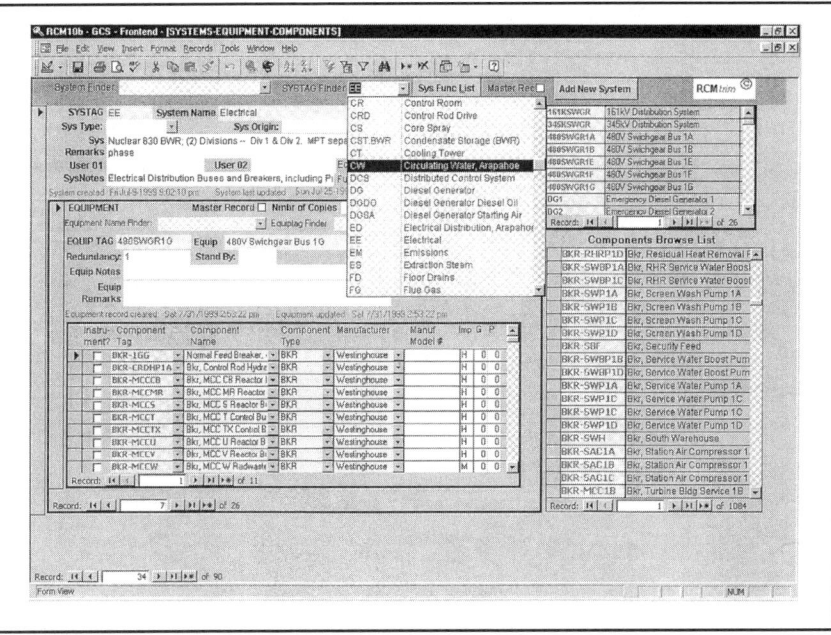

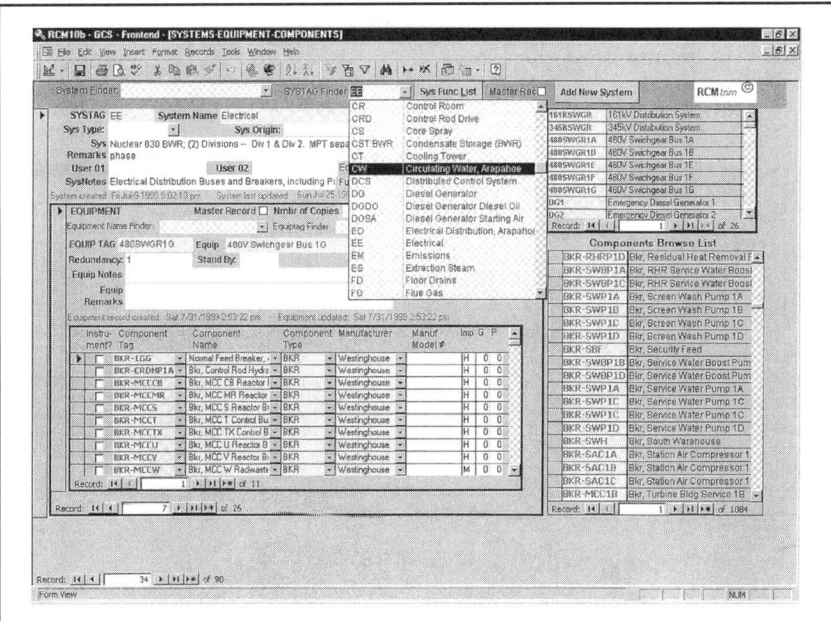

Pull-down Menus and Finders: Pull Down Selection

Pull-down Menus & Finders. Pull-down menus should provide pre-defined selection values and a category based on lookup tables that standardize and speed user data entry. All basic system types — service, production, safety, etc.—should be included for easy reference with the capability to quickly add more. Pull-down data selection and characteristics for many data field attributes such as manufacturers, component types, failure modes and symptoms, basis data source, and secondary failures should be provided. Users should be able to quickly scan options and extract and enter data. System, equipment, and component finders speed locating relevant equipment, information, and location for modeled equipment in the database. Finders should also facilitate jumping to other analysis data source areas and back to the home information for detailed review and selection—quickly.

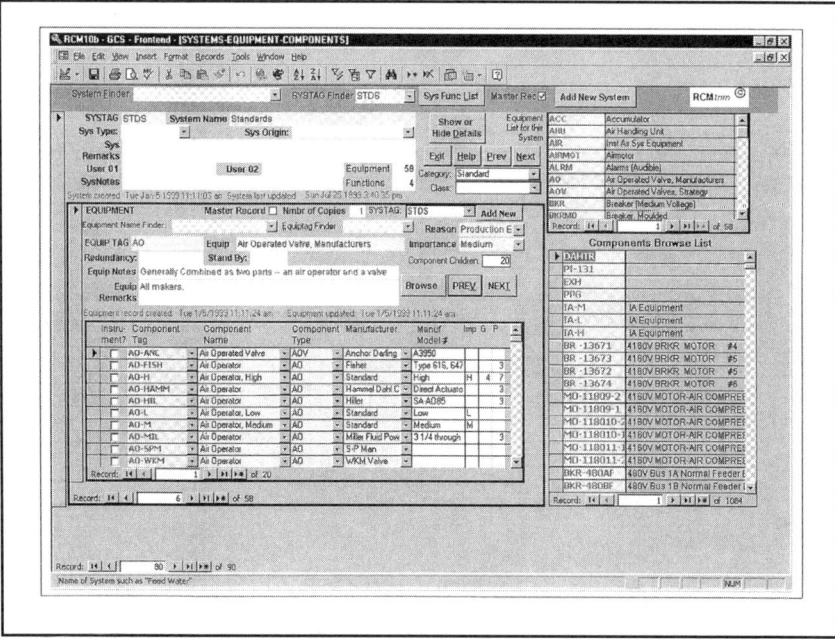

Standards: Air Operated Valves and Suppliers

System Functions & Instruments Hierarchy. "Functions" should be provided beginning at the system level. The user should be able to select a system to model or review, drill down to the system's functions, identify their importance, the instruments and controls used to monitor

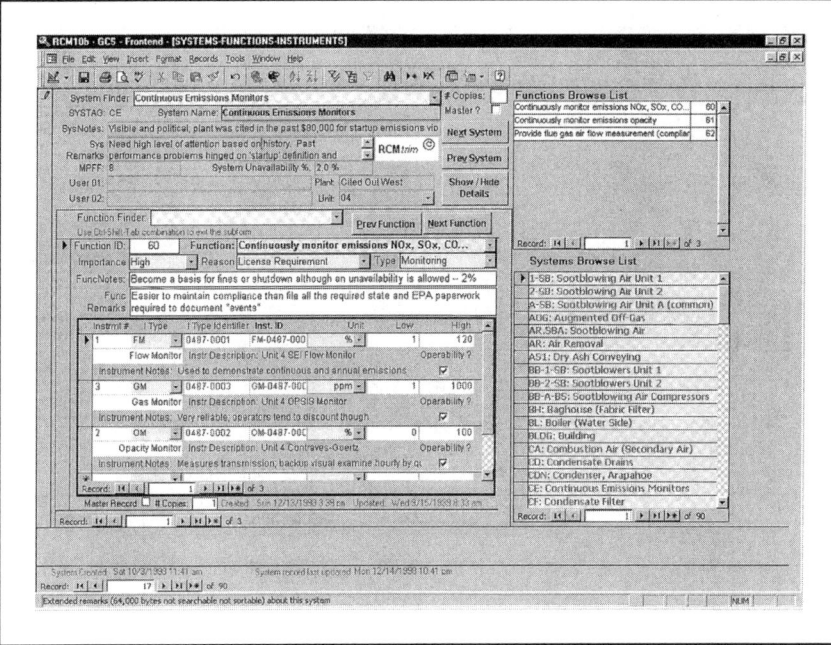

System Functions and Instruments Hierarchy

Systems-Functions-Instruments

RCM*trim* ©

SYSTAG	Function	Instrument	Low-Hi	Unit
CRD	Provide CRDM cooling & motive water	FT-212	30-47	gpm
		PI-220	-	
		PI-236A	11.9-10.9	in Hg
		TSC-243	-250	F
	Provide flow control	FC-304	20-40	gpm
		FI-206	-	
		FI-214	20-40	gpm
	Provide Reactor Recirc pump seal water	FI-45A	5-10	gpm
		FI-45B	5-10	gpm
		PI-48	-	psig
	Provide RWCU pumps seal water	FREG-1A	10-30	gpm
		FREG-1B	10-30	gpm
	Provide safe shutdown reactivity control	DPI-224	-	
		PI-131 (02-19)	-	
		PI-219	-	
		PI-226	-	
		PI-228	-	
		PI-229	-	
		PI-233	-	

Copyright 1998 by Golden Consulting Services, Operations Maintenance Engineering, Inc. and Compelling Operations Reliability Engineering, Inc.

Thursday, September 16, 1999

Page 2 of 3

System Functions and Instruments Hierarchy: Key Instruments and Limits

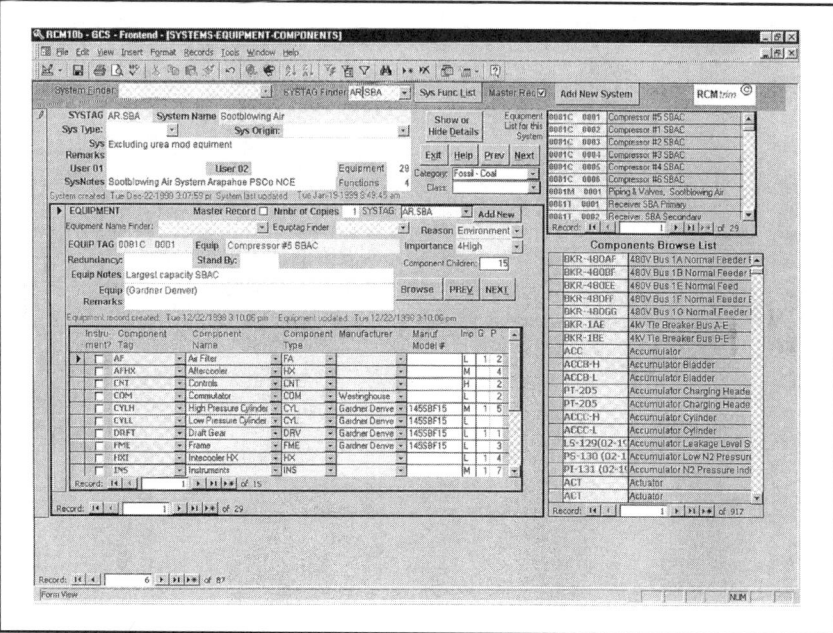

Pull-down Menus and Finders: Browse Lists and Finders System Equipment Computers Hierarchy

these functions, and identify instrument range limits. Function importance rank should be standardized with pull-down menus and scaled based on the relative rank of each category. Functions should further identify type. All systems in the database are readily available for quick development reference, and the functions can be categorized based upon RCM-based attributes such as safety, environment, production, and cost.

Functions Zoom. Instruments that provide key monitoring and safety information should allow further development for operations impact and requirements. These important instruments require calibrations, channel, or alarm checks. Lists of similar instruments and requirements based on functions should be available to be developed as instrument calibration programs based upon importance.

System-Equipment-Components Hierarchy. Systems are broken down by major pieces of equipment such as a sootblowing air compressor assembly or a startup boiler feedpump. "Equipment" can vary from

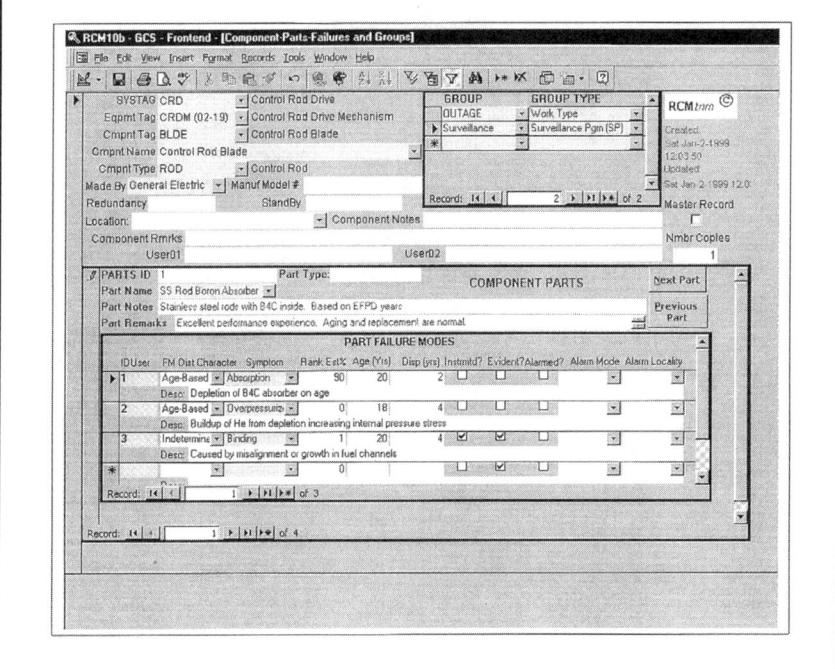

Components Parts Failure Modes

simple skids and associated components on a skid to complex trains and subsystems. Abstract associations of equipment based on proximity, interfaces, and joint functionality should also be allowed to develop as the user's needs require. A feedwater pump train, for example, could be treated as a skid. User-defined fields allow special requirements to be added for general purposes, such as nuclear "environmentally qualified" equipment. Hot links embedded in peripheral Browse Lists speed the identification and selection of useful development material from similar systems. The model should emphasize similarity of equipment and systems, standardizing plans implicitly as the plant model is built.

Components-Parts-Failure Modes. For each component, parts can be identified and described in terms of failure information. Failure modes, causes, system impact (based on redundancy and standby/running applications), location, notes, and user-defined fields can uniquely describe the parts and their affect on the component and systems based on each failure. Fundamental failure characteristics can be classed sta-

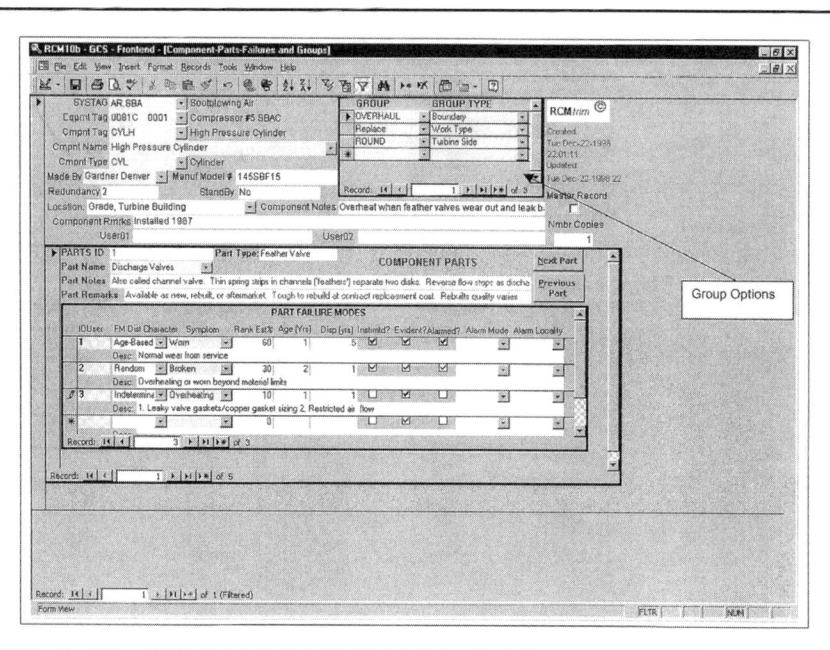

Groups and Routes

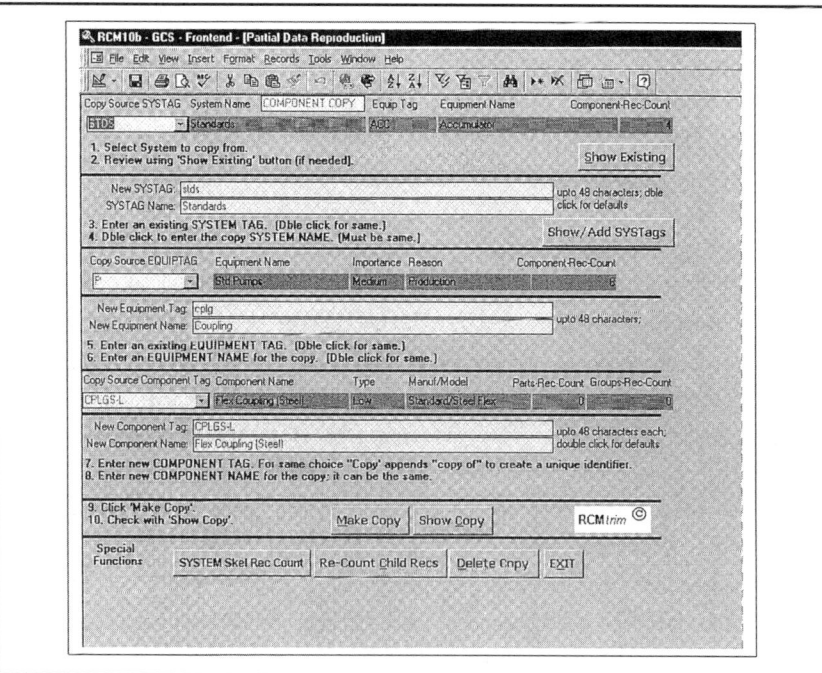

Data Copying: Components

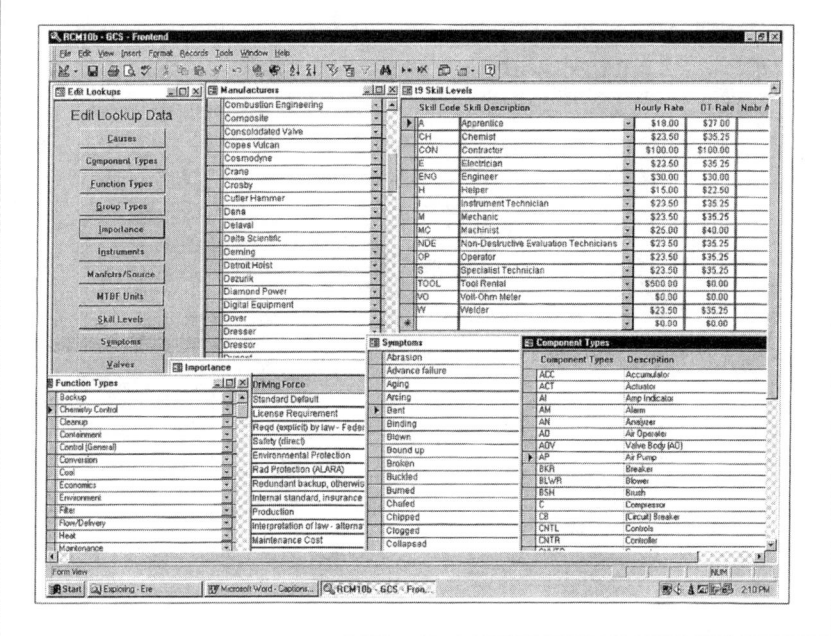

Pull-down Menus and Finders: Lookup Tables

tistically (random, age-based,) and by symptom, occurrence frequency, age estimate, and description. Failures can be ordered by importance rank to emphasize high-risk failures. Basic part failure information can be quickly scanned and validated by workers, or reviewed and used by operators or engineering. Secondary failures and failure causes are also identified. Models are available at every level — Systems, Equipment, Component, and Parts—and their failures are available for immediate review using browse and hot link features.

Applicable models and their subordinate information can be extracted from any level, and be incorporated or recreated as new systems and subordinate equipment conveniently. Users can capitalize on what they already know, rather than bog down in endless creation of new documents and files. Menus allow users to summarize all-important monitoring equipment in the database. Operators can quickly locate an instrument and determine its importance to plant functions. "Loss of function" consequences are clearly identified, allowing the operator to evaluate impact on unit operations. Support instrumenta-

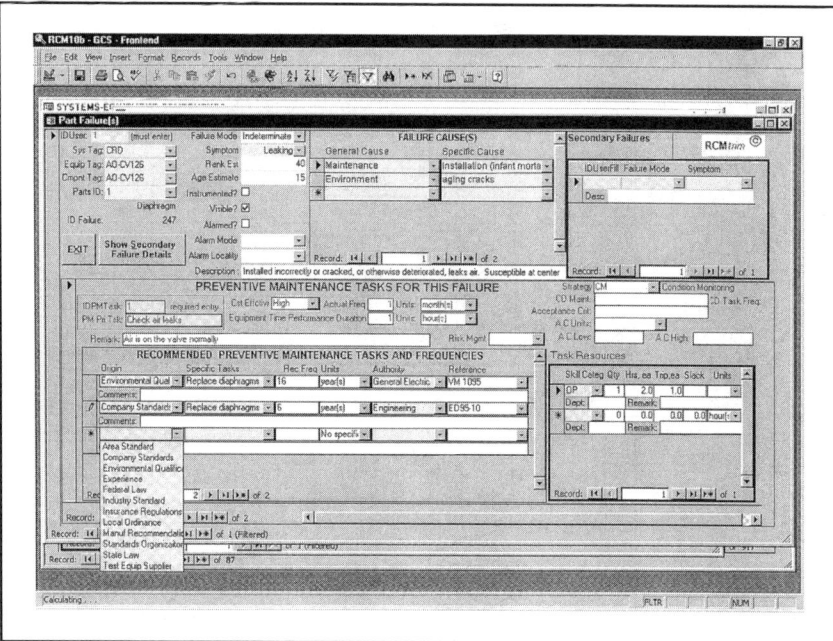

Detailed Information Part Failure Preventative Maintenance Tasks-Reccomended PM Tasks.

tion maintenance managers and crews can quickly review instruments, ascertain intended operational role, and review failures and maintenance plans. Failure risks can quickly be determined based on operator instrument usage, design functionality, and maintenance strategy. The requirements of operator training programs can be identified to support cost-effective, failure-based development of formal and informal on-the-job training.

Part Failure-Preventive Maintenance Tasks-Recommended PM Tasks. From the failure information (especially causes), applicable and effective "PM tasks" can be identified. Tasks can be uniquely identified by number, labeled for RCM-type (time-based, on-condition, condition-directed, condition-monitored), be ranked by cost-effectiveness and risk management value, and provide performance information (frequency interval and units). Where the primary maintenance activity requires condition-directed maintenance on failure, the failure definition limits and CDM tasks can be clearly identified. Where condition-

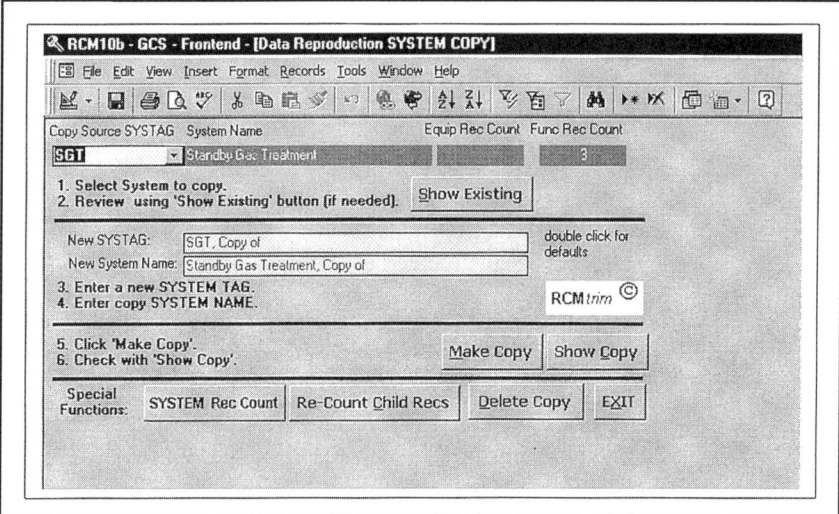

Data Copying: Systems

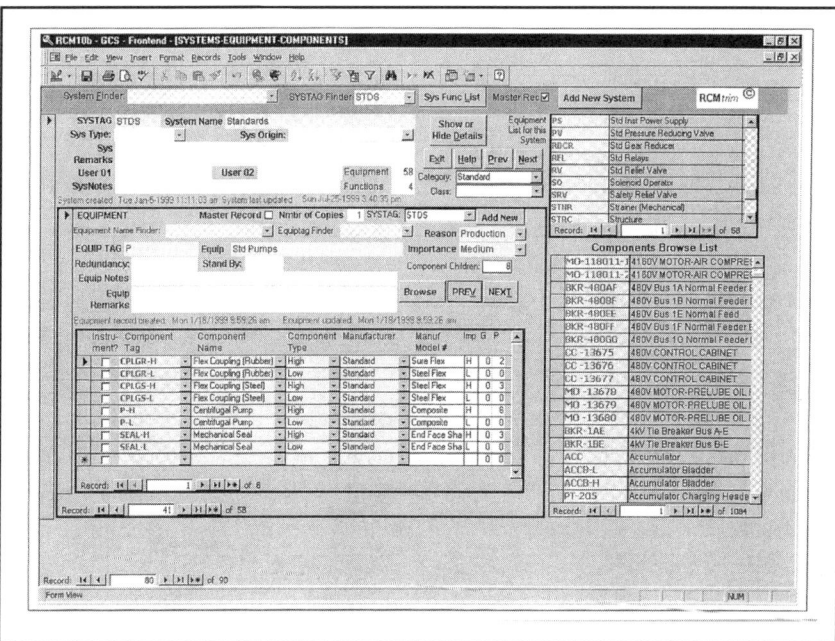

System Equipment Components Hierarchy

monitoring is the strategy selected, any condition-directed maintenance requirements may also be specified — with or without limits. When no

Operator Rounds Instrument List   SBA

SYSTAG	SBA
Sys Name	Sootblowing Air

Function	Maintain boiler differential pressure
Function ID	141 **Importance** i3Medium
Reason	Production
Func Type	Flow/Delivery

INSTRUMENTS

ID	Instrument	Description	Low	High	Unit	Oper
55	DPI-FG1		0	28	inches	☑
56	NI-ID1		0	1500	rpm	☐
57	NI-ID2		0	1500	rpm	☐

Function	Provide backup air to instrument/control air
Function ID	139 **Importance** Low
Reason	Redundant backup, otherwise "High"
Func Type	Control

INSTRUMENTS

ID	Instrument	Description	Low	High	Unit	Oper
53	PI-SBA-PI-20		60	120	psig	☐

Function	Provide baghouse bag cleaning air
Function ID	140 **Importance** Medium
Reason	Production economics
Func Type	Economics

INSTRUMENTS

ID	Instrument	Description	Low	High	Unit	Oper
54	DPI-1		12	20	psid	☐

Operator's Rounds

scheduled maintenance is selected, the factual review of a failure mode and its impact can be documented with no scheduled maintenance program as the planned strategy. The review is available for engineering, diagnostics or reassessment at a later time.

Recommended Preventive Maintenance Tasks and Frequencies may also be identified based on origin, task(s), frequency, authority, and references. Users that require detailed bases for all work done — such as nuclear power plants—will be able to provide this information. All users will be able to explicitly document requirements such as ASME

PM Task Frequency

SYSTAG		Skill Level	PM Task Name	Actual Frequency
			Check vibration during startup	2 year(s)
	SEAL-H	Mechanical Seal		
		Operator		
			Check leak rate	1 month(s)
PS				
	PS-HIGH	Power Supply		
		Instrument Technician		
			Check voltage	5 year(s)
			Check ripple voltage	
PV				
	PV-HIGH	Pressure Reducing Valve		
		Operator		
			Check control range	1 year(s)
RDCR				
	RDCR-H	Gear Reducer		
		Operator		
			Check/trend noise	1 week(s)
			Check area	month(s)
			Sample lube oil/analysis	year(s)
REL				
	REL-HIGH	Relay		
		Instrument Technician		
			Check operation	1 year(s)
		Operator		
			Check chattering relays	1 shift(s)
RV				

Thursday, September 16, 1999	RCM*tnm* ©	*Page 47 of 49*

Reports

codes, EPA Title 5 emissions, and other requirements should they so desire. When documented, users can see those aspects of their scheduled maintenance program that are based on the force of law. Every PM task has task resources necessary to perform the task identified. This includes work classification, department, work hours, travel and slack time. PM tasks can be grouped at the component, and part failure level to arrange packaged activity that can be performed as convenient work packages.

Data Copying. Data copying subroutines allow the user to select, extract, and apply failure data at many levels and copy large chunks of pre-existing systems — up to and including the entire system itself— into new or existing models. A process that avoids recreating basic engineering and failure data available elsewhere for the multitude of replicated components, equipment, and even systems present in a large industrial facility can be used to develop plans quickly. Data copying allows rapid similarity modeling at multiple levels.

PM Strategy for Air Removal

RCM*trim* ©

PartName (Strategy)	(PM Task)	Failure Description	Symptom(s) (Actual PM Freq)	Rank Estimate %	Age Estimate (years)
SYSTEM: AR	*Air Removal*				
EQUIPMENT: GSCEA	*Gland Steam Condensing Exhauster A*				
COMPONENTTAG: MO-163MV	*Gland Exhauster A Discharge*				
Motor Operator		*Drives past torque switch onto seat and binds*	Seized	20	10
OCM	Test MO setup		3 year(s)		
Motor Operator		*Times out before reaching position*	Drift	60	5
TBM	Reset valve toque/limit		6 year(s)		
COMPONENTTAG: MOT-163MV	*Gland Ex A Discharge MO Motor*				
Stator		*dust air particles cause insulation erosion*	Abrasion	20	8
OCM	Visually inspect end turns		5 year(s)		
Bearings		*surface fatigue/overload*	Galled	60	15
OCM	Lube		36 mile(s)		
Bearings		*contamination/debris/seal wear*	Damaged	30	8
CM	Check loading		1 start(s)		
Rotor		*asyncronous loading*	Broken	80	15
OCM	Check vibration level		3 month(s)		
Rotor		*asyncronous loading*	Broken	80	15
CDM	Motor analysis		12 month(s)		

Copyright 1998 by Operations Maintenance Engineering, Inc. and Economic Reliability Engineering, Inc.

Thursday, September 16, 1999

Page 1 of 32

Part Failure—Preventative Maintenance Tasks—Reccomended PM Tasks

Annualized Work Hours

RCM*trim* ©

SYSTAG	Hierarchy	SkillLevel	Strategy
CRD		**Control Rod Drive**	

Control Rod Drive Mechanism/Collet Locking Mechanism/Collet Mechanism

				Actual Freq	Quantity	Hours	TripTime	SlackTime	Units
OP	Operator								
CM									
	PM Task: **Manipulate HCU manual valve to flush wit**			20	2	2	1		
		Skill Level Subtotals:				2	1		

Control Rod Drive Mechanism/Collet Locking Mechanism/Collet Spring

				Actual Freq	Quantity	Hours	TripTime	SlackTime	Units
M	Mechanic								
OCM									
	PM Task: **Check spring constant**			30	1	0	0		
TBM									
	PM Task: **Replace cracked springs**			15	1	0	0		
		Skill Level Subtotals:				0	0		

Control Rod Drive Mechanism/Control Rod Blade/SS Rod Boron Absorber

				Actual Freq	Quantity	Hours	TripTime	SlackTime	Units
ENG	Engineer								
OCM									
	PM Task: **Check time (friction test)**			2	1	2	0		

Reports

Standards. A special set of equipment, component, or even system standards is set up for one special purpose — providing a library of existing components and their failure modes, applicable PM tasks, suppliers, and supplier or other recommended programs. The standards should be able to jump-start the creation of a new system model from scratch, providing a ready source of component failure data. Further, the standards should provide the option for a facility to tailor their maintenance strategy to reflect their predominant failure modes and experience in their facility, based on their particular methods of operation and environment.

Detailed Information. The database ideally offers the opportunity to capture and reuse detailed equipment information such as manufacturer-specific failure and maintenance strategies without starting over for each new component type. Characteristics of different equipment types should be readily available to easy application. Design similarities with fleets or product evolutionary cycles can be captured.

Operator's Rounds. Operator rounds can be developed as hard-copy printouts or exportable files for the performance of rounds. Operator rounds provide information about the monitoring strategy that helps operators perform effective rounds and understand rational behind monitoring intervals.

Reports. A variety of useful management, maintenance, operations, and training reports should be available based upon systems, their functions, work classifications, and risk. Cost management reports should support maintenance strategy adjustments to reflect experience and costs. Management should easily be able to document, maintain and use the basis rational for the operating maintenance strategy. The information should provide a useful tool to perform comparison analysis for parts, manufacturers, and applications to derive future cost reductions, performance improvements, and other benefits. The plan should integrate operations monitoring, maintenance rework and replace tasks, and engineering technical assessment. Useful derivative products — such as the calibration schedule, time-based maintenance plan, and lubrication schedules—should be available from the basic plant model.

Export/Import Routines. Predefined routines should allow export of completed maintenance and monitoring strategies to CMMS and

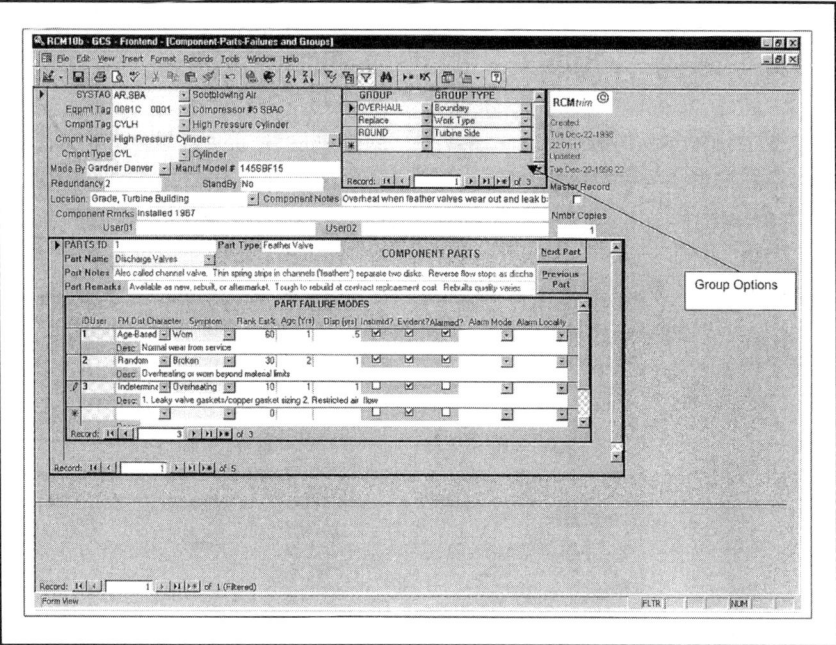

Groups and Routes

operating rounds management systems. Hardcopy results of the same information should also be available to support training and review. Existing facilities should be able to download existing equipment files and PM plans as "delimited" files, have these loaded into the PMO database to provide a starting point, and have finished plan results "up-loadable" to the database as complete, organized tasks in file format. Little or no data re-entry should be required to accomplish this transfer.

Groups & Routes. Work should be easily organized into groups such as overhauls, surveillance tests, or special work activity that can be issued at any time. Those planning work should have the option to deconstruct and re-plan maintenance activity in many different ways based on work that mitigates failure. The software should be a flexible work-planning tool. The activities the plant would like to download for their Project Management Software, such as outage work, should be available in exportable file format. The software should facilitate outage vs. online work and risk analysis. Routes of repetitive "light" PM activity that can be performed together should be available.

Lubrication, calibrations, and even operating inspections should conveniently package activity for streamlined performance. Since a key to successful route performance is grouping and performing many brief PM's together in sequence reducing trip time, the software should facilitate this use.

References

1. F.S. Nolan, H.L. Heap, et al, *Reliability-Centered Maintenance,* United Airlines, San Francisco, CA, Dec. 1978 NTIS AD/A066579
2. *Reliability Centered Maintenance* (edited summary of Nolan & Heap), R. Keith Young, MQS, PdMA , Millersville, MD, 1996
3. Smith, A.M., *Reliability-Centered Maintenance,* McGraw-Hill, New York, NY, 1993
4. Scherkenbach, W.W., *The Deming Route,* CEE Press Books, George Washington University, Washington, DC, 1990
5. Ishikawa, Ki., *What is Total Quality Control? The Japanese Way* (Translated by D.L. Lu), Prentice-Hall, Inc., Englewood, NJ, 1985
6. Bloch, H.P., Geitner, F.K. *Machine Reliability Assessment*, Van Nostrand Reinhard, New York, NY, 1990
7. Tajiri, M., & Gotoh, *Total Productive Maintenance Implementation—A Japanese Approach*, McGraw-Hill, Inc., New York, NY, 1992
8. Rao, S., *Reliability-Based Design*, McGraw-Hill, New York, NY, 1992
9. Kececioglu, D, *Reliability Engineering Handbook Vol. 1-2*, Prentice Hall, Englewood Cliffs, NJ, 1991
10. Ireson, W. Grant, et al, *Handbook of Reliability Engineering and Management*, McGraw-Hill, 1988
11. *Equipment Maintenance Optimization Group Meeting Minutes*, ERPI Boston, MA, 1995
12. *RCM Handbook*, EPRI, 1994
13. *Reliability Centered Maintenance Implementation*, EPRI NDE Center, Charlotte, NC, RCM Maintenance Training, Nov. 1993, (NUS)
14. *RCM for Substations Technical Reference*, EPRI NUS Gaithersburg, MD, June 1996
15. *RCM Proceedings RCM for Substations Conference*, EPRI Cambias & Associates, August 1996
16. *Predictive Maintenance Primer*, NMAC EPRI (NUS), Palo Alto,

CA, April 1991

17. *RCM Generic Applications Guide,* EPRI, Feb. 1991

18. *Demonstration of RCM, Project Description & Results, VI,* Schwan, et al, EPRI NUS, April 1991

19. *Demonstration of RCM, First Annual Progress Report, V3.,* Anderson, J, et al, EPRI Southern California Edison/ERIN Engineering, April 1991

20. *Demonstration of RCM, Final Report of San Onofre NGS, V3.,* Betros, et al, EPRI Southern California Edison/ERIN Engineering, April 1991

21. *Proactive Operations and Maintenance Workshops*

22. *Root Cause Analysis Workshops,* PSCo Improvement Technology Group, 1996, Denver, CO

23. *Process Improvement Technology,* Brunetti, Wayne, 1994

24. *Reliability Centered Maintenance,* MQS, R. Keith Young, 1996, PdMA, Millersville, MD

25. *Code of Federal Regulations 10CFR50.65, Part 10 Section 50.65* "The Maintenance Rule"

26. *Industry Guideline for Monitoring the Effectiveness of Maintenance at Nuclear Power Plants,* Nuclear Management and Resource Council, Inc. (NUMARC), May 1993

27. *Inspection Procedure 62002, 50.65 Maintenance Rule Inspection, NRC Inspection Manual.* US Nuclear Regulatory Commission

28. *Inspection of Maintenance Rule Implementation at Nuclear Power Plants Inspection Procedure 62706, Implementation, NRC Inspection Manual,* US Nuclear Regulatory Commission

29. *Inspection of Structures, Passive Components, and Civil Engineering Features at Nuclear Power Plants Inspection Procedure 62007, Maintenance Observation, NRC Inspection Manual,* US Nuclear Regulatory Commission

30. *NRC Information Notice 97-18: Problems Identified During Maintenance Rule Baseline Inspections,* US Nuclear Regulatory Commission, April 14, 1997

31. *"The Maintenance Rule" Presentation and Interpretation,* American Nuclear Society Winter Annual Meeting, Albuquerque, NM, Nov 16, 1997

32. *Understanding Reliability Centered Maintenance: A Practical Guide to Maintenance, Second Edition 1998*, Jack Nicholas, R. Keith Young, MQS, Millersville, MD 21108

33. Moubray, J. *Reliability Centred Maintenance*, Industrial Press, New York, NY, 1992

34. *In Search of Excellence*, Peters, T.J., Waterman, R.H., Harper and Row, Cambridge, MA, 1982

35. *Faults & Failures: The Auckland Outage*, Sweet, W., IEEE Spectrum, April 1998, p.72

36. *Auckland Unplugged: The Story of a Blackout*, Ackermann, T. & Muller, D., Electric Light and Power, Nov 1998, p. 20-23

37. *Coal Handling Maintenance Optimization at Pawnee*, Rohde, S. & August, J., P/PM Technology, October 1996

38. *Damn the Torpedoes! Hit or Miss* (American Submarines entered WWII armed with dangerously unreliable torpedoes that took almost two years to fix), Murphy, D., American Heritage of Invention and Technology, Spring 1998, Vol. 13, Num. 4, p. 56-63

39. *Auckland: City in Crisis,* Internet, www.nzwires/crisis, 2/09/98-4/28/98

40. *Proceedings*, Predictive Maintenance Technology National Conferences, 1996, 1998

41. *Notes, Poke Yoke,* Fury Enterprises, Dallas, TX

42. Metteson, Gene, *The Air Transportation Industry-Birthplace of RCM*, (presentation & paper), RCM for Substations Conference, Dec 4, 1995, Newport Beach, CA

43. Netherton, Dana, *Standard to Define RCM*, Maintenance Technology, pp. 17-24, June 1999

44. *The Deacon's Masterpiece*, Oliver Wendell Holmes, McGraw-Hill, 1965

Index

A

Accuracy (instrumentation), 243-248
Acronyms, xi-xiv
Age exploration, 6, 178-180, 199-205, 307, 338, 426-427:
   definition, 199-201;
   value, 201-204;
   systematic application, 204-205
Aging analysis, 47, 270, 338, 405-407
Alarms, 242-243
Alternative solutions, 349-350
Ambiguity, 146-147
Analysis software, 312-315
Applicability criterion, 31-32, 122-124
Applications software, 304
Applications (RCM), 12-14, 161-193, 344-350:
   overview, 161-165;
   engineering, 165-173;
   integration of functions, 173-185;
   safety, 185-187;
   case histories, 187-193;
   statistical maintenance, 346-349;
   alternatives, 349-350
Area checks, 238-240
Areas not worked online, 292-293
As low as reasonably achievable, 396-397
Assessing programs, 122-125:
   applicability, 122-124;
   cost effectiveness, 124-125
Assumptions, 31-37:
   applicability, 31-32;
   effectiveness, 32-33;

CMMS, 218-221, 256-258, 304-307, 318, 441-459:
barriers, 218-220;
software, 304-307;
integration, 318; example, 441-459
CMMS example (Power FM), 441-459
CMMS software, 304-307:
maintenance process, 304-305;
custom products, 305-306;
system level measurement, 306;
age exploration, 307
Coal belt fire, 69-70
Coded components, 141
Coding levels (software), 303-304
Common mode failures, 144-147:
keep it simple stupid, 146;
ambiguity, 146-147
Company roles, 129-130
Comparison analysis, 104-105
Complex failures, 142-144
Complexity (failure), 141-144, 228-229
Component failure, 156
Components hierarchy, 464-465
Condition monitoring, 6, 153-154, 415-420
Condition-based maintenance, 6, 42, 153-154, 226-228, 392-393, 401-405, 415-420
Condition-directed maintenance, 226-228, 418-420
Configurations simulation, 319
Confusion implications, 234-235
Consequences, 185-187
Conservatism, 134-139:
over-conservatism, 135-139
Consistency (maintenance), 29-37:
statistics, 30-37;
assumptions, 31-37
Consistency (parts), 181-182
Corporate maintenance management system. SEE CMMS.

D

E

F

O

Q

R

S

T